Differential Geometry and Kinematics of Continua

Differential Geometry and Kinematics of Continua

John D Clayton
US Army Research Laboratory, USA

World Scientific
NEW JERSEY • LONDON • SINGAPORE • BEIJING • SHANGHAI • HONG KONG • TAIPEI • CHENNAI

Published by

World Scientific Publishing Co. Pte. Ltd.
5 Toh Tuck Link, Singapore 596224
USA office: 27 Warren Street, Suite 401-402, Hackensack, NJ 07601
UK office: 57 Shelton Street, Covent Garden, London WC2H 9HE

Library of Congress Cataloging-in-Publication Data
Clayton, John D., 1976–
Differential geometry and kinematics of continua / John D Clayton, US Army Research Laboratory, USA.
pages cm
Includes bibliographical references and index.
ISBN 978-981-4616-03-4 (hardcover : alk. paper)
1. Geometry, Differential. 2. Calculus of tensors. 3. Continuum mechanics. 4. Field theory (Physics) I. Title.
QC20.7.D52C53 2014
531'.1130151636--dc23

2014014295

British Library Cataloguing-in-Publication Data
A catalogue record for this book is available from the British Library.

Printed in Singapore

Preface

The primary purpose of this work is to provide precise definitions and complete derivations of fundamental mathematical relationships of tensor analysis encountered in nonlinear continuum mechanics and continuum physics, with a focus on classical differential geometry applied towards kinematics of continua. Of key interest are anholonomic aspects arising from a multiplicative decomposition of the gradient of a certain vector-valued field (*i.e.*, the material gradient of the motion of points of a continuous body through space-time) into a product of two distinct two-point tensor fields, neither of which in isolation necessarily obeys the integrability conditions for the gradient of a smooth vector field.

As deemed appropriate for a research monograph that may be used for repeated reference and instruction, detailed step-by-step derivations of key results are included throughout. These include generic mathematical identities as well as their operation in the context of deformation of continua of a general nature, *e.g.*, derived quantities are not restricted to description of a particular class of solid or fluid bodies. In this regard, the style of presentation differs from that of conventional research articles that often omit line-by-line derivations and leave critical steps to either be obtained by the reader or accepted without proof. By working through and thus verifying the ample derivations in this text, researchers and advanced students will gain valuable practice in analytical tensor calculations of finite deformation kinematics in a classical geometric setting, applicable in generalized systems of curvilinear coordinates. Such fundamental practice is complementary and conducive to theoretical and applied research in such diverse topics as nonlinear elasticity of generic substances, finite plasticity of crystalline solids, mechanics of fluids and liquid crystals, and field theories of lattice defects.

The technical content can be divided into two broad subject areas: geometry and kinematics of holonomic deformation, and geometry and kinematics of anholonomic deformation. The former, dealing with holonomic spaces, is covered in Chapters 2 and 3. These chapters offer a new and comprehensive style of presentation of fundamental topics, with a scope of identities and derivations not given in complete form in any other known single publication. However, much of the content in Chapters 2 and 3 needed for completeness, comparison, and/or derivation of subsequent new results has been necessarily collected from important works by others. The latter broad subject area, dealing with anholonomic spaces, is addressed in Chapters 4 and 5. These two chapters contain a number of new derivations and results of tensor calculus not published elsewhere, in addition to thorough treatment of collected previously known ideas. The scope of these chapters is thought more comprehensive than previously available from any other known single work dealing with anholonomic geometry and anholonomic deformation kinematics.

The systematic format of this book emphasizes clarity and ease of reference, with minimal, if any, superfluous text. This work is self-contained and presumably will be a useful desktop reference and educational supplement for researchers and advanced graduate students in various disciplines of applied mathematics, theoretical and mathematical physics, condensed matter physics, engineering mechanics, and materials science. The reader is assumed to have background knowledge of tensor calculus and nonlinear continuum mechanics, the latter most important for identifying particular applications of the examples considered herein. Most of the text is directed towards general mathematical definitions and identities rather than specific problems peculiar to any particular branch of physics or engineering.

John D. Clayton
Aberdeen, Maryland, USA
2014

Contents

Chapter 1

Introduction

In this chapter, an overview of motivation, objectives, and contents of each chapter of this book is given. Historical remarks provide context for the contents of this book. The general scheme of notation used in later chapters is described. Unique features of this book are summarized.

1.1 Motivation, Objectives, and Scope

The main objective of this book is to provide a somewhat complete, yet still concise, collection of mathematical relationships associated with differential geometry in the context of deformable continua. Bodies whose points may undergo relative motion but do not physically separate or break, and thus can be described kinematically by smooth fields of continuum theory, are examples of such deformable continua. These mathematical relationships are of fundamental importance in historic and modern research activities in nonlinear continuum mechanics and continuum physics of solids, and analogous concepts from differential geometry and field theory exist among governing equations of various other branches of continuum physics such as fluid mechanics, biomechanics, electromagnetics, and gravitation.

As formally discussed in later chapters, the total deformation gradient $\boldsymbol{F}$ for an element of a material body is decomposed multiplicatively as $\boldsymbol{F} = \bar{\boldsymbol{F}}\tilde{\boldsymbol{F}}$, where $\tilde{\boldsymbol{F}}$ and $\bar{\boldsymbol{F}}$ are generic two-point mappings from the reference to intermediate configuration and intermediate to current configuration, respectively. In general curvilinear coordinates, $F^a_{.A} = \partial\varphi^a/\partial X^A$ is the tangent map from the reference to current configuration, where spatial coordinates $x^a = \varphi^a(X, t)$ follow the motion φ that may depend on time t and material particle X.

The aforementioned multiplicative decomposition is encountered frequently in mechanics and physics literature. More specifically, some version of such a decomposition is often considered the most physically realistic and mathematically sound ingredient of geometrically nonlinear kinematic and constitutive models of many kinds of materials, including various kinds of solid crystals (metals, ceramics, minerals, molecular crystals, *etc.*), polymers, and biological materials. Thorough accounts of applications of multiplicative decompositions in the context of nonlinear elastic, plastic, and/or anelastic crystalline solids can be found in several monographs [3,7,23,30]. Solutions to boundary value problems corresponding to discrete lattice defects—crystal dislocations, disclinations, and point defects—embedded in otherwise nonlinear elastic media have been derived [9,10,30,33,34]. Applications towards other classes of matter (*e.g.*, biological materials, viscous solids and fluids) have likewise been described geometrically [13,27]. Several other general theoretical and mathematically rigorous treatments of particular relevance to the present work are collected in [25]; see also [16,29] for review articles of differential geometric fundamentals in the context of continuum mechanics and continuum physics.

In the terminology adopted throughout this book, the deformation gradient $\boldsymbol{F}$ is said to be integrable or "holonomic" since the differentiable one-to-one mapping $\varphi(X,t)$ exists between referential and spatial positions of material particles at any given time t. Conversely, $\tilde{F}^{\alpha}_{.A}(X,t)$ and $\bar{F}^{-1\alpha}_{.a}(x,t)$ need not be integrable functions of X^A and x^a, respectively; in such cases, these mappings are said to be "anholonomic" [28]. When $\tilde{\boldsymbol{F}}$ ($\bar{\boldsymbol{F}}^{-1}$) is not integrable at a given time instant, coordinates $\tilde{x}^{\alpha}$ that are differentiable one-to-one functions of X^A (x^a) do not exist, and the resulting intermediate configuration is labeled an anholonomic space [8] or anholonomic reference [21].

The first half of this book deals with geometry and kinematics of holonomic deformation. Chapter 2 introduces requisite notation and definitions associated with holonomic coordinate systems and various kinds of coordinate differentiation such as covariant differentiation. Identities from classical differential geometry associated with torsion and curvature, for example, are given. Chapter 3 introduces the deformation gradient and other fundamental derived quantities and identities, and it includes specific descriptions and examples in Cartesian, cylindrical, spherical, and convected coordinate systems.

Certain material in Chapters 2 and 3—which is necessary for comparison with or derivation of new results in later chapters—has been collected from the vast respected literature on continuum mechanics and nonlinear elasticity of solids [15,22,24,31] and thus may already be familiar to many applied mathematicians and scholars of these disciplines. However, the organization, scope, stepwise derivations, and examples in these two chapters differ from those of any known earlier works and will thus presumably benefit the individual researcher (as a desktop reference), advanced student (as a practice guide for detailed tensor calculations), and instructor (as supplemental educational material). Furthermore, many of the example calculations expressed in indicial form in general curvilinear coordinates have not been presented elsewhere, to the author's knowledge.

The second half of this book describes geometry and kinematics of anholonomic deformation. Chapter 4 develops rules for differentiation and derives numerous differential-geometric identities associated with a multiplicative decomposition of the deformation gradient into two generally anholonomic mappings. Extensive treatments of integrability conditions and choices of possibly curvilinear anholonomic coordinate systems are included. Chapter 5 contains important results pertaining to derived kinematic quantities and integral theorems in the context of anholonomic deformation. The treatment in this book for mathematical operations in anholonomic space associated with a multiplicative decomposition of the total deformation gradient is thought to be more comprehensive than any given elsewhere, at least in the context of literature known to the author. Although certain classical geometric identities [16,28] are often used, much of Chapters 4 and 5 is new, with a number of identities, derivations, and example calculations not published elsewhere. The content of Chapters 4 and 5 of this book significantly extends a related study of anholonomic deformation, geometry, and differentiation published two years ago in a more brief research paper by the author [8].

A comprehensive list of symbols used throughout this book is provided in Appendix A, immediately after Chapter 5. The bibliography and subject index then follow.

1.2 Historical Remarks

The history of anholonomic geometry is reviewed concisely here. In the context of classical differential geometry, analysis of anholonomic spaces began with work of G. Vrânceanu [32] and É. Cartan [5] in the 1920s. The notion of moving, possibly non-coordinate (*i.e.*, non-holonomic), sets of director vectors or frames is also credited to Cartan [4]. With regard to construction of a unified field theory in physics, A. Einstein [12] considered geometric spaces with teleparallelism, meaning space-time described by a manifold equipped with a connection with vanishing curvature and non-vanishing torsion, the latter feature being related to anholonomicity. With regard to thermomechanical behavior of condensed matter, C. Eckart [11] formally conceived the notion of a locally relaxed configuration in anelasticity, corresponding geometrically to a non-Euclidean material manifold. Not long thereafter in the 1950s–1960s, fundamental linkages between anholonomicity associated with Cartan's torsion tensor of a certain affine connection and the presence of dislocations in crystalline solids were established by B. Bilby [1, 2], E. Kröner [19, 20], and K. Kondo [17, 18]. These works established the geometric foundations for numerous more recent applications of the theory such as those cited in Section 1.1.

The history of the study of holonomic geometry and kinematics of continuous bodies is much older and much larger in scope. A comprehensive bibliography is contained in the 1960 treatise of C. Truesdell and R. Toupin [31], wherein proper credit is given to historical works of renowned European scholars dating back several centuries. In particular, correspondence between satisfaction of the local compatibility equations, for example those appearing in nonlinear elasticity theory, and vanishing of the curvature tensor in Riemannian geometry has been known since the middle of the 19th century [31].

1.3 Notation

Notations of both classical and modern tensor analysis and differential geometry typically encountered in the context of continuum mechanics/physics are used. Direct (*i.e.*, coordinate-free) and indicial (*i.e.*, coordinate-based) notations are used interchangeably, though the latter are preferred here for clarity in many derivations involving classical differential geometry. Vectors and tensors of higher order are written in bold italic font, while scalars and components of vectors and higher-order tensors are written in normal italic font.

Unless noted otherwise, coordinate-based expressions are valid in general (*i.e.*, curvilinear) coordinate systems. When the index notation is used, Einstein's summation convention is invoked over repeated contravariant-covariant indices. The following conventions are followed for coordinate indices. Indices referred to a referential configuration are written in capital Roman font. Indices referred to a spatial configuration are written in lower case Roman font. Indices referred to an intermediate configuration are written in lower case Greek font. Accordingly, Roman fonts necessarily denote holonomic coordinates, whereas Greek fonts denote possibly anholonomic coordinates. For example, a vector $\boldsymbol{V}$ has components V^A in a reference coordinate system, components V^a in a spatial coordinate system, and components V^α in an intermediate coordinate system.

As will become clear later, most developments require some degree of smoothness of field quantities. Often, continuity and differentiability requirements are stated explicitly; otherwise, field variables are assumed to be "sufficiently smooth" to enable requisite mathematical operations (*e.g.*, partial coordinate differentiation or time differentiation) as they appear, without introduction of singularities. Let f denote a generic function or field quantity depending on independent variable X. Notation $f(X)$ is used interchangeably to denote either the field f (*i.e.*, defining f to be a function of X) or the value of f at a specific value of X, with the particular meaning implied by context.

1.4 Summary

Unique or key features of this work can be summarized succinctly as follows:

- Presentation of mathematical operations and examples in anholonomic space associated with a multiplicative decomposition (*e.g.*, of the gradient of motion) is considered both more general and more comprehensive than any given elsewhere and contains many original ideas and new results.
- Line-by-line derivations are frequent and exhaustive, to facilitate practice and enable verification of final results.
- Indicial and direct notations of tensor calculus enable connections with historic and modern literature, respectively.
- General analysis is given in generic curvilinear coordinates; particular sections deal with applications and examples in Cartesian, cylindrical, spherical, and convected coordinates.
- A comprehensive list of symbols and notation—with pointers to equations wherein symbols first appear—is included at the end of the text.

Chapter 2

Geometric Fundamentals

In this chapter, fundamental definitions and derivations of differential geometry and tensor analysis are provided. The style of presentation most often follows that of classical differential geometry as applied in various branches of finite deformation continuum mechanics and physics.

2.1 Configurations, Coordinates, and Metrics

Described consecutively in what follows are motion, coordinates, metric tensors, shifter tensors, and associated mathematical operations. Kronecker delta symbols and permutation symbols are also introduced.

2.1.1 *Motion and coordinates*

A material point or particle in reference configuration B_0 is labeled X. The corresponding point in the current or spatial configuration B is labeled x. Time is denoted by t. The motion and its inverse, respectively, are described at any given time by the one-to-one mappings

$$x^a = \varphi^a(X,t) = x^a(X,t), \qquad X^A = \Phi^A(x,t) = X^A(x,t). \tag{2.1}$$

Unless noted otherwise, spatial coordinates x^a and reference coordinates X^A are assumed sufficiently differentiable functions of their arguments in subsequent derivations. In particular, later derivations require that x^a be at least three times differentiable with respect to X^A and at least twice differentiable with respect to t. Such coordinates are assumed to fully cover simply connected, open region(s) of a material body embedded in corresponding configurations. Partial differentiation with respect to coordinates is written alternatively as follows:

$$\frac{\partial(\cdot)}{\partial X^A} = \partial_A(\cdot) = (\cdot)_{,A}, \qquad \frac{\partial(\cdot)}{\partial x^a} = \partial_a(\cdot) = (\cdot)_{,a}. \tag{2.2}$$

The following identities are used frequently:

$$\partial_A[\partial_B(\cdot)] = \partial_B[\partial_A(\cdot)], \qquad \partial_a[\partial_b(\cdot)] = \partial_b[\partial_a(\cdot)]. \tag{2.3}$$

Let $\boldsymbol{X} \in B_0$ and $\boldsymbol{x} \in B$ denote position vectors in Euclidean space. Referential and spatial basis vectors are then written, respectively, as

$$\boldsymbol{G}_A(X) = \partial_A \boldsymbol{X}, \qquad \boldsymbol{g}_a(x) = \partial_a \boldsymbol{x}. \tag{2.4}$$

Reciprocal basis vectors are $\boldsymbol{G}^A(X)$ and $\boldsymbol{g}^a(x)$. Scalar products of basis vectors and their reciprocals are

$$\langle \boldsymbol{G}^A, \boldsymbol{G}_B \rangle = \delta^A_B, \qquad \langle \boldsymbol{g}^a, \boldsymbol{g}_b \rangle = \delta^a_b. \tag{2.5}$$

Kronecker delta symbols obey

$$\begin{aligned} &\delta^A_B = 1 \quad \forall A = B, \qquad &\delta^A_B = 0 \quad \forall A \neq B; \\ &\delta^a_b = 1 \quad \forall a = b, \qquad &\delta^a_b = 0 \quad \forall a \neq b. \end{aligned} \tag{2.6}$$

Einstein's summation convention applies throughout this text, whereby summation is enforced over duplicate contra-covariant indices. Applying this convention to Kronecker delta symbols in three dimensions,

$$\delta^A_A = \delta^a_a = \delta^1_1 + \delta^2_2 + \delta^3_3 = 3. \tag{2.7}$$

Third-order permutation symbols are now introduced, and will be used frequently later:

$$\begin{aligned} e^{ABC} &= e_{ABC} = 0 \quad &&\text{when any two indices are equal,} \\ e^{ABC} &= e_{ABC} = +1 \quad &&\text{for } ABC = 123, 231, 312, \\ e^{ABC} &= e_{ABC} = -1 \quad &&\text{for } ABC = 132, 213, 321; \end{aligned} \tag{2.8}$$

$$\begin{aligned} e^{abc} &= e_{abc} = 0 \quad &&\text{when any two indices are equal,} \\ e^{abc} &= e_{abc} = +1 \quad &&\text{for } abc = 123, 231, 312, \\ e^{abc} &= e_{abc} = -1 \quad &&\text{for } abc = 132, 213, 321. \end{aligned} \tag{2.9}$$

Definitions and identities involving third-order permutation symbols implicitly assume a three-dimensional ambient space. Identities for permutation symbols are listed below:

$$\begin{aligned} e^{ABC}e_{DEF} = \det \begin{bmatrix} \delta^A_D & \delta^A_E & \delta^A_F \\ \delta^B_D & \delta^B_E & \delta^B_F \\ \delta^C_D & \delta^C_E & \delta^C_F \end{bmatrix}, \qquad e^{ABC}e_{ADE} = \delta^B_D\delta^C_E - \delta^B_E\delta^C_D; \\ e^{ABC}e_{ABE} = 2\delta^C_E, \qquad e^{ABC}e_{ABC} = 2\delta^A_A = 6. \end{aligned} \tag{2.10}$$

Analogous identities apply for symbols with spatial (lower-case) indices. Also,

$$e^{(AB)C} = e^{A(BC)} = e^{(A|B|C)} = e^{(ABC)} = 0,$$
$$e^{[AB]C} = e^{A[BC]} = e^{[A|B|C]} = e^{[ABC]} = e^{ABC}, \tag{2.11}$$

where indices in parentheses are symmetric, indices in square brackets are anti-symmetric, and indices between vertical bars are excluded from (anti)symmetry operations, as indicated below for arbitrary second- and third-order quantities:

$$A_{(AB)} = \frac{1}{2}(A_{AB} + A_{BA}), \tag{2.12}$$

$$A_{[AB]} = \frac{1}{2}(A_{AB} - A_{BA}); \tag{2.13}$$

$$A_{(ABC)} = \frac{1}{6}(A_{ABC} + A_{ACB} + A_{BCA} + A_{BAC} + A_{CAB} + A_{CBA}), \tag{2.14}$$

$$A_{[ABC]} = \frac{1}{6}(A_{ABC} - A_{ACB} + A_{BCA} - A_{BAC} + A_{CAB} - A_{CBA}), \tag{2.15}$$

$$A_{(A|B|C)} = \frac{1}{2}(A_{ABC} + A_{CBA}), \quad A_{[A|B|C]} = \frac{1}{2}(A_{ABC} - A_{CBA}). \tag{2.16}$$

From (2.3) and (2.13),

$$\partial_{[A}[\partial_{B]}(\cdot)] = 0, \qquad \partial_{[a}[\partial_{b]}(\cdot)] = 0. \tag{2.17}$$

Using (2.5), the scalar product of a vector $\boldsymbol{V} = V^A \boldsymbol{G}_A$ and a covector $\boldsymbol{\alpha} = \alpha_B \boldsymbol{G}^B$ is

$$\langle \boldsymbol{V}, \boldsymbol{\alpha} \rangle = V^A \alpha_B \langle \boldsymbol{G}_A, \boldsymbol{G}^B \rangle = V^A \alpha_B \delta_A^B = V^A \alpha_A. \tag{2.18}$$

The tensor product or outer product $\otimes$ obeys

$$(\boldsymbol{G}^A \otimes \boldsymbol{G}^B)\boldsymbol{G}_C = \boldsymbol{G}^A \langle \boldsymbol{G}^B, \boldsymbol{G}_C \rangle. \tag{2.19}$$

In modern treatments of differential geometry, the following perhaps more intuitive notation is often introduced for basis vectors and their reciprocals:

$$\boldsymbol{G}_A = \frac{\boldsymbol{\partial}}{\boldsymbol{\partial} X^A} = \boldsymbol{\partial}_A, \qquad \boldsymbol{G}^A = \boldsymbol{d}X^A. \tag{2.20}$$

The above notation is used sparingly in the present work. Often in mathematical literature, bold font is omitted for $\boldsymbol{\partial}_A$ and $\boldsymbol{d}X^A$, but it is retained herein for clarity. With this notation, (2.5) becomes

$$\langle \boldsymbol{\partial}_B, \boldsymbol{d}X^A \rangle = \delta_B^A. \tag{2.21}$$

Analogous expressions can be written for spatial quantities.

2.1.2 Metric tensors

Symmetric metric tensors $\boldsymbol{G}(X)$ and $\boldsymbol{g}(x)$ are introduced for respective configurations B_0 and B:

$$\boldsymbol{G}(X) = G_{AB}\boldsymbol{G}^A \otimes \boldsymbol{G}^B = (\boldsymbol{G}_A \cdot \boldsymbol{G}_B)\boldsymbol{G}^A \otimes \boldsymbol{G}^B, \quad G_{AB} = G_{(AB)}; \tag{2.22}$$

$$\boldsymbol{g}(x) = g_{ab}\boldsymbol{g}^a \otimes \boldsymbol{g}^b = (\boldsymbol{g}_a \cdot \boldsymbol{g}_b)\boldsymbol{g}^a \otimes \boldsymbol{g}^b, \quad g_{ab} = g_{(ab)}; \tag{2.23}$$

where the dot product of vectors $\boldsymbol{V}$ and $\boldsymbol{W}$ is

$$\begin{aligned}\boldsymbol{V} \cdot \boldsymbol{W} &= V^A\boldsymbol{G}_A \cdot W^B\boldsymbol{G}_B \\ &= V^A W^B(\boldsymbol{G}_A \cdot \boldsymbol{G}_B) \\ &= V^A G_{AB} W^B \\ &= V^A W_A \\ &= W^A V_A \\ &= \boldsymbol{W} \cdot \boldsymbol{V}.\end{aligned} \tag{2.24}$$

As indicated, metric tensors can be used to lower contravariant indices:

$$V_A = V^B G_{AB}, \qquad \boldsymbol{G}_A = G_{AB}\boldsymbol{G}^B. \tag{2.25}$$

Inverses of metric tensors are

$$\boldsymbol{G}^{-1} = G^{AB}\boldsymbol{G}_A \otimes \boldsymbol{G}_B = (\boldsymbol{G}^A \cdot \boldsymbol{G}^B)\boldsymbol{G}_A \otimes \boldsymbol{G}_B, \quad G^{AB} = G^{(AB)}; \tag{2.26}$$

$$\boldsymbol{g}^{-1} = g^{ab}\boldsymbol{g}_a \otimes \boldsymbol{g}_b = (\boldsymbol{g}^a \cdot \boldsymbol{g}^b)\boldsymbol{g}_a \otimes \boldsymbol{g}_b, \quad g^{ab} = g^{(ab)}; \tag{2.27}$$

where the dot product of covectors $\boldsymbol{\alpha}$ and $\boldsymbol{\beta}$ is

$$\begin{aligned}\boldsymbol{\alpha} \cdot \boldsymbol{\beta} &= \alpha_A\boldsymbol{G}^A \cdot \beta_B\boldsymbol{G}^B \\ &= \alpha_A\beta_B(\boldsymbol{G}^A \cdot \boldsymbol{G}^B) \\ &= \alpha_A G^{AB}\beta_B \\ &= \alpha^A\beta_A \\ &= \beta^A\alpha_A \\ &= \boldsymbol{\beta} \cdot \boldsymbol{\alpha}.\end{aligned} \tag{2.28}$$

As indicated, inverse metric tensors can be used to raise covariant indices:

$$V^A = V_B G^{AB}, \qquad \boldsymbol{G}^A = G^{AB}\boldsymbol{G}_B. \tag{2.29}$$

Furthermore, from the definition of the inverse operation,

$$G^{AC}G_{CB} = \delta^A_B, \qquad g^{ac}g_{cb} = \delta^a_b. \tag{2.30}$$

Determinants of metric tensors and their inverses are written as

$$G = \det \boldsymbol{G} = \det (G_{AB}), \tag{2.31}$$

$$g = \det \boldsymbol{g} = \det (g_{ab}); \tag{2.32}$$

$$G^{-1} = \det \boldsymbol{G}^{-1} = \det (G^{AB}) = 1/G, \tag{2.33}$$

$$g^{-1} = \det \boldsymbol{g}^{-1} = \det (g^{ab}) = 1/g. \tag{2.34}$$

In terms of permutation symbols, determinants obey

$$G = \frac{1}{6} e^{ABC} e^{DEF} G_{AD} G_{BE} G_{CF}, \tag{2.35}$$

$$g = \frac{1}{6} e^{abc} e^{def} g_{ad} g_{be} g_{cf}; \tag{2.36}$$

$$G^{-1} = \frac{1}{6} e_{ABC} e_{DEF} G^{AD} G^{BE} G^{CF}, \tag{2.37}$$

$$g^{-1} = \frac{1}{6} e_{abc} e_{def} g^{ad} g^{be} g^{cf}. \tag{2.38}$$

Since $G_{AB} = G_{BA}$, the determinant of $\boldsymbol{G}$ is, explicitly,

$$\begin{aligned} G &= \det(G_{AB}) \\ &= \det(\partial_A \boldsymbol{X} \cdot \partial_B \boldsymbol{X}) \\ &= \det \begin{bmatrix} G_{11} & G_{12} & G_{13} \\ G_{21} & G_{22} & G_{23} \\ G_{31} & G_{32} & G_{33} \end{bmatrix} \\ &= \det \begin{bmatrix} G_{11} & G_{12} & G_{13} \\ G_{12} & G_{22} & G_{23} \\ G_{13} & G_{23} & G_{33} \end{bmatrix} \\ &= G_{11}(G_{22}G_{33} - G_{32}G_{23}) \\ &\quad - G_{12}(G_{21}G_{33} - G_{31}G_{23}) \\ &\quad + G_{13}(G_{21}G_{32} - G_{31}G_{22}) \\ &= G_{11}G_{22}G_{33} + G_{12}G_{23}G_{31} + G_{13}G_{21}G_{32} \\ &\quad - G_{13}G_{22}G_{31} - G_{12}G_{21}G_{33} - G_{11}G_{23}G_{32} \\ &= G_{11}G_{22}G_{33} + 2G_{12}G_{13}G_{23} \\ &\quad - G_{11}(G_{23})^2 - G_{22}(G_{13})^2 - G_{33}(G_{12})^2; \end{aligned} \tag{2.39}$$

similar equations apply for determinants of $\boldsymbol{G}^{-1}$, $\boldsymbol{g}$, and $\boldsymbol{g}^{-1}$.

Permutation tensors are defined as

$$\epsilon^{ABC} = \frac{1}{\sqrt{G}}\, e^{ABC}, \quad \epsilon_{ABC} = \sqrt{G}\, e_{ABC}; \tag{2.40}$$

$$\epsilon^{abc} = \frac{1}{\sqrt{g}}\, e^{abc}, \quad \epsilon_{abc} = \sqrt{g}\, e_{abc}. \tag{2.41}$$

Permutation tensors obey the same identities as permutation symbols in (2.10) and (2.11). Herein, metric tensors are assumed positive definite over any volume [15, 22] such that

$$G(X) > 0, \quad G^{-1}(X) > 0, \quad g(x) > 0, \quad g^{-1}(x) > 0. \tag{2.42}$$

In certain coordinate systems, determinants of metric tensors or their inverses may be zero or undefined at certain points, lines, or surfaces; for example, at points or lines along null radial coordinates in spherical or cylindrical systems. Throughout this book it is assumed that coordinate systems are positively oriented; *e.g.*, a right-handed Cartesian frame is positively oriented, while a left-handed Cartesian frame is not.

Permutation tensors are used to determine cross products. The vector cross product $\times$ obeys, for two vectors $\boldsymbol{V}$ and $\boldsymbol{W}$,

$$\begin{aligned} \boldsymbol{V} \times \boldsymbol{W} &= \epsilon_{ABC} V^B W^C \boldsymbol{G}^A \\ &= \epsilon_{ABC} V^A W^B \boldsymbol{G}^C \\ &= \epsilon_{ABC} V^C W^A \boldsymbol{G}^B, \end{aligned} \tag{2.43}$$

and for two covectors $\boldsymbol{\alpha}$ and $\boldsymbol{\beta}$,

$$\begin{aligned} \boldsymbol{\alpha} \times \boldsymbol{\beta} &= \epsilon^{ABC} \alpha_B \beta_C \boldsymbol{G}_A \\ &= \epsilon^{ABC} \alpha_A \beta_B \boldsymbol{G}_C \\ &= \epsilon^{ABC} \alpha_C \beta_A \boldsymbol{G}_B. \end{aligned} \tag{2.44}$$

The following identities relate basis vectors and permutation tensors:

$$\begin{aligned} \boldsymbol{G}_A \cdot (\boldsymbol{G}_B \times \boldsymbol{G}_C) &= \boldsymbol{G}_A \cdot (\epsilon_{BCD} \boldsymbol{G}^D) \\ &= \epsilon_{BCD} \langle \boldsymbol{G}_A, \boldsymbol{G}^D \rangle \\ &= \epsilon_{BCD} \delta^D_A \\ &= \epsilon_{BCA} \\ &= \epsilon_{CAB} \\ &= \epsilon_{ABC}, \end{aligned} \tag{2.45}$$

and analogously for reciprocal basis vectors,

$$\begin{aligned}\boldsymbol{G}^A\cdot(\boldsymbol{G}^B\times\boldsymbol{G}^C)&=\boldsymbol{G}^A\cdot(\epsilon^{BCD}\boldsymbol{G}_D)\\&=\epsilon^{BCD}\langle\boldsymbol{G}^A,\boldsymbol{G}_D\rangle\\&=\epsilon^{BCD}\delta^A_D\\&=\epsilon^{BCA}\\&=\epsilon^{CAB}\\&=\epsilon^{ABC}.\end{aligned}\tag{2.46}$$

2.1.3 *Shifter tensors*

Shifter tensors, which are examples of two-point tensors, can be introduced in Euclidean space [15]:

$$g^a_A(x,X)=\langle\boldsymbol{g}^a,\boldsymbol{G}_A\rangle,\qquad g^A_a(x,X)=\langle\boldsymbol{g}_a,\boldsymbol{G}^A\rangle;\tag{2.47}$$

$$g^{aA}(x,X)=\boldsymbol{g}^a\cdot\boldsymbol{G}^A=g^{Aa},\qquad g_{aA}(x,X)=\boldsymbol{g}_a\cdot\boldsymbol{G}_A=g_{Aa};\tag{2.48}$$

$$g^a_Bg^B_b=\delta^a_b,\qquad g^A_bg^b_B=\delta^A_B;\tag{2.49}$$

$$g_{aA}=g_{ab}g^b_A=G_{AB}g^B_a=g_{ab}G_{AB}g^{Bb}.\tag{2.50}$$

Noting that

$$\det(g^{aA})=\det(g^a_BG^{AB})=G^{-1}\det(g^a_B),\tag{2.51}$$

$$\det(g_{aA})=\det(g^b_Ag_{ab})=g\det(g^b_A),\tag{2.52}$$

it follows from the product rule for determinants that

$$1=\det(\delta^a_b)=\det(g^{aA}g_{bA})=\frac{g}{G}[\det(g^a_A)]^2,\quad 1=\det(g^a_A)\det(g^A_b).\tag{2.53}$$

Therefore, determinants of shifter tensors obey

$$\det(g^A_a)=1/\det(g^a_A)=\sqrt{\det(g_{ab})/\det(G_{AB})}=\sqrt{g/G}.\tag{2.54}$$

Vector $\boldsymbol{V}$ and covector $\boldsymbol{\alpha}$ can be written as follows in terms of basis vectors in either configuration:

$$\boldsymbol{V}=V^A\boldsymbol{G}_A=(V^ag^A_a)(g^b_A\boldsymbol{g}_b)=V^a\delta^b_a\boldsymbol{g}_b=V^a\boldsymbol{g}_a;\tag{2.55}$$

$$\boldsymbol{\alpha}=\alpha_A\boldsymbol{G}^A=(\alpha_ag^a_A)(g^A_b\boldsymbol{g}^b)=\alpha_a\delta^a_b\boldsymbol{g}^b=\alpha_a\boldsymbol{g}^a.\tag{2.56}$$

The following rules for shifting of basis vectors are implied:

$$\boldsymbol{g}_a=g^A_a\boldsymbol{G}_A,\quad\boldsymbol{g}^a=g^a_A\boldsymbol{G}^A;\qquad\boldsymbol{G}_A=g^a_A\boldsymbol{g}_a,\quad\boldsymbol{G}^A=g^A_a\boldsymbol{g}^a.\tag{2.57}$$

2.2 Linear Connections

Discussed in what follows are connection coefficients and covariant differentiation, torsion and curvature tensors, Riemannian geometry, and features of Euclidean space.

2.2.1 *Connection coefficients and covariant derivatives*

Associated with a linear connection in configuration B_0 is the covariant derivative operator ∇. The covariant derivative assigns to two differentiable vector fields $\boldsymbol{V}$, $\boldsymbol{W}$ a third vector field $\nabla_{\boldsymbol{V}}\boldsymbol{W}$, called the covariant derivative of $\boldsymbol{W}$ along $\boldsymbol{V}$, that obeys the following conditions [24, 28]:

$$\begin{aligned}&\text{(i) } \nabla_{\boldsymbol{V}}\boldsymbol{W} \text{ is linear in both } \boldsymbol{V} \text{ and } \boldsymbol{W};\\ &\text{(ii) } \nabla_{f\boldsymbol{V}}\boldsymbol{W} = f\nabla_{\boldsymbol{V}}\boldsymbol{W} \text{ for scalar function } f;\\ &\text{(iii) } \nabla_{\boldsymbol{V}}(f\boldsymbol{W}) = f\nabla_{\boldsymbol{V}}\boldsymbol{W} + \langle \boldsymbol{V}, \boldsymbol{D}f\rangle\,\boldsymbol{W}.\end{aligned} \tag{2.58}$$

In reference coordinates X^A, the derivative of differentiable function f in the direction of $\boldsymbol{V}$ is $\langle \boldsymbol{V}, \boldsymbol{D}f\rangle = V^A\partial_A f$. In reference coordinates, the covariant derivative of $\boldsymbol{W}$ along $\boldsymbol{V}$ is

$$\nabla_{\boldsymbol{V}}\boldsymbol{W} = (V^B\partial_B W^A + \Gamma^{\cdot\cdot A}_{BC}W^C V^B)\boldsymbol{G}_A. \tag{2.59}$$

In n-dimensional space, n^3 entries of the object $\Gamma^{\cdot\cdot A}_{BC}$ are called connection coefficients. Let $X^A(X)$ and $\hat{X}^{\hat{A}}(X)$ denote coordinate systems obeying

$$\frac{\partial\hat{X}^{\hat{A}}}{\partial X^B}\frac{\partial X^B}{\partial\hat{X}^{\hat{C}}} = (\partial_B\hat{X}^{\hat{A}})(\partial_{\hat{C}}\partial X^B) = \delta^{\hat{A}}_{\hat{C}}, \tag{2.60}$$

$$\partial_{\hat{B}}\left(\frac{\partial\hat{X}^{\hat{A}}}{\partial X^B}\frac{\partial X^B}{\partial\hat{X}^{\hat{C}}}\right) = \partial_{\hat{B}}(\delta^{\hat{A}}_{\hat{C}}) = 0. \tag{2.61}$$

Connection coefficients transform between these two coordinate systems as [7, 28]

$$\begin{aligned}\hat{\Gamma}^{\cdot\cdot\hat{A}}_{\hat{B}\hat{C}} &= \partial_{\hat{B}}X^B\,\partial_{\hat{C}}X^C\,\partial_A\hat{X}^{\hat{A}}\,\Gamma^{\cdot\cdot A}_{BC} + \partial_C\hat{X}^{\hat{A}}\,\partial_{\hat{B}}(\partial_{\hat{C}}X^C)\\ &= \partial_{\hat{B}}X^B\,\partial_{\hat{C}}X^C\left[\partial_A\hat{X}^{\hat{A}}\,\Gamma^{\cdot\cdot A}_{BC} - \partial_B(\partial_C\hat{X}^{\hat{A}})\right],\end{aligned} \tag{2.62}$$

where the second equality follows from (2.60) and (2.61).

The covariant derivative of a differentiable vector field $\boldsymbol{V}$ is the second-order tensor field

$$\nabla\boldsymbol{V} = \nabla_B V^A\boldsymbol{G}_A\otimes\boldsymbol{G}^B = (\partial_B V^A + \Gamma^{\cdot\cdot A}_{BC}V^C)\boldsymbol{G}_A\otimes\boldsymbol{G}^B. \tag{2.63}$$

Let $\boldsymbol{A}(X)$ be a vector or higher-order tensor field. The covariant derivative is applied to components of vectors and tensors of higher order as [7, 28]

$$\begin{aligned}\nabla_N A^{A\ldots F}_{G\ldots M} = \partial_N A^{A\ldots F}_{G\ldots M} \\ + \Gamma^{\cdot\cdot A}_{NR} A^{R\ldots F}_{G\ldots M} + \cdots + \Gamma^{\cdot\cdot F}_{NR} A^{A\ldots R}_{G\ldots M} \\ - \Gamma^{\cdot\cdot R}_{NG} A^{A\ldots F}_{R\ldots M} - \cdots - \Gamma^{\cdot\cdot R}_{NM} A^{A\ldots F}_{G\ldots R}.\end{aligned} \quad (2.64)$$

Indices of covariant differentition on the left correspond to those of partial differentiation and the first covariant index of the connection coefficients on the right.

Additional identities following from (2.58) and (2.64) are

$$\nabla_N(V^A + W^A) = \nabla_N V^A + \nabla_N W^A, \quad (2.65)$$

$$\nabla_N(V^A W^B) = V^A \nabla_N W^B + W^B \nabla_N V^A, \quad (2.66)$$

$$\nabla_N \delta^A_B = \partial_N \delta^A_B + \Gamma^{\cdot\cdot A}_{NC}\delta^C_B - \Gamma^{\cdot\cdot C}_{NB}\delta^A_C = \partial_N \delta^A_B + \Gamma^{\cdot\cdot A}_{NB} - \Gamma^{\cdot\cdot A}_{NB} = 0. \quad (2.67)$$

The covariant derivative of a scalar field $A(X)$ is equivalent to its partial derivative:

$$\nabla_N A = \partial_N A. \quad (2.68)$$

A linear connection can alternatively be called an affine connection.

Using (2.58) and modern notation of (2.20), connection coefficients associated with ∇ can be defined as the following gradients of basis vectors:

$$\Gamma^{\cdot\cdot A}_{BC}\boldsymbol{\partial}_A = \nabla_{\boldsymbol{\partial}_B}\boldsymbol{\partial}_C. \quad (2.69)$$

Application of this definition to (2.21) leads to

$$\begin{aligned}0 &= \nabla_{\boldsymbol{\partial}_A}\delta^B_C \\ &= \nabla_{\boldsymbol{\partial}_A}\langle\boldsymbol{\partial}_C, \boldsymbol{d}X^B\rangle \\ &= \langle\nabla_{\boldsymbol{\partial}_A}\boldsymbol{\partial}_C, \boldsymbol{d}X^B\rangle + \langle\boldsymbol{\partial}_C, \nabla_{\boldsymbol{\partial}_A}\boldsymbol{d}X^B\rangle \\ &= \Gamma^{\cdot\cdot D}_{AC}\langle\boldsymbol{\partial}_D, \boldsymbol{d}X^B\rangle + \langle\boldsymbol{\partial}_C, \nabla_{\boldsymbol{\partial}_A}\boldsymbol{d}X^B\rangle \\ &= \Gamma^{\cdot\cdot B}_{AC} + \langle\boldsymbol{\partial}_C, \nabla_{\boldsymbol{\partial}_A}\boldsymbol{d}X^B\rangle,\end{aligned} \quad (2.70)$$

such that gradients of reciprocal basis vectors obey

$$\nabla_{\boldsymbol{\partial}_A}\boldsymbol{d}X^B = -\Gamma^{\cdot\cdot B}_{AC}\boldsymbol{d}X^C. \quad (2.71)$$

The covariant derivative of a differentiable vector field $\boldsymbol{V}$ can then be computed as

$$\begin{aligned}\nabla\boldsymbol{V} &= \nabla_{\boldsymbol{\partial}_B}\boldsymbol{V}\otimes\boldsymbol{d}X^B \\ &= \nabla_{\boldsymbol{\partial}_B}(V^A\boldsymbol{\partial}_A)\otimes\boldsymbol{d}X^B \\ &= \nabla_{\boldsymbol{\partial}_B}(V^A)\boldsymbol{\partial}_A\otimes\boldsymbol{d}X^B + V^A\nabla_{\boldsymbol{\partial}_B}(\boldsymbol{\partial}_A)\otimes\boldsymbol{d}X^B \\ &= \partial_B V^A\boldsymbol{\partial}_A\otimes\boldsymbol{d}X^B + V^A\Gamma^{\cdot\cdot C}_{BA}\boldsymbol{\partial}_C\otimes\boldsymbol{d}X^B \\ &= (\partial_B V^A + V^C\Gamma^{\cdot\cdot A}_{BC})\boldsymbol{\partial}_A\otimes\boldsymbol{d}X^B,\end{aligned} \quad (2.72)$$

where the notation $\nabla_{\boldsymbol{\partial}_A}(\cdot) = \partial_A(\cdot)$ when the argument is a scalar or scalar component of a vector or tensor. End results in (2.63) and (2.72) are fully consistent.

All definitions and identities in Section 2.2.1 can be applied analogously to current configuration B with spatial coordinates x^a.

2.2.2 *Torsion and curvature*

The torsion tensor of an arbitrary affine connection with coefficients $\Gamma_{BC}^{\cdot\cdot A}$ is defined as

$$\boldsymbol{T} = T_{BC}^{\cdot\cdot A} \boldsymbol{G}^B \otimes \boldsymbol{G}^C \otimes \boldsymbol{G}_A = \Gamma_{[BC]}^{\cdot\cdot A} \boldsymbol{G}^B \otimes \boldsymbol{G}^C \otimes \boldsymbol{G}_A. \tag{2.73}$$

Components of the torsion tensor are the anti-symmetric covariant components of its corresponding connection:

$$T_{BC}^{\cdot\cdot A} = \Gamma_{[BC]}^{\cdot\cdot A}. \tag{2.74}$$

Components of the Riemann-Christoffel curvature tensor of this arbitrary connection are

$$\begin{aligned} R_{BCD}^{\cdot\cdot\cdot A} &= \frac{\partial \Gamma_{CD}^{\cdot\cdot A}}{\partial X^B} - \frac{\partial \Gamma_{BD}^{\cdot\cdot A}}{\partial X^C} + \Gamma_{BE}^{\cdot\cdot A}\Gamma_{CD}^{\cdot\cdot E} - \Gamma_{CE}^{\cdot\cdot A}\Gamma_{BD}^{\cdot\cdot E} \\ &= 2\partial_{[B}\Gamma_{C]D}^{\cdot\cdot A} + 2\Gamma_{[B|E|}^{\cdot\cdot A}\Gamma_{C]D}^{\cdot\cdot E}. \end{aligned} \tag{2.75}$$

By definition, $R_{BCD}^{\cdot\cdot\cdot A} = R_{[BC]D}^{\cdot\cdot\cdot A}$. Both the torsion and curvature transform under a conventional change of basis as true tensors. Different definitions are widespread in the literature for torsion and curvature; those listed here follow [7, 28]. Skew second covariant derivatives of a contravariant vector $\boldsymbol{V}$ and a covector $\boldsymbol{\alpha}$ can be expressed as [28]

$$\begin{aligned} \nabla_{[B}\nabla_{C]}V^A &= \nabla_{[B}(\partial_{C]}V^A + \Gamma_{C]D}^{\cdot\cdot A}V^D) \\ &= \partial_{[B}(\partial_{C]}V^A) + \partial_{[B}\Gamma_{C]D}^{\cdot\cdot A}V^D + \Gamma_{[B|D|}^{\cdot\cdot A}\partial_{C]}V^D \\ &\quad + \Gamma_{[C|D|}^{\cdot\cdot A}\partial_{B]}V^D + \Gamma_{[B|D|}^{\cdot\cdot A}\Gamma_{C]E}^{\cdot\cdot D}V^E \\ &\quad - \Gamma_{[BC]}^{\cdot\cdot D}\partial_D V^A - \Gamma_{[BC]}^{\cdot\cdot E}\Gamma_{ED}^{\cdot\cdot A}V^D \\ &= \partial_{[B}\Gamma_{C]D}^{\cdot\cdot A}V^D + \Gamma_{[B|D|}^{\cdot\cdot A}\Gamma_{C]E}^{\cdot\cdot D}V^E \\ &\quad - \Gamma_{[BC]}^{\cdot\cdot D}\partial_D V^A - \Gamma_{[BC]}^{\cdot\cdot D}\Gamma_{DE}^{\cdot\cdot A}V^E \\ &= \frac{1}{2}R_{BCD}^{\cdot\cdot\cdot A}V^D - T_{BC}^{\cdot\cdot D}\nabla_D V^A, \end{aligned} \tag{2.76}$$

$$\begin{aligned}
\nabla_{[B}\nabla_{C]}\alpha_D &= \nabla_{[B}(\partial_{C]}\alpha_D - \Gamma^{\cdot\cdot A}_{C]D}\alpha_A) \\
&= \partial_{[B}(\partial_{C]}\alpha_D) - \partial_{[B}\Gamma^{\cdot\cdot A}_{C]D}\alpha_A - \Gamma^{\cdot\cdot A}_{[B|D|}\partial_{C]}\alpha_A \\
&\quad - \Gamma^{\cdot\cdot A}_{[C|D|}\partial_{B]}\alpha_A - \Gamma^{\cdot\cdot A}_{[BC]}\partial_A\alpha_D \\
&\quad + \Gamma^{\cdot\cdot A}_{[BC]}\Gamma^{\cdot\cdot E}_{AD}\alpha_E + \Gamma^{\cdot\cdot A}_{[B|D|}\Gamma^{\cdot\cdot E}_{C]A}\alpha_E \\
&= -\partial_{[B}\Gamma^{\cdot\cdot A}_{C]D}\alpha_A - \Gamma^{\cdot\cdot A}_{[B|E|}\Gamma^{\cdot\cdot E}_{C]D}\alpha_A \\
&\quad - (\Gamma^{\cdot\cdot A}_{[BC]}\partial_A\alpha_D - \Gamma^{\cdot\cdot A}_{[BC]}\Gamma^{\cdot\cdot E}_{AD}\alpha_E) \\
&= -\frac{1}{2}R^{\cdot\cdot\cdot A}_{BCD}\alpha_A - T^{\cdot\cdot A}_{BC}\nabla_A\alpha_D. \qquad (2.77)
\end{aligned}$$

Skew second covariant derivatives of higher-order tensors also involve torsion and curvature; *e.g.*, for a mixed second-order tensor $\boldsymbol{A} = A^{A}_{\cdot B}\boldsymbol{G}_A \otimes \boldsymbol{G}^B$ [16],

$$\nabla_{[B}\nabla_{C]}A^{A}_{\cdot D} = \frac{1}{2}R^{\cdot\cdot\cdot A}_{BCE}A^{E}_{\cdot D} - \frac{1}{2}R^{\cdot\cdot\cdot E}_{BCD}A^{A}_{\cdot E} - T^{\cdot\cdot E}_{BC}\nabla_E A^{A}_{\cdot D}. \qquad (2.78)$$

The following additional definitions apply for the curvature:

$$R_{ABCD} = R^{\cdot\cdot\cdot E}_{ABC}G_{DE} \qquad \text{(covariant curvature);} \quad (2.79)$$

$$R_{BC} = R^{\cdot\cdot\cdot A}_{ABC} \qquad \text{(Ricci curvature);} \quad (2.80)$$

$$\theta^{AB} = \frac{1}{4}\epsilon^{ACD}\epsilon^{BEF}R_{CDEF} \qquad \text{(Einstein tensor);} \quad (2.81)$$

$$\kappa = \frac{1}{n(n-1)}R_{AB}G^{AB} = \frac{1}{n(n-1)}R \ \text{(scalar curvature).} \quad (2.82)$$

Recall that n is the dimension of the space. Furthermore, contraction over the final two indices of the curvature tensor can be expressed simply as

$$\begin{aligned}
R^{\cdot\cdot\cdot A}_{BCA} &= 2\partial_{[B}\Gamma^{\cdot\cdot A}_{C]A} + 2\Gamma^{\cdot\cdot A}_{[B|E|}\Gamma^{\cdot\cdot E}_{C]A} \\
&= 2\partial_{[B}\Gamma^{\cdot\cdot A}_{C]A} + \Gamma^{\cdot\cdot A}_{BE}\Gamma^{\cdot\cdot E}_{CA} - \Gamma^{\cdot\cdot A}_{CE}\Gamma^{\cdot\cdot E}_{BA} \\
&= 2\partial_{[B}\Gamma^{\cdot\cdot A}_{C]A} + \Gamma^{\cdot\cdot A}_{BE}\Gamma^{\cdot\cdot E}_{CA} - \Gamma^{\cdot\cdot E}_{CA}\Gamma^{\cdot\cdot A}_{BE} \\
&= 2\partial_{[B}\Gamma^{\cdot\cdot A}_{C]A}. \qquad (2.83)
\end{aligned}$$

2.2.3 *Identities for connection coefficients and curvature*

Coefficients of an arbitrary affine connection can be written [28]

$$\Gamma^{\cdot\cdot A}_{BC} = \{^{\cdot\cdot A}_{BC}\} + T^{\cdot\cdot A}_{BC} - T^{\cdot A\cdot}_{C\cdot B} + T^{A\cdot\cdot}_{\cdot BC} + \frac{1}{2}(M^{\cdot\cdot A}_{BC} + M^{\cdot A\cdot}_{C\cdot B} - M^{A\cdot\cdot}_{\cdot BC}). \quad (2.84)$$

Here, G_{AB} and its inverse G^{AB} are symmetric, three times differentiable, invertible, but otherwise arbitrary second-order tensors. Christoffel symbols of the first and second kind for the tensor G_{AB} are, respectively,

$$\{_{BC,A}\} = \frac{1}{2}(\partial_B G_{AC} + \partial_C G_{AB} - \partial_A G_{BC}) \quad \text{(first kind)}; \tag{2.85}$$

$$\{^{..A}_{BC}\} = \frac{1}{2}G^{AD}(\partial_B G_{CD} + \partial_C G_{BD} - \partial_D G_{BC}) \text{ (second kind)}. \tag{2.86}$$

By definition, Christoffel symbols are related by

$$\{_{BC,A}\} = G_{AD}\{^{..D}_{BC}\} \tag{2.87}$$

and have the following symmetry:

$$\{^{..A}_{BC}\} = \{^{..A}_{CB}\}, \qquad \{_{BC,A}\} = \{_{CB,A}\}. \tag{2.88}$$

The third-order object in parentheses on the right of (2.84) obeys

$$M^{..A}_{BC} = G^{AD}M_{BCD} = -G^{AD}\nabla_B G_{CD} = G_{CD}\nabla_B G^{AD}, \tag{2.89}$$

where the final equality follows from

$$\nabla_B(G_{CD}G^{AD}) = \nabla_B \delta^A_C = 0. \tag{2.90}$$

The covariant derivative of G_{AB} follows as

$$\begin{aligned} \nabla_A G_{BC} &= \partial_A G_{BC} - \Gamma^{..D}_{AB}G_{DC} - \Gamma^{..D}_{AC}G_{BD} \\ &= -M_{ABC} \\ &= -M_{A(BC)}. \end{aligned} \tag{2.91}$$

When $\nabla_A G_{BC} = 0$, or equivalently, when $M_{ABC} = 0$, the connection is said to be metric with respect to G_{AB}, *i.e.*, a metric connection. For a metric connection, covariant differentiation via ∇ and lowering (raising) indices by G_{AB} (G^{AB}) commute.

Relation (2.84) can be written in a more compact way upon introduction of the following notation for arbitrary third-order object A_{ABC} [28]:

$$A_{\{ABC\}} = A_{ABC} - A_{BCA} + A_{CAB}. \tag{2.92}$$

Dividing the partial derivative of G_{BC} by two as

$$\chi_{ABC} = \frac{1}{2}\partial_A G_{BC} = \chi_{ACB}, \tag{2.93}$$

arbitrary connection (2.84) in holonomic coordinates can be expressed as

$$\Gamma^{..A}_{BC} = \frac{1}{2}G^{AD}(2\chi_{\{BDC\}} - 2T_{\{BDC\}} + M_{\{BDC\}}). \tag{2.94}$$

Covariant components of the torsion tensor are defined as

$$T_{BCD} = T_{BC}^{\cdot\cdot A} G_{AD} = -T_{CBD}. \tag{2.95}$$

The Riemann-Christoffel curvature tensor obeys the following identities:

$$R_{(BC)D}^{\cdot\cdot\cdot A} = 0; \tag{2.96}$$

$$R_{[BCD]}^{\cdot\cdot\cdot A} = 2\nabla_{[B} T_{CD]}^{\cdot\cdot A} - 4T_{[BC}^{\cdot\cdot E} T_{D]E}^{\cdot\cdot A}; \tag{2.97}$$

$$R_{AB(CD)} = \nabla_{[A} M_{B]CD} + T_{AB}^{\cdot\cdot E} M_{ECD}; \tag{2.98}$$

$$\nabla_{[E} R_{BC]D}^{\cdot\cdot\cdot A} = 2T_{[EB}^{\cdot\cdot F} R_{C]FD}^{\cdot\cdot\cdot A}. \tag{2.99}$$

Relation (2.99) is known as Bianchi's identity. Relation (2.98) can be rearranged as

$$\nabla_{[A}\nabla_{B]} G_{CD} = -\nabla_{[A} M_{B]CD} = -R_{AB(CD)} - T_{AB}^{\cdot\cdot E} \nabla_E G_{CD}. \tag{2.100}$$

Let an arbitrary linear connection with coefficients $\Gamma_{BC}^{\cdot\cdot A}$ be decomposed additively as

$$\Gamma_{BC}^{\cdot\cdot A} = \Xi_{BC}^{\cdot\cdot A} + \Upsilon_{BC}^{\cdot\cdot A}, \tag{2.101}$$

where $\Xi_{BC}^{\cdot\cdot A}$ are also connection coefficents and $\Upsilon_{BC}^{\cdot\cdot A}$ are components of a differentiable third-order tensor field [28]. The torsion of this connection is, from (2.74),

$$T_{BC}^{\cdot\cdot A} = \Gamma_{[BC]}^{\cdot\cdot A} = \Xi_{[BC]}^{\cdot\cdot A} + \Upsilon_{[BC]}^{\cdot\cdot A} = \tau_{BC}^{\cdot\cdot A} + \Upsilon_{[BC]}^{\cdot\cdot A}, \tag{2.102}$$

where $\tau_{BC}^{\cdot\cdot A}$ is the torsion of the first term on the right of (2.101). Denote by $\nabla(\cdot)$ the covariant derivative with respect to the total connection $\Gamma_{BC}^{\cdot\cdot A}$, and denote by $\bar{\nabla}(\cdot)$ the covariant derivative with respect to $\Xi_{BC}^{\cdot\cdot A}$. The total Riemann-Christoffel curvature tensor is then

$$\begin{aligned}
R_{BCD}^{\cdot\cdot\cdot A} &= 2\partial_{[B}\Gamma_{C]D}^{\cdot\cdot A} + 2\Gamma_{[B|E|}^{\cdot\cdot A}\Gamma_{C]D}^{\cdot\cdot E} \\
&= 2\partial_{[B}\Xi_{C]D}^{\cdot\cdot A} + 2\partial_{[B}\Upsilon_{C]D}^{\cdot\cdot A} \\
&\quad + 2(\Xi_{[B|E|}^{\cdot\cdot A} + \Upsilon_{[B|E|}^{\cdot\cdot A})(\Xi_{C]D}^{\cdot\cdot E} + \Upsilon_{C]D}^{\cdot\cdot E}) \\
&= 2\partial_{[B}\Xi_{C]D}^{\cdot\cdot A} + 2\Xi_{[B|E|}^{\cdot\cdot A}\Xi_{C]D}^{\cdot\cdot E} \\
&\quad + 2(\partial_{[B}\Upsilon_{C]D}^{\cdot\cdot A} + \Upsilon_{[B|E|}^{\cdot\cdot A}\Upsilon_{C]D}^{\cdot\cdot E} \\
&\qquad + \Xi_{[B|E|}^{\cdot\cdot A}\Upsilon_{C]D}^{\cdot\cdot E} + \Upsilon_{[B|E|}^{\cdot\cdot A}\Xi_{C]D}^{\cdot\cdot E}) \\
&= \bar{R}_{BCD}^{\cdot\cdot\cdot A} + 2\nabla_{[B}\Upsilon_{C]D}^{\cdot\cdot A} + 2\Gamma_{[BC]}^{\cdot\cdot E}\Upsilon_{ED}^{\cdot\cdot A} - 2\Upsilon_{[C|D|}^{\cdot\cdot E}\Upsilon_{B]E}^{\cdot\cdot A} \\
&= \bar{R}_{BCD}^{\cdot\cdot\cdot A} + 2\nabla_{[B}\Upsilon_{C]D}^{\cdot\cdot A} + 2T_{BC}^{\cdot\cdot E}\Upsilon_{ED}^{\cdot\cdot A} + 2\Upsilon_{[B|D|}^{\cdot\cdot E}\Upsilon_{C]E}^{\cdot\cdot A},
\end{aligned} \tag{2.103}$$

where $\bar{R}^{\cdot\cdot\cdot A}_{BCD}$ is the curvature tensor of $\Xi^{\cdot\cdot A}_{BC}$ and

$$\nabla_B \Upsilon^{\cdot\cdot A}_{CD} = \partial_B \Upsilon^{\cdot\cdot A}_{CD} + \Gamma^{\cdot\cdot A}_{BE} \Upsilon^{\cdot\cdot E}_{CD} - \Gamma^{\cdot\cdot E}_{BC} \Upsilon^{\cdot\cdot A}_{ED} - \Gamma^{\cdot\cdot E}_{BD} \Upsilon^{\cdot\cdot A}_{CE}. \tag{2.104}$$

Total curvature also obeys

$$\begin{aligned} R^{\cdot\cdot\cdot A}_{BCD} &= \bar{R}^{\cdot\cdot\cdot A}_{BCD} + 2(\partial_{[B} \Upsilon^{\cdot\cdot A}_{C]D} + \Upsilon^{\cdot\cdot A}_{[B|E|} \Upsilon^{\cdot\cdot E}_{C]D} \\ &\qquad + \Xi^{\cdot\cdot A}_{[B|E|} \Upsilon^{\cdot\cdot E}_{C]D} + \Upsilon^{\cdot\cdot A}_{[B|E|} \Xi^{\cdot\cdot E}_{C]D}) \\ &= \bar{R}^{\cdot\cdot\cdot A}_{BCD} + 2\bar{\nabla}_{[B} \Upsilon^{\cdot\cdot A}_{C]D} + 2\Xi^{\cdot\cdot E}_{[BC]} \Upsilon^{\cdot\cdot A}_{ED} + 2\Upsilon^{\cdot\cdot E}_{[C|D|} \Upsilon^{\cdot\cdot A}_{B]E} \\ &= \bar{R}^{\cdot\cdot\cdot A}_{BCD} + 2\bar{\nabla}_{[B} \Upsilon^{\cdot\cdot A}_{C]D} + 2\tau^{\cdot\cdot E}_{BC} \Upsilon^{\cdot\cdot A}_{ED} - 2\Upsilon^{\cdot\cdot E}_{[B|D|} \Upsilon^{\cdot\cdot A}_{C]E}, \end{aligned} \tag{2.105}$$

where

$$\bar{\nabla}_B \Upsilon^{\cdot\cdot A}_{CD} = \partial_B \Upsilon^{\cdot\cdot A}_{CD} + \Xi^{\cdot\cdot A}_{BE} \Upsilon^{\cdot\cdot E}_{CD} - \Xi^{\cdot\cdot E}_{BC} \Upsilon^{\cdot\cdot A}_{ED} - \Xi^{\cdot\cdot E}_{BD} \Upsilon^{\cdot\cdot A}_{CE}. \tag{2.106}$$

2.2.4 *Riemannian geometry*

In Riemannian geometry, by definition, the torsion vanishes and the connection is metric:

$$\Gamma^{\cdot\cdot A}_{BC} = \{^{\cdot\cdot A}_{BC}\} = \frac{1}{2} G^{AD} (\partial_B G_{CD} + \partial_C G_{BD} - \partial_D G_{BC}), \tag{2.107}$$

$$T^{\cdot\cdot A}_{BC} = 0, \tag{2.108}$$

$$M_{ABC} = 0. \tag{2.109}$$

The Riemann-Christoffel curvature tensor of (2.75) becomes

$$R^{\cdot\cdot\cdot A}_{BCD} = 2\partial_{[B} \{^{\cdot\cdot A}_{C]D}\} + 2\{^{\cdot\cdot A}_{[B|E|}\}\{^{\cdot\cdot E}_{C]D}\}. \tag{2.110}$$

Using (2.107), this tensor or its covariant form can be expressed in terms of the metric tensor and first and second partial derivatives of the metric tensor [16]:

$$\begin{aligned} R_{ABCD} &= R^{\cdot\cdot\cdot E}_{ABC} G_{DE} \\ &= \frac{1}{2} (\partial_A \partial_C G_{BD} + \partial_B \partial_D G_{AC} - \partial_A \partial_D G_{BC} - \partial_B \partial_C G_{AD}) \\ &\qquad + G^{EF} (\{_{AC,E}\}\{_{BD,F}\} - \{_{BC,E}\}\{_{AD,F}\}). \end{aligned} \tag{2.111}$$

Identities (2.96)–(2.99) reduce to, respectively,

$$R_{(AB)CD} = 0, \tag{2.112}$$

$$R_{[ABC]D} = 0, \tag{2.113}$$

$$R_{AB(CD)} = 0, \tag{2.114}$$

$$\nabla_{[E} R_{AB]CD} = 0. \tag{2.115}$$

Furthermore, in Riemannian geometry,

$$R_{ABCD} = R_{CDAB}. \tag{2.116}$$

The number of independent components of the Riemann-Christoffel curvature tensor is $n^2(n^2 - 1)/12$; for example, one component for two-dimensional space and six components for three-dimensional space. Einstein's tensor and Ricci's tensor are both symmetric and are related by

$$\theta_{AB} = R_{AB} - \frac{1}{2} R G_{AB}, \tag{2.117}$$

and the following identity holds:

$$\nabla_A \theta^A_{.B} = \nabla_A \left(R_{BC} G^{CA} - \frac{1}{2} R \delta^A_B \right) = 0. \tag{2.118}$$

When $n = 2$ or $n = 3$, the curvature tensor can be reconstructed from the Ricci curvature as

$$R_{ABCD} = 2(R_{B[C} G_{D]A} - R_{A[C} G_{D]B}) - R G_{B[C} G_{D]A}. \tag{2.119}$$

Recall from (2.82) that

$$R = R^A_{.A} = R_{AB} G^{AB} = R_{CABD} G^{AB} G^{CD}. \tag{2.120}$$

When $n = 2$, $R = 2\kappa$ is the Gaussian curvature, and

$$R_{AB} = R G_{AB} = 2\kappa G_{AB}. \tag{2.121}$$

2.2.5 *Euclidean space*

Let $G_{AB}(X) = \boldsymbol{G}_A \cdot \boldsymbol{G}_B$ be the metric tensor of the space. The Levi-Civita connection coefficients of G_{AB}, written as $\overset{G}{\Gamma}{}^{..A}_{BC}$, are the associated metric and torsion-free connection coefficients of (2.107):

$$\overset{G}{\Gamma}{}^{..A}_{BC} = \frac{1}{2} G^{AD} (\partial_B G_{CD} + \partial_C G_{BD} - \partial_D G_{BC}) = \overset{G}{\Gamma}{}^{..A}_{CB}. \tag{2.122}$$

Notice that the superposed G is a descriptive label rather than a free index and is exempt from the summation convention. In Euclidean space, the Riemann-Christoffel curvature tensor of the Levi-Civita connection vanishes identically:

$$\overset{G}{R}{}^{...A}_{BCD} = 2 \left(\partial_{[B} \overset{G}{\Gamma}{}^{..A}_{C]D} + \overset{G}{\Gamma}{}^{..A}_{[B|E|} \overset{G}{\Gamma}{}^{..E}_{C]D} \right) = 0. \tag{2.123}$$

In n-dimensional Euclidean space, a transformation to an n-dimensional Cartesian coordinate system is available at each point X such that $G_{AB}(X) \to \delta_{AB}$, where δ_{AB} are covariant Kronecker delta symbols. When the curvature tensor of the connection vanishes, the space is said to be intrinsically flat; otherwise, the space is said to be intrinsically curved. In two dimensions, a cylindrical shell is intrinsically flat, while a spherical shell is intrinsically curved.

Henceforward, reference configuration B_0 will be treated as a three-dimensional Euclidean space. More precisely, herein, when a configuration is identified with n-dimensional Euclidean space, a simply connected deformable body of finite size in that configuration is assumed to occupy an open region of infinitely extended n-dimensional Euclidean vector space. In the interest of brevity, such a configuration is simply labeled a Euclidean space. Embedding of each point $X \in B_0$ in a region of Euclidean space has already been implied if the existence of global position vector $\boldsymbol{X}(X)$ is presumed as in (2.4).

The covariant derivative associated with the Levi-Civita connection on B_0 is written as

$$\overset{G}{\nabla}_A(\cdot) = (\cdot)_{;A}. \tag{2.124}$$

According to the notation scheme used here, symbolic covariant derivatives of basis vectors and their reciprocals vanish, leading to

$$\boldsymbol{G}_{A;B} = \partial_B \boldsymbol{G}_A - \overset{G}{\Gamma}{}^{..C}_{BA} \boldsymbol{G}_C = 0 \Leftrightarrow \partial_B \boldsymbol{G}_A = \overset{G}{\Gamma}{}^{..C}_{BA} \boldsymbol{G}_C, \tag{2.125}$$

$$\boldsymbol{G}^A_{;B} = \partial_B \boldsymbol{G}^A + \overset{G}{\Gamma}{}^{..A}_{BC} \boldsymbol{G}^C = 0 \Leftrightarrow \partial_B \boldsymbol{G}^A = -\overset{G}{\Gamma}{}^{..A}_{BC} \boldsymbol{G}^C. \tag{2.126}$$

It follows that Christoffel symbols can be computed as

$$\begin{aligned}
\overset{G}{\Gamma}{}^{..A}_{BC} &= \overset{G}{\Gamma}{}^{..D}_{BC} \delta^A_D \\
&= \overset{G}{\Gamma}{}^{..D}_{BC} \langle \boldsymbol{G}_D, \boldsymbol{G}^A \rangle \\
&= \langle \partial_B \boldsymbol{G}_C, \boldsymbol{G}^A \rangle \\
&= \langle \partial_B (\partial_C \boldsymbol{X}), \boldsymbol{G}_D G^{-1DA} \rangle. \\
&= \langle \partial_B \partial_C \boldsymbol{X}, \partial_D \boldsymbol{X} (\partial_A \boldsymbol{X} \cdot \partial_D \boldsymbol{X})^{-1} \rangle.
\end{aligned} \tag{2.127}$$

The above notation for covariantly constant basis vectors is also used in [16], equations (4.4.13)–(4.4.16).

In the context of notation in (2.69), setting

$$\nabla \to \overset{G}{\nabla} \Leftrightarrow \Gamma^{..A}_{BC} \to \overset{G}{\Gamma}{}^{..A}_{BC} \tag{2.128}$$

leads to equivalence between partial derivatives and gradients of basis vectors:

$$\partial_A \boldsymbol{G}_B = \partial_A \partial_B \boldsymbol{X} = \overset{G}{\Gamma}{}^{..C}_{AB} \boldsymbol{G}_C = \overset{G}{\Gamma}{}^{..C}_{AB} \boldsymbol{\partial}_C = \overset{G}{\nabla}_{\boldsymbol{\partial}_A} \boldsymbol{\partial}_B. \tag{2.129}$$

Therefore,

$$\nabla = \overset{G}{\nabla} \Rightarrow \nabla_{\boldsymbol{\partial}_A} \boldsymbol{\partial}_B = \nabla_{\boldsymbol{\partial}_B} \boldsymbol{\partial}_A = \partial_B \boldsymbol{\partial}_A = \partial_A \boldsymbol{\partial}_B, \tag{2.130}$$

but the same identity does not always apply for an arbitrary connection ∇; *i.e.*, the reverse of (2.130) need not hold. Notice the difference in notation and terminology: symbolic covariant derivatives of basis vectors in (2.125) always vanish, while covariant derivatives (*i.e.*, gradients) of basis vectors in (2.129) do not except for certain choices of coordinate systems (*e.g.*, a Cartesian basis with necessarily vanishing Christoffel symbols).

Returning to classical notation, the partial derivative of the metric tensor is

$$\begin{aligned}
\partial_C G_{AB} &= \partial_C (\boldsymbol{G}_A \cdot \boldsymbol{G}_B) \\
&= \partial_C \boldsymbol{G}_A \cdot \boldsymbol{G}_B + \boldsymbol{G}_A \cdot \partial_C \boldsymbol{G}_B \\
&= \overset{G}{\Gamma}{}^{..D}_{CA} \boldsymbol{G}_D \cdot \boldsymbol{G}_B + \overset{G}{\Gamma}{}^{..D}_{CB} \boldsymbol{G}_D \cdot \boldsymbol{G}_A \\
&= \overset{G}{\Gamma}{}^{..D}_{CA} G_{BD} + \overset{G}{\Gamma}{}^{..D}_{CB} G_{AD} \\
&= 2 \overset{G}{\Gamma}{}^{..D}_{C(A} G_{B)D}.
\end{aligned} \tag{2.131}$$

Similarly, for the inverse of the covariant metric tensor,

$$\begin{aligned}
\partial_C G^{AB} &= \partial_C (\boldsymbol{G}^A \cdot \boldsymbol{G}^B) \\
&= \partial_C \boldsymbol{G}^A \cdot \boldsymbol{G}^B + \boldsymbol{G}^A \cdot \partial_C \boldsymbol{G}^B \\
&= -\overset{G}{\Gamma}{}^{..A}_{CD} \boldsymbol{G}^D \cdot \boldsymbol{G}^B - \overset{G}{\Gamma}{}^{..B}_{CD} \boldsymbol{G}^D \cdot \boldsymbol{G}^A \\
&= -\overset{G}{\Gamma}{}^{..A}_{CD} G^{BD} - \overset{G}{\Gamma}{}^{..B}_{CD} G^{AD} \\
&= -2 \overset{G}{\Gamma}{}^{..(A}_{CD} G^{B)D}.
\end{aligned} \tag{2.132}$$

As the Levi-Civita connection is a metric connection,

$$G_{AB;C} = \partial_C G_{AB} - \overset{G}{\Gamma}{}^{..D}_{CA} G_{DB} - \overset{G}{\Gamma}{}^{..D}_{CB} G_{AD} = 0, \tag{2.133}$$

$$G^{AB}{}_{;C} = \partial_C G^{AB} + \overset{G}{\Gamma}{}^{..A}_{CD} G^{DB} + \overset{G}{\Gamma}{}^{..B}_{CD} G^{AD} = 0. \tag{2.134}$$

It follows immediately from (2.133), (2.134), and the product rule that covariant differentiation with respect to the Levi-Civita connection commutes with raising and lowering indices via the corresponding metric. For example, for a differentiable vector field $\boldsymbol{V} = V^A \boldsymbol{G}_A = V_A \boldsymbol{G}^A$,

$$V^A{}_{;B} = (G^{AC} V_C)_{;B} = G^{AC} V_{C;B} + G^{AC}{}_{;B} V_C = G^{AC} V_{C;B}, \tag{2.135}$$

$$V_{A;B} = (G_{AC}V^C)_{;B} = G_{AC}V^C_{;B} + G_{AC;B}V^C = G_{AC}V^C_{;B}. \tag{2.136}$$

Partial coordinate differentiation generally does not commute with raising or lowering indices unless the connection coefficients vanish:

$$\begin{aligned}\partial_B V_A &= \partial_B(G_{AC}V^C) \\ &= G_{AC}\partial_B V^C + V^C \partial_B G_{AC} \\ &= G_{AC}\partial_B V^C + 2V^C \overset{G}{\Gamma}{}_{B(A}^{..D} G_{C)D}.\end{aligned} \tag{2.137}$$

From the symmetry of the Levi-Civita connection [or from (2.1) and (2.17)] skew partial derivatives of basis vectors vanish in Euclidean space:

$$\partial_{[B}\boldsymbol{G}_{A]} = \overset{G}{\Gamma}{}_{[BA]}^{..C}\boldsymbol{G}_C = \partial_{[B}\partial_{A]}\boldsymbol{X} = 0. \tag{2.138}$$

From (2.76), the skew second covariant derivative of a differentiable vector field with components $V^A(X)$ vanishes in Euclidean space:

$$\overset{G}{\nabla}_{[B}\overset{G}{\nabla}_{C]}V^A = V^A_{;[CB]} = \frac{1}{2}\overset{G}{R}{}_{BCD}^{...A}V^D - \overset{G}{\Gamma}{}_{[BC]}^{..D}\overset{G}{\nabla}_D V^A = 0. \tag{2.139}$$

Similarly, from (2.77), for a differentiable covariant vector field with components $\alpha_D(X)$:

$$\overset{G}{\nabla}_{[B}\overset{G}{\nabla}_{C]}\alpha_D = \alpha_{D;[CB]} = -\frac{1}{2}\overset{G}{R}{}_{BCD}^{...A}\alpha_A - \overset{G}{\Gamma}{}_{[BC]}^{..A}\overset{G}{\nabla}_A\alpha_D = 0. \tag{2.140}$$

Therefore, (2.139) and (2.140) lead to

$$V^A_{;BC} = V^A_{;CB}, \qquad \alpha_{A;BC} = \alpha_{A;CB}. \tag{2.141}$$

The following identity will often be used for the derivative of the determinant of a non-singular but otherwise arbitrary second-order tensor $\boldsymbol{A}$ [14]:

$$\frac{\partial \det \boldsymbol{A}}{\partial A^A_{.B}} = A^{-1B}_{\ \ .A} \det \boldsymbol{A}. \tag{2.142}$$

The quantity in (2.142) is often referred to as the cofactor of $\boldsymbol{A}$. It follows from the chain rule that

$$\begin{aligned}\partial_A[\ln(\det \boldsymbol{A})] &= \frac{1}{\det \boldsymbol{A}}\partial_A \det \boldsymbol{A} \\ &= \frac{1}{\det \boldsymbol{A}}\frac{\partial \det \boldsymbol{A}}{\partial A^B_{.C}}\partial_A A^B_{.C} \\ &= A^{-1C}_{\ \ .B}\partial_A A^B_{.C}.\end{aligned} \tag{2.143}$$

Applying this identity to the determinant of the metric tensor,

$$\begin{aligned}
\partial_A(\ln\sqrt{G}) &= \frac{1}{\sqrt{G}}\partial_A(\sqrt{G}) \\
&= \frac{1}{2G}\partial_A G \\
&= \frac{1}{2G}\frac{\partial \det \boldsymbol{G}}{\partial G_{BC}}\partial_A G_{BC} \\
&= \frac{1}{2G} G G^{CB}\partial_A G_{BC} \\
&= \frac{1}{2} G^{CB}\left(\overset{G}{\Gamma}{}^{\cdot\cdot D}_{AC} G_{BD} + \overset{G}{\Gamma}{}^{\cdot\cdot D}_{AB} G_{CD}\right) \\
&= \frac{1}{2}\left(\overset{G}{\Gamma}{}^{\cdot\cdot D}_{AC}\delta^C_D + \overset{G}{\Gamma}{}^{\cdot\cdot D}_{AB}\delta^B_D\right) \\
&= \overset{G}{\Gamma}{}^{\cdot\cdot D}_{AD} \\
&= \overset{G}{\Gamma}{}^{\cdot\cdot D}_{DA}.
\end{aligned} \tag{2.144}$$

Notice also that

$$\begin{aligned}
\partial_A G &= \frac{\partial G}{\partial G_{BC}}\partial_A G_{BC} \\
&= G G^{CB}\left(\overset{G}{\Gamma}{}^{\cdot\cdot D}_{AC} G_{BD} + \overset{G}{\Gamma}{}^{\cdot\cdot D}_{AB} G_{CD}\right) \\
&= 2G\overset{G}{\Gamma}{}^{\cdot\cdot D}_{AD} \\
&= 2G\overset{G}{\Gamma}{}^{\cdot\cdot D}_{DA}.
\end{aligned} \tag{2.145}$$

For a metric connection, the covariant derivative of the permutation tensor vanishes:

$$\begin{aligned}
\epsilon_{ABC;D} &= \partial_D\epsilon_{ABC} - \overset{G}{\Gamma}{}^{\cdot\cdot E}_{DA}\epsilon_{EBC} - \overset{G}{\Gamma}{}^{\cdot\cdot E}_{DB}\epsilon_{AEC} - \overset{G}{\Gamma}{}^{\cdot\cdot E}_{DC}\epsilon_{ABE} \\
&= \partial_D\epsilon_{ABC} - \overset{G}{\Gamma}{}^{\cdot\cdot E}_{DE}\epsilon_{ABC} \\
&= \partial_D(\sqrt{G})e_{ABC} - \partial_D(\ln\sqrt{G})\epsilon_{ABC} \\
&= \partial_D(\ln\sqrt{G})\epsilon_{ABC} - \partial_D(\ln\sqrt{G})\epsilon_{ABC} \\
&= 0.
\end{aligned} \tag{2.146}$$

Similarly, for the contravariant version,

$$\begin{aligned}
\epsilon^{ABC}{}_{;D} &= \partial_D\epsilon^{ABC} + \overset{G}{\Gamma}{}^{\cdot\cdot A}_{DE}\epsilon^{EBC} + \overset{G}{\Gamma}{}^{\cdot\cdot B}_{DE}\epsilon^{AEC} + \overset{G}{\Gamma}{}^{\cdot\cdot C}_{DE}\epsilon^{ABE} \\
&= \partial_D\epsilon^{ABC} + \overset{G}{\Gamma}{}^{\cdot\cdot E}_{DE}\epsilon^{ABC} \\
&= \partial_D(1/\sqrt{G})e^{ABC} + \partial_D(\ln\sqrt{G})\epsilon^{ABC} \\
&= -\partial_D(\ln\sqrt{G})\epsilon^{ABC} + \partial_D(\ln\sqrt{G})\epsilon^{ABC} \\
&= 0.
\end{aligned} \tag{2.147}$$

If the covariant derivative of the determinant $G[G_{AB}(X)] = \det[G_{AB}(X)]$ is defined by application of the chain rule as [7]

$$G_{;A} = \frac{\partial G}{\partial G_{BC}} G_{BC;A} = G G^{CB} G_{BC;A}, \tag{2.148}$$

then $G_{;A} = 0$ follows from (2.133). This notation is also used in [16], (2.5.17). Finally, contravariant reference basis vectors can be written symbolically as gradients of coordinates in Euclidean space:

$$\overset{G}{\nabla} X^A = \partial_B X^A \boldsymbol{G}^B = \delta^A_B \boldsymbol{G}^B = \boldsymbol{G}^A. \tag{2.149}$$

Now consider spatial configuration B, which is also henceforward assumed a Euclidean space, with metric tensor components $g_{ab}(x) = \boldsymbol{g}_a \cdot \boldsymbol{g}_b$. The spatial Levi-Civita connection is

$$\overset{g}{\Gamma}{}_{bc}^{\cdot\cdot a} = \frac{1}{2} g^{ad}(\partial_b g_{cd} + \partial_c g_{bd} - \partial_d g_{bc}) = \overset{g}{\Gamma}{}_{cb}^{\cdot\cdot a}. \tag{2.150}$$

Superposed g is a descriptive label rather than a free index and is exempt from the summation convention. In Euclidean space, the Riemann-Christoffel curvature tensor vanishes:

$$\overset{g}{R}{}_{bcd}^{\cdot\cdot\cdot a} = 2\left(\partial_{[b}\overset{g}{\Gamma}{}_{c]d}^{\cdot\cdot a} + \overset{g}{\Gamma}{}_{[b|e|}^{\cdot\cdot a}\overset{g}{\Gamma}{}_{c]d}^{\cdot\cdot e}\right) = 0. \tag{2.151}$$

The covariant derivative associated with the Levi-Civita connection on B is written as

$$\overset{g}{\nabla}_a(\cdot) = (\cdot)_{;a}. \tag{2.152}$$

Analogously to identities for the reference configuration,

$$\boldsymbol{g}_{a;b} = 0 \Leftrightarrow \partial_b \boldsymbol{g}_a = \overset{g}{\Gamma}{}_{ba}^{\cdot\cdot c} \boldsymbol{g}_c, \tag{2.153}$$

$$\boldsymbol{g}^a_{;b} = 0 \Leftrightarrow \partial_b \boldsymbol{g}^a = -\overset{g}{\Gamma}{}_{bc}^{\cdot\cdot a} \boldsymbol{g}^c, \tag{2.154}$$

$$\overset{g}{\Gamma}{}_{bc}^{\cdot\cdot a} = \langle \partial_b \boldsymbol{g}_c, \boldsymbol{g}^a \rangle, \tag{2.155}$$

$$\partial_{[b}\boldsymbol{g}_{a]} = \overset{g}{\Gamma}{}_{[ba]}^{\cdot\cdot c} \boldsymbol{g}_c = \partial_{[b}\partial_{a]}\boldsymbol{x} = 0, \tag{2.156}$$

$$\partial_c g_{ab} = 2\overset{g}{\Gamma}{}_{c(a}^{\cdot\cdot d} g_{b)d}, \tag{2.157}$$

$$\partial_c g^{ab} = -2\overset{g}{\Gamma}{}_{cd}^{\cdot\cdot(a} g^{b)d}, \tag{2.158}$$

$$\partial_a(\ln\sqrt{g}) = \overset{g}{\Gamma}{}_{ad}^{\cdot\cdot d} = \overset{g}{\Gamma}{}_{da}^{\cdot\cdot d}, \tag{2.159}$$

$$g_{ab;c} = 0, \tag{2.160}$$

$$\epsilon_{abc;d} = 0, \tag{2.161}$$

$$\epsilon^{abc}{}_{;d} = 0. \tag{2.162}$$

From (2.76), the skew second covariant derivative of a vector field with components $V^a(x)$ vanishes in Euclidean space:

$$\overset{g}{\nabla}_{[b}\overset{g}{\nabla}_{c]}V^a = V^a_{;[cb]} = \frac{1}{2}\overset{g}{R}{}_{bcd}^{\ldots a}V^d - \overset{g}{\Gamma}{}_{[bc]}^{\ldots d}\overset{g}{\nabla}_d V^a = 0. \tag{2.163}$$

Similarly, from (2.77), for a differentiable covariant vector field $\boldsymbol{\alpha}(x)$,

$$\overset{g}{\nabla}_{[b}\overset{g}{\nabla}_{c]}\alpha_d = \alpha_{d\,;[cb]} = -\frac{1}{2}\overset{g}{R}{}_{bcd}^{\ldots a}\alpha_a - \overset{g}{\Gamma}{}_{[bc]}^{\ldots a}\overset{g}{\nabla}_a\alpha_d = 0. \tag{2.164}$$

Therefore,

$$V^a_{;\,bc} = V^a_{;\,cb}, \qquad \alpha_{a\,;bc} = \alpha_{a\,;cb}. \tag{2.165}$$

Contravariant spatial basis vectors can be written symbolically as gradients of coordinates:

$$\overset{g}{\nabla}x^a = \partial_b x^a \boldsymbol{g}^b = \delta^a_b \boldsymbol{g}^b = \boldsymbol{g}^a. \tag{2.166}$$

2.3 Differential Operators and Related Notation

Differential operators—specifically the gradient, divergence, curl, and Laplacian—are defined. Partial and total covariant derivatives of two-point tensors are introduced. Generalized scalar products for second- and higher-order tensors are defined.

2.3.1 *Gradient, divergence, curl, and Laplacian*

The description that follows in Section 2.3.1 is framed in reference configuration B_0 covered by material coordinates X^A. Analogous definitions and identities apply for a description in spatial coordinates x^a covering the current configuration B.

The gradient of a differentiable scalar function $f(X)$ is equivalent to its partial derivative:

$$\overset{G}{\nabla}f = \overset{G}{\nabla}_A f\,\boldsymbol{G}^A = f_{;\,A}\boldsymbol{G}^A = \partial_A f\,\boldsymbol{G}^A. \tag{2.167}$$

The gradient of a differentiable vector field $\boldsymbol{V}(X) = V^A\boldsymbol{G}_A$ is

$$\begin{aligned}
\overset{G}{\nabla}\boldsymbol{V} &= \partial_B\boldsymbol{V}\otimes\boldsymbol{G}^B \\
&= \partial_B V^A\boldsymbol{G}_A\otimes\boldsymbol{G}^B + V^A\partial_B\boldsymbol{G}_A\otimes\boldsymbol{G}^B \\
&= \partial_B V^A\boldsymbol{G}_A\otimes\boldsymbol{G}^B + V^A\overset{G}{\Gamma}{}_{BA}^{\ldots C}\boldsymbol{G}_C\otimes\boldsymbol{G}^B \\
&= \left(\partial_B V^A + \overset{G}{\Gamma}{}_{BC}^{\ldots A}V^C\right)\boldsymbol{G}_A\otimes\boldsymbol{G}^B \\
&= V^A_{;\,B}\boldsymbol{G}_A\otimes\boldsymbol{G}^B \\
&= \overset{G}{\nabla}_B V^A\boldsymbol{G}_A\otimes\boldsymbol{G}^B.
\end{aligned} \tag{2.168}$$

In the present notation, the gradient can also be found directly using (2.125) as

$$\begin{aligned}
\overset{G}{\nabla} \boldsymbol{V} &= \overset{G}{\nabla}_B V^A \boldsymbol{G}_A \otimes \boldsymbol{G}^B \\
&= V^A_{;B} \boldsymbol{G}_A \otimes \boldsymbol{G}^B \\
&= V^A_{;B} \boldsymbol{G}_A \otimes \boldsymbol{G}^B + V^A \boldsymbol{G}_{A;B} \otimes \boldsymbol{G}^B \\
&= (V^A \boldsymbol{G}_A)_{;B} \otimes \boldsymbol{G}^B \\
&= \boldsymbol{V}_{;B} \otimes \boldsymbol{G}^B \\
&= \overset{G}{\nabla}_B \boldsymbol{V} \otimes \boldsymbol{G}^B. \qquad (2.169)
\end{aligned}$$

Let $\boldsymbol{V}(X)$ and $\boldsymbol{W}(X)$ denote two differentiable vector fields. The following calculation demonstrates that the gradient operation with respect to the Levi-Civita connection commutes with the dot (scalar) product of vectors, as implied already in (2.131) and (2.132):

$$\begin{aligned}
\partial_A(\boldsymbol{V} \cdot \boldsymbol{W}) &= \partial_A(V^B W_B) \\
&= V^B \partial_A W_B + W_B \partial_A V^B \\
&= V^B(\partial_A W_B - \overset{G}{\Gamma}{}^{..C}_{AB} W_C) + W_C(\partial_A V^C + \overset{G}{\Gamma}{}^{..C}_{AB} V^B) \\
&= V^B(\partial_A W_B - \overset{G}{\Gamma}{}^{..C}_{AB} W_C) + W_B(\partial_A V^B + \overset{G}{\Gamma}{}^{..B}_{AC} V^C) \\
&= V^B W_{B;A} + W_B V^B_{;A} \\
&= \boldsymbol{V} \cdot \partial_A \boldsymbol{W} + \boldsymbol{W} \cdot \partial_A \boldsymbol{V}. \qquad (2.170)
\end{aligned}$$

The trace of a second-order tensor $\boldsymbol{A}$ is

$$\mathrm{tr}\boldsymbol{A} = A^A_{.A} = A^{AB} G_{AB} = A_{AB} G^{AB}. \qquad (2.171)$$

The divergence of a differentiable vector field $\boldsymbol{V}(X) = V^A \boldsymbol{G}_A$ is

$$\begin{aligned}
\langle \overset{G}{\nabla}, \boldsymbol{V} \rangle &= \mathrm{tr}\,(\overset{G}{\nabla} \boldsymbol{V}) \\
&= \langle \partial_A \boldsymbol{V}, \boldsymbol{G}^A \rangle \\
&= \langle \partial_A (V^B \boldsymbol{G}_B), \boldsymbol{G}^A \rangle \\
&= \langle V^B_{;A} \boldsymbol{G}_B, \boldsymbol{G}^A \rangle \\
&= V^A_{;A} \\
&= \partial_A V^A + \overset{G}{\Gamma}{}^{..A}_{AB} V^B \\
&= \frac{1}{\sqrt{G}} \partial_A (\sqrt{G} V^A), \qquad (2.172)
\end{aligned}$$

where the final equality follows from (2.144). Recall from (2.43) and (2.44) that the vector cross product $\times$ obeys, for two vectors $\boldsymbol{V}$ and $\boldsymbol{W}$ and two covectors $\boldsymbol{\alpha}$ and $\boldsymbol{\beta}$,

$$\boldsymbol{V} \times \boldsymbol{W} = \epsilon_{ABC} V^B W^C \boldsymbol{G}^A, \qquad (2.173)$$

$$\boldsymbol{\alpha} \times \boldsymbol{\beta} = \epsilon^{ABC} \alpha_B \beta_C \boldsymbol{G}_A. \tag{2.174}$$

The curl of a differentiable covariant vector field $\boldsymbol{\alpha}(X)$ obeys

$$\begin{aligned}\overset{G}{\nabla} \times \boldsymbol{\alpha} &= \boldsymbol{G}^A \times \partial_A (\alpha_B \boldsymbol{G}^B) \\ &= \boldsymbol{G}^A \times \boldsymbol{G}^B \alpha_{B;\,A} \\ &= \epsilon^{ABC} \alpha_{B;A} \boldsymbol{G}_C \\ &= \epsilon^{ABC} \alpha_{C;B} \boldsymbol{G}_A \\ &= \epsilon^{ABC} \partial_B \alpha_C \boldsymbol{G}_A,\end{aligned} \tag{2.175}$$

where the final equality follows from the symmetry of the Levi-Civita connection:

$$\epsilon^{ABC} \alpha_{C;\,B} = \epsilon^{ABC} \partial_B \alpha_C - \epsilon^{ABC} \alpha_D \overset{G}{\Gamma}{}^{..D}_{[BC]} = \epsilon^{ABC} \partial_B \alpha_C. \tag{2.176}$$

The Laplacian of a twice differentiable scalar field $f(X)$ is

$$\overset{G}{\nabla}{}^2 f = G^{AB} f_{;\,AB} = (G^{AB} \partial_A f)_{;\,B} = \frac{1}{\sqrt{G}} \partial_B (\sqrt{G} G^{AB} \partial_A f). \tag{2.177}$$

The divergence of the curl of a twice differentiable (co)vector field vanishes identically:

$$\langle \overset{G}{\nabla}, \overset{G}{\nabla} \times \boldsymbol{\alpha} \rangle = \epsilon^{ABC} \alpha_{C\,;\,BA} = \epsilon^{C[AB]} \alpha_{C\,;\,(AB)} = 0, \tag{2.178}$$

as does the curl of the gradient of a twice differentiable scalar field:

$$\overset{G}{\nabla} \times \overset{G}{\nabla} f = \epsilon^{ABC} f_{;\,CB} \boldsymbol{G}_A = \epsilon^{A[BC]} f_{;\,(CB)} \boldsymbol{G}_A = 0. \tag{2.179}$$

2.3.2 *Partial and total covariant derivatives*

Consider a two-point tensor (*i.e.*, double tensor) $\boldsymbol{A}(x, X)$ of order two:

$$\boldsymbol{A}(x, X) = A^a_{\cdot A}(x, X) \boldsymbol{g}_a(x) \otimes \boldsymbol{G}^A(X). \tag{2.180}$$

Components of the total covariant derivative of $A^a_{\cdot A}$ are defined as

$$\begin{aligned}A^a_{\cdot A:B} &= (A^a_{\cdot A;B}) + (A^a_{\cdot A;b}) \partial_B x^b \\ &= \left(\frac{\partial A^a_{\cdot A}}{\partial X^B}\bigg|_x - \overset{G}{\Gamma}{}^{..C}_{BA} A^a_{\cdot C} \right) + \left(\frac{\partial A^a_{\cdot A}}{\partial x^b}\bigg|_X + \overset{g}{\Gamma}{}^{..a}_{bc} A^c_{\cdot A} \right) \frac{\partial x^b}{\partial X^B} \\ &= \left(\frac{\partial A^a_{\cdot A}}{\partial X^B}\bigg|_x - \overset{G}{\Gamma}{}^{..C}_{BA} A^a_{\cdot C} \right) + \left(\frac{\partial A^a_{\cdot A}}{\partial x^b}\bigg|_X + \overset{g}{\Gamma}{}^{..a}_{bc} A^c_{\cdot A} \right) \partial_B x^b.\end{aligned} \tag{2.181}$$

Quantities $\partial_B x^b$ will be identified with components of the deformation gradient in Section 3.1.1. Partial covariant derivatives $A^a_{\cdot A;B}$ and $A^a_{\cdot A;b}$ are defined as usual covariant derivatives with respect to indices in one configuration, with those of the other configuration held fixed:

$$A^{a}_{\cdot A;B} = \frac{\partial A^{a}_{\cdot A}}{\partial X^{B}}\bigg|_{x} - \overset{G}{\Gamma}{}^{..C}_{BA} A^{a}_{\cdot C}, \tag{2.182}$$

$$A^{a}_{\cdot A;b} = \frac{\partial A^{a}_{\cdot A}}{\partial x^{b}}\bigg|_{X} + \overset{g}{\Gamma}{}^{..a}_{bc} A^{c}_{\cdot A}. \tag{2.183}$$

Now recall from (2.1) that $x^a = x^a(X, t)$. Writing $A^{a}_{\cdot A}[X, x(X, t)] = A^{a}_{\cdot A}(X, t)$, the partial derivative of $A^{a}_{\cdot A}$ at fixed t is

$$\frac{\partial A^{a}_{\cdot A}}{\partial X^{B}} = \frac{\partial A^{a}_{\cdot A}}{\partial X^{B}}\bigg|_{t} = \frac{\partial A^{a}_{\cdot A}}{\partial X^{B}}\bigg|_{x} + \frac{\partial A^{a}_{\cdot A}}{\partial x^{b}}\bigg|_{X} \partial_B x^b = \partial_B A^{a}_{\cdot A}. \tag{2.184}$$

With this notation, (2.181) becomes

$$A^{a}_{\cdot A:B} = \frac{\partial A^{a}_{\cdot A}}{\partial X^{B}} - \overset{G}{\Gamma}{}^{..C}_{BA} A^{a}_{\cdot C} + \overset{g}{\Gamma}{}^{..a}_{bc} A^{c}_{\cdot A} \partial_B x^b. \tag{2.185}$$

The total covariant derivative can also be obtained directly by taking the (material) gradient of $\boldsymbol{A}$:

$$\begin{aligned}
\overset{G}{\nabla} \boldsymbol{A} &= \partial_B \boldsymbol{A} \otimes \boldsymbol{G}^B \\
&= \partial_B (A^{a}_{\cdot A} \boldsymbol{g}_a \otimes \boldsymbol{G}^A) \otimes \boldsymbol{G}^B \\
&= \partial_B A^{a}_{\cdot A} \boldsymbol{g}_a \otimes \boldsymbol{G}^A \otimes \boldsymbol{G}^B \\
&\quad + A^{c}_{\cdot A} \partial_B x^b (\partial_b \boldsymbol{g}_c) \otimes \boldsymbol{G}^A \otimes \boldsymbol{G}^B \\
&\quad + A^{a}_{\cdot C} \boldsymbol{g}_a \otimes (\partial_B \boldsymbol{G}^C) \otimes \boldsymbol{G}^B \\
&= \partial_B A^{a}_{\cdot A} \boldsymbol{g}_a \otimes \boldsymbol{G}^A \otimes \boldsymbol{G}^B \\
&\quad + A^{c}_{\cdot A} \partial_B x^b (\overset{g}{\Gamma}{}^{..a}_{bc} \boldsymbol{g}_a) \otimes \boldsymbol{G}^A \otimes \boldsymbol{G}^B \\
&\quad - A^{a}_{\cdot C} \boldsymbol{g}_a \otimes (\overset{G}{\Gamma}{}^{..C}_{BA} \boldsymbol{G}^A) \otimes \boldsymbol{G}^B \\
&= \left(\partial_B A^{a}_{\cdot A} + \overset{g}{\Gamma}{}^{..a}_{bc} A^{c}_{\cdot A} \partial_B x^b - \overset{G}{\Gamma}{}^{..C}_{BA} A^{a}_{\cdot C} \right) \boldsymbol{g}_a \otimes \boldsymbol{G}^A \otimes \boldsymbol{G}^B \\
&= A^{a}_{\cdot A:B} \boldsymbol{g}_a \otimes \boldsymbol{G}^A \otimes \boldsymbol{G}^B.
\end{aligned} \tag{2.186}$$

The total covariant derivative can be extended to two-point tensors of arbitrary order as [14]

$$(A^{a\ldots e}_{A\ldots E})_{:K} = (A^{a\ldots e}_{A\ldots E})_{;K} + (A^{a\ldots e}_{A\ldots E})_{;k} \partial_K x^k, \tag{2.187}$$

where partial covariant derivatives are taken with respect to indices referred to a single configuration, as in (2.182) and (2.183). From the definition of the total covariant derivative, noting that $\partial_K x^k \, \partial_l X^K = \delta^k_l$,

$$\begin{aligned}
(A^{a\ldots e}_{A\ldots E})_{:K} \partial_k X^K &= (A^{a\ldots e}_{A\ldots E})_{;K} \partial_k X^K + (A^{a\ldots e}_{A\ldots E})_{;l} \partial_K x^l \, \partial_k X^K \\
&= (A^{a\ldots e}_{A\ldots E})_{:k}.
\end{aligned} \tag{2.188}$$

Thus, total covariant derivatives map between configurations like partial derivatives:

$$(A^{a\ldots e}_{A\ldots E})_{:k} = (A^{a\ldots e}_{A\ldots E})_{:K}\partial_k X^K, \quad (A^{a\ldots e}_{A\ldots E})_{:K} = (A^{a\ldots e}_{A\ldots E})_{:k}\partial_K x^k. \tag{2.189}$$

As an example of a two-point tensor field, consider a shifter of (2.47):

$$g^a_A(x, X) = \langle \boldsymbol{g}^a(x), \boldsymbol{G}_A(X)\rangle. \tag{2.190}$$

Treating x^a and X^A as independent variables, partial derivatives are

$$\left.\frac{\partial g^a_A}{\partial X^B}\right|_x = \langle \boldsymbol{g}^a, \partial_B \boldsymbol{G}_A\rangle = \langle \boldsymbol{g}^a, \overset{G}{\Gamma}{}^{\cdot\cdot C}_{BA}\boldsymbol{G}_C\rangle = \overset{G}{\Gamma}{}^{\cdot\cdot C}_{BA}\, g^a_C, \tag{2.191}$$

$$\left.\frac{\partial g^a_A}{\partial x^b}\right|_X = \langle \partial_b \boldsymbol{g}^a, \boldsymbol{G}_A\rangle = \langle -\overset{g}{\Gamma}{}^{\cdot\cdot a}_{bc}\boldsymbol{g}^c, \boldsymbol{G}_A\rangle = -\overset{g}{\Gamma}{}^{\cdot\cdot a}_{bc}\, g^c_A. \tag{2.192}$$

It follows that partial covariant derivatives of the shifter vanish identically [7, 14]:

$$g^a_{A;B} = \left.\frac{\partial g^a_A}{\partial X^B}\right|_x - \overset{G}{\Gamma}{}^{\cdot\cdot C}_{BA}\, g^a_C = \langle \boldsymbol{g}^a, \boldsymbol{G}_{A;B}\rangle = 0, \tag{2.193}$$

$$g^a_{A;b} = \left.\frac{\partial g^a_A}{\partial x^b}\right|_X + \overset{g}{\Gamma}{}^{\cdot\cdot a}_{bc}\, g^c_A = \langle \boldsymbol{g}^a_{;b}, \boldsymbol{G}_A\rangle = 0. \tag{2.194}$$

Consistent with the above developments, the total covariant derivative of the shifter also vanishes:

$$\begin{aligned}
g^a_{A:B} &= g^a_{A;B} + g^a_{A;b}\,\partial_B x^b \\
&= \partial_B g^a_A[x(X,t),t] - \overset{G}{\Gamma}{}^{\cdot\cdot C}_{BA}\, g^a_C + \overset{g}{\Gamma}{}^{\cdot\cdot a}_{bc}\, g^c_A\,\partial_B x^b \\
&= \left.\frac{\partial g^a_A}{\partial X^B}\right|_x + \left.\frac{\partial g^a_A}{\partial x^b}\right|_X \partial_B x^b - \overset{G}{\Gamma}{}^{\cdot\cdot C}_{BA}\, g^a_C + \overset{g}{\Gamma}{}^{\cdot\cdot a}_{bc}\, g^c_A\,\partial_B x^b \\
&= g^a_{A:b}\partial_B x^b \\
&= 0.
\end{aligned} \tag{2.195}$$

It follows that shifting of indices commutes with covariant differentiation.

2.3.3 *Generalized scalar products*

In mathematics and mechanics literature, various notations are often introduced to denote summation over multiple indices of higher-order tensors. Notation listed in what follows next is not used often in the present text, but is included here because it is encountered frequently elsewhere (*e.g.*, [7] and references therein).

Let $\boldsymbol{A}$ and $\boldsymbol{B}$ be second-order tensors. The following notation is introduced for their scalar products:

$$\langle \boldsymbol{A}, \boldsymbol{B} \rangle = \operatorname{tr}(\boldsymbol{A}\boldsymbol{B}) = A^{AB} B_{BA}, \quad \boldsymbol{A} : \boldsymbol{B} = \operatorname{tr}(\boldsymbol{A}\boldsymbol{B}^{\mathrm{T}}) = A^{AB} B_{AB}. \tag{2.196}$$

The transpose of a second-order tensor denotes a horizontal switch of indices:

$$A^{\mathrm{T}}_{AB} = A_{BA}, \qquad (A^{\mathrm{T}})^{AB} = A^{BA}. \tag{2.197}$$

A second-order tensor is symmetric if and only if it is equal to its transpose. Similar definitions apply when $\boldsymbol{A}$ and $\boldsymbol{B}$ are two-point tensors:

$$\langle \boldsymbol{A}, \boldsymbol{B} \rangle = A^{a}_{\cdot A} B^{A}_{\cdot a}, \qquad \boldsymbol{A} : \boldsymbol{B} = A^{a}_{\cdot A} B_{a}^{\cdot A}. \tag{2.198}$$

The colon is also used to denote summation over two indices with tensors of order higher than two, *e.g.*,

$$(\boldsymbol{C} : \boldsymbol{D})^{AB} = C^{ABCD} D_{CD}. \tag{2.199}$$

Chapter 3

Kinematics of Integrable Deformation

In this chapter, a mathematical framework describing kinematics of continuous bodies is presented. The description is directed towards bodies undergoing possibly large holonomic (*i.e.*, integrable) deformations.

3.1 The Deformation Gradient and Derived Quantities

The deformation gradient and its inverse are introduced. The Jacobian determinant associated with volume changes is defined, and Piola's identities are derived. Stretch, rotation, and strain tensors are defined. Compatibility conditions are derived. The Piola transform for a differential area element (*i.e.*, Nanson's formula) is derived.

3.1.1 *Deformation gradient*

The deformation gradient $\boldsymbol{F}$ is the two-point tensor

$$\boldsymbol{F} = F^a_{.A}\boldsymbol{g}_a \otimes \boldsymbol{G}^A, \tag{3.1}$$

with components, from (2.1),

$$F^a_{.A}(X,t) = \frac{\partial \varphi^a(X,t)}{\partial X^A} = \frac{\partial x^a(X,t)}{\partial X^A} = x^a_{,A}(X,t) = \partial_A x^a(X,t). \tag{3.2}$$

Similarly, the inverse deformation gradient and its components are

$$\boldsymbol{F}^{-1} = F^{-1A}_{\ .a}\boldsymbol{G}_A \otimes \boldsymbol{g}^a, \tag{3.3}$$

$$F^{-1A}_{\ .a}(x,t) = \frac{\partial \Phi^A(x,t)}{\partial x^a} = \frac{\partial X^A(x,t)}{\partial x^a} = X^A_{,a}(x,t) = \partial_a X^A(x,t). \tag{3.4}$$

By definition,

$$F^a_{.A}F^{-1A}_{\ .b} = \delta^a_b, \qquad F^{-1A}_{\ .a}F^a_{.B} = \delta^A_B. \tag{3.5}$$

Figure 3.1 illustrates, conceptually, the role of the deformation gradient and its inverse with regard to mappings between reference and current configurations.

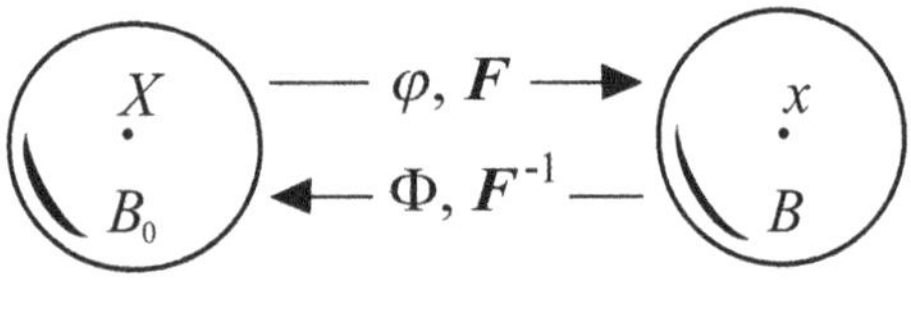

Fig. 3.1 Mappings between reference and current configurations of a deformable body.

The deformation gradient and its inverse are necessarily non-singular from invertibility properties of one-to-one mappings (at fixed t) φ and Φ; stronger conditions usually imposed in the context of continuum mechanics are that their determinants remain positive:

$$\det \boldsymbol{F} = \frac{1}{\det \boldsymbol{F}^{-1}} = \frac{1}{6} e^{ABC} e_{abc} F^{a}_{.A} F^{b}_{.B} F^{c}_{.C} > 0, \tag{3.6}$$

$$\det \boldsymbol{F}^{-1} = \frac{1}{\det \boldsymbol{F}} = \frac{1}{6} e^{abc} e_{ABC} F^{-1A}_{.a} F^{-1B}_{.b} F^{-1C}_{.c} > 0. \tag{3.7}$$

As will be demonstrated later in Section 3.1.2, such positive determinants ensure that differential volumes remain positive during deformation.

Partial differentiation (holding time t fixed) proceeds as

$$\partial_A(\cdot) = \frac{\partial(\cdot)}{\partial X^A} = \frac{\partial(\cdot)}{\partial x^a}\frac{\partial x^a}{\partial X^A} = \partial_a(\cdot)\partial_A x^a = \partial_a(\cdot) F^{a}_{.A}, \tag{3.8}$$

$$\partial_a(\cdot) = \frac{\partial(\cdot)}{\partial x^a} = \frac{\partial(\cdot)}{\partial X^A}\frac{\partial X^A}{\partial x^a} = \partial_A(\cdot)\partial_a X^A = \partial_A(\cdot) F^{-1A}_{.a}. \tag{3.9}$$

It follows from these identities and (2.4) that the deformation gradient and its inverse can be written

$$\boldsymbol{F} = \frac{\partial x^a}{\partial X^A}\partial_a \boldsymbol{x} \otimes \boldsymbol{G}^A = \partial_A \boldsymbol{x} \otimes \boldsymbol{G}^A, \tag{3.10}$$

$$\boldsymbol{F}^{-1} = \frac{\partial X^A}{\partial x^a}\partial_A \boldsymbol{X} \otimes \boldsymbol{g}^a = \partial_a \boldsymbol{X} \otimes \boldsymbol{g}^a. \tag{3.11}$$

Taking derivatives of Kronecker delta symbols with respect to deformation gradient components or inverse deformation gradient components leads to the following identities:

$$\begin{aligned}
\frac{\partial \delta^a_b}{\partial F^{-1C}_{\ .c}} &= \frac{\partial (F^a_{.A} F^{-1A}_{\ .b})}{\partial F^{-1C}_{\ .c}} \\
&= \frac{\partial F^a_{.A}}{\partial F^{-1C}_{\ .c}} F^{-1A}_{\ .b} + \frac{\partial F^{-1A}_{\ .b}}{\partial F^{-1C}_{\ .c}} F^a_{.A} \\
&= \frac{\partial F^a_{.A}}{\partial F^{-1C}_{\ .c}} F^{-1A}_{\ .b} + F^a_{.C} \delta^c_b \\
&= 0 \\
&\Rightarrow \frac{\partial F^a_{.A}}{\partial F^{-1C}_{\ .c}} = \frac{\partial}{\partial \partial_c X^C} \left(\frac{\partial x^a}{\partial X^A} \right) = -F^a_{.C} F^c_{.A},
\end{aligned} \tag{3.12}$$

$$\begin{aligned}
\frac{\partial \delta^A_B}{\partial F^c_{.C}} &= \frac{\partial (F^{-1A}_{\ .a} F^a_{.B})}{\partial F^c_{.C}} \\
&= \frac{\partial F^{-1A}_{\ .a}}{\partial F^c_{.C}} F^a_{.B} + \frac{\partial F^a_{.B}}{\partial F^c_{.C}} F^{-1A}_{\ .a} \\
&= \frac{\partial F^{-1A}_{\ .a}}{\partial F^c_{.C}} F^a_{.B} + F^{-1A}_{\ .c} \delta^C_B \\
&= 0 \\
&\Rightarrow \frac{\partial F^{-1A}_{\ .a}}{\partial F^c_{.C}} = \frac{\partial}{\partial \partial_C x^c} \left(\frac{\partial X^A}{\partial x^a} \right) = -F^{-1A}_{\ .c} F^{-1C}_{\ .a}.
\end{aligned} \tag{3.13}$$

Consider a differential line element $\mathrm{d}\boldsymbol{X}$ in the reference configuration. At a given time instant, such an element is mapped to its representation in the current configuration $\mathrm{d}\boldsymbol{x}$ via the Taylor series [7]

$$\begin{aligned}
\mathrm{d}x^a(X) &= (\partial_A x^a)\Big|_X \mathrm{d}X^A + \frac{1}{2!}(x^a_{:AB})\Big|_X \mathrm{d}X^A \mathrm{d}X^B \\
&\quad + \frac{1}{3!}(x^a_{:ABC})\Big|_X \mathrm{d}X^A \mathrm{d}X^B \mathrm{d}X^C + \cdots \\
&= (F^a_{.A})\Big|_X \mathrm{d}X^A + \frac{1}{2!}(F^a_{.A:B})\Big|_X \mathrm{d}X^A \mathrm{d}X^B \\
&\quad + \frac{1}{3!}(F^a_{.A:BC})\Big|_X \mathrm{d}X^A \mathrm{d}X^B \mathrm{d}X^C + \cdots ,
\end{aligned} \tag{3.14}$$

where components of the total covariant derivative of $\boldsymbol{F}$ are defined as in (2.185):

$$\begin{aligned}
F^a_{.A:B} &= \frac{\partial F^a_{.A}}{\partial X^B} - \overset{G}{\Gamma}{}^{..C}_{BA} F^a_{.C} + \overset{g}{\Gamma}{}^{..a}_{bc} F^c_{.A} F^b_{.B} \\
&= \partial_B(\partial_A x^a) - \overset{G}{\Gamma}{}^{..C}_{BA} \partial_C x^a + \overset{g}{\Gamma}{}^{..a}_{bc} \partial_A x^c \partial_B x^b \\
&= x^a_{:AB}.
\end{aligned} \tag{3.15}$$

Analogously, for the total covariant derivative of the inverse deformation gradient,

$$\begin{aligned}
F^{-1A}_{\ \ \ .a:B} &= F^{b}_{.B} F^{-1A}_{\ \ \ .a:b} \\
&= F^{b}_{.B}(\partial_b F^{-1A}_{\ \ \ .a} - \overset{g}{\Gamma}{}^{..c}_{ba} F^{-1A}_{\ \ \ .c} + \overset{G}{\Gamma}{}^{..A}_{DC} F^{-1C}_{\ \ \ .a} F^{-1D}_{\ \ \ .b}) \\
&= \partial_B x^b (\partial_b \partial_a X^A - \overset{g}{\Gamma}{}^{..c}_{ba} \partial_c X^A + \overset{G}{\Gamma}{}^{..A}_{DC} \partial_a X^C \partial_b X^D) \\
&= \partial_B x^b \, X^A_{:ab} \\
&= X^A_{:aB}.
\end{aligned} \tag{3.16}$$

The following notation scheme is evident [31]:

$$F^{a}_{.A} = \partial_A x^a = x^a_{,A} = x^a_{:A}, \qquad F^{-1A}_{\ \ \ .a} = \partial_a X^A = X^A_{,a} = X^A_{:a}. \tag{3.17}$$

From the identity

$$0 = (\delta^B_A)_{:C} = (F^a_{.A} F^{-1B}_{\ \ \ .a})_{:C} = F^a_{.A:C} F^{-1B}_{\ \ \ .a} + F^a_{.A} F^{-1B}_{\ \ \ .a:C}, \tag{3.18}$$

it follows that quantities in (3.15) and (3.16) are related by

$$\begin{aligned}
F^{-1A}_{\ \ \ .a:b} &= -F^{-1A}_{\ \ \ .c} F^{-1B}_{\ \ \ .a} F^{-1C}_{\ \ \ .b} F^{c}_{.B:C} \\
&\Leftrightarrow X^A_{:ab} = -\partial_c X^A \partial_a X^B \partial_b X^C x^c_{:BC}.
\end{aligned} \tag{3.19}$$

The third-order position gradient follows likewise as

$$F^a_{.A:BC} = (F^a_{.A:B})_{:C} = (x^a_{:AB})_{:C} = x^a_{:ABC} = \partial_C[\partial_B(\partial_A x^a)] + \cdots. \tag{3.20}$$

In direct notation,

$$F^a_{.A:BC} \boldsymbol{g}_a \otimes \boldsymbol{G}^A \otimes \boldsymbol{G}^B \otimes \boldsymbol{G}^C = \partial_C(\partial_B \boldsymbol{F} \otimes \boldsymbol{G}^B) \otimes \boldsymbol{G}^C. \tag{3.21}$$

Applying the total covariant derivative operation twice in succession to $F^a_{.A}$ gives the following full 15-term expression:

$$\begin{aligned}
F^a_{.A:BC} &= \partial_C(F^a_{.A:B}) - \overset{G}{\Gamma}{}^{..D}_{CA} F^a_{.D:B} \\
&\quad - \overset{G}{\Gamma}{}^{..D}_{CB} F^a_{.A:D} + \overset{g}{\Gamma}{}^{..a}_{cd} F^d_{.A:B} F^c_{.C} \\
&= \partial_C \partial_B F^a_{.A} - F^a_{.D} \partial_C \overset{G}{\Gamma}{}^{..D}_{BA} + F^e_{.C} F^d_{.A} F^b_{.B} \partial_e \overset{g}{\Gamma}{}^{..a}_{bd} \\
&\quad - \overset{G}{\Gamma}{}^{..D}_{BA} \partial_C F^a_{.D} - \overset{G}{\Gamma}{}^{..D}_{CA} \partial_B F^a_{.D} - \overset{G}{\Gamma}{}^{..D}_{CB} \partial_D F^a_{.A} \\
&\quad + \overset{g}{\Gamma}{}^{..a}_{bd} F^b_{.B} \partial_C F^d_{.A} + \overset{g}{\Gamma}{}^{..a}_{bd} F^d_{.A} \partial_C F^b_{.B} + \overset{g}{\Gamma}{}^{..a}_{cd} F^c_{.C} \partial_B F^d_{.A} \\
&\quad - \overset{G}{\Gamma}{}^{..D}_{CA} \overset{g}{\Gamma}{}^{..a}_{bd} F^d_{.D} F^b_{.B} - \overset{G}{\Gamma}{}^{..D}_{CB} \overset{g}{\Gamma}{}^{..a}_{db} F^b_{.A} F^d_{.D} \\
&\quad - \overset{G}{\Gamma}{}^{..E}_{BA} \overset{g}{\Gamma}{}^{..a}_{cd} F^c_{.C} F^d_{.E} + \overset{G}{\Gamma}{}^{..D}_{CA} \overset{G}{\Gamma}{}^{..E}_{BD} F^a_{.E} \\
&\quad + \overset{G}{\Gamma}{}^{..D}_{CB} \overset{G}{\Gamma}{}^{..E}_{DA} F^a_{.E} + \overset{g}{\Gamma}{}^{..a}_{cd} \overset{g}{\Gamma}{}^{..d}_{be} F^c_{.C} F^e_{.A} F^b_{.B}.
\end{aligned} \tag{3.22}$$

From the identity $\partial_A[\partial_B(\cdot)] = \partial_B[\partial_A(\cdot)]$ of (2.3),

$$\partial_B F^a_{.A} = \partial_B \partial_A x^a = \partial_A \partial_B x^a = \partial_A F^a_{.B} = \partial_{(B} F^a_{.A)}, \tag{3.23}$$

$$\partial_C \partial_B F^a_{.A} = \partial_C \partial_B \partial_A x^a = \partial_{(C} \partial_B \partial_{A)} x^a = \partial_{(C} \partial_B F^a_{.A)}. \tag{3.24}$$

From symmetry of the (torsion-free) Levi-Civita connection coefficients in both reference and current configurations, it also follows that

$$F^a_{.A:B} = F^a_{.B:A} = F^a_{.(A:B)}, \qquad F^a_{.A:BC} = F^a_{.(A:BC)}. \tag{3.25}$$

The first of (3.25) is evident from direct inspection of (3.15). The second is evident from (2.141), (2.165), and expansion of the total covariant derivative in terms of partial covariant derivatives:

$$\begin{aligned} F^a_{.A:BC} &= (F^a_{.A;B})_{:C} + (F^a_{.A;b} F^b_{.B})_{:C} \\ &= F^a_{.A;BC} + (F^a_{.A;B})_{;c} F^c_{.C} + (F^a_{.A;b})_{:C} F^b_{.B} + F^a_{.A;b} F^b_{.B:C} \\ &= F^a_{.A;BC} + (F^a_{.A;B})_{;c} F^c_{.C} + F^a_{.A;bc} F^b_{.B} F^c_{.C} \\ &\qquad + (F^a_{.A;b})_{;C} F^b_{.B} + F^a_{.A;b} F^b_{.B:C} \\ &= F^a_{.A;BC} + (F^a_{.A;b})_{;B} F^b_{.C} + F^a_{.A;bc} F^b_{.B} F^c_{.C} \\ &\qquad + (F^a_{.A;b})_{;C} F^b_{.B} + F^a_{.A;b} F^b_{.B:C} \\ &= F^a_{.A:CB}. \end{aligned} \tag{3.26}$$

The second total covariant derivative of $\boldsymbol{F}$ maps between configurations as

$$\begin{aligned} F^a_{.A:bc} &= (F^a_{.A:b})_{:C} F^{-1C}_{.c} \\ &= (F^a_{.A:B} F^{-1B}_{.b})_{:C} F^{-1C}_{.c} \\ &= F^a_{.A:BC} F^{-1B}_{.b} F^{-1C}_{.c} + F^a_{.A:B} F^{-1B}_{.b:c}. \end{aligned} \tag{3.27}$$

To first order in $\mathrm{d}\boldsymbol{X}$, (3.14) is

$$\mathrm{d}x^a(X) = (\partial_A x^a)\Big|_X \mathrm{d}X^A = (F^a_{.A})\Big|_X \mathrm{d}X^A, \qquad \mathrm{d}\boldsymbol{x} = \boldsymbol{F}\mathrm{d}\boldsymbol{X}. \tag{3.28}$$

Relationship (3.28) is the usual assumption from classical continuum field theory [31] and will often be used in subsequent derivations.

Covariant positions in referential and spatial coordinates are symbolically defined, respectively, as

$$\begin{aligned} x_a(X,t) &= x^b(X,t) g_{ab}[x(X,t)], \\ X_A(x,t) &= X^B(x,t) G_{AB}[X(x,t)]. \end{aligned} \tag{3.29}$$

Covariant deformation gradient components and covariant inverse deformation gradient components obey, respectively,

$$F_{aA} = g_{ab} F^b_{.A} = g_{ab} \partial_A x^b, \qquad F^{-1}_{Aa} = G_{AB} F^{-1B}_{.a} = G_{AB} \partial_a X^B. \tag{3.30}$$

The following identity then arises from (2.157) and (3.8):

$$\begin{aligned}\partial_A x_a &= \partial_A(x^b g_{ab}) \\ &= g_{ab}\partial_A x^b + x^b \partial_A g_{ab} \\ &= g_{ab} F^b_{\cdot A} + x^b \partial_A x^c \, \partial_c g_{ab} \\ &= F_{aA} + 2x^b F^c_{\cdot A} \overset{g}{\Gamma}{}^{\cdot\cdot d}_{c(a} g_{b)d}.\end{aligned} \tag{3.31}$$

Similarly, for the inverse motion, the following identity arises from (2.131) and (3.9):

$$\begin{aligned}\partial_a X_A &= \partial_a(X^B G_{AB}) \\ &= G_{AB}\partial_a X^B + X^B \partial_a G_{AB} \\ &= G_{AB} F^{-1B}_{\cdot a} + X^B \partial_a X^C \, \partial_C G_{AB} \\ &= F^{-1}_{Aa} + 2X^B F^{-1C}_{\cdot a} \overset{G}{\Gamma}{}^{\cdot\cdot D}_{C(A} G_{B)D}.\end{aligned} \tag{3.32}$$

It follows that, in general, $F_{aA} \neq \partial_A x_a$ unless metric components g_{ab} are constant, and likewise $F^{-1}_{Aa} \neq \partial_a X_A$ unless G_{AB} are constants.

Covariant components of differential line elements in configurations B_0 and B are defined, respectively, as

$$\mathrm{d}X_A(X) = \mathrm{d}X^B(X)\, G_{AB}(X), \qquad \mathrm{d}x_a(x) = \mathrm{d}x^b(x)\, g_{ab}(x), \tag{3.33}$$

such that squared lengths of line elements obey

$$|\mathrm{d}\boldsymbol{X}|^2 = \mathrm{d}\boldsymbol{X} \cdot \mathrm{d}\boldsymbol{X} = \mathrm{d}X^A G_{AB} \mathrm{d}X^B = \mathrm{d}X^A \mathrm{d}X_A, \tag{3.34}$$

$$|\mathrm{d}\boldsymbol{x}|^2 = \mathrm{d}\boldsymbol{x} \cdot \mathrm{d}\boldsymbol{x} = \mathrm{d}x^a g_{ab} \mathrm{d}x^b = \mathrm{d}x^a \mathrm{d}x_a. \tag{3.35}$$

Covariant quantities $\mathrm{d}X_A$ and $\mathrm{d}x_a$ defined in this way are not exact differentials of X_A and x_a defined in (3.29) unless respective metric components G_{AB} and g_{ab} are constant.

3.1.2 *Jacobian determinant*

The Jacobian determinant J provides the relationship between differential volume elements in reference and current configurations. Differential volume elements in the reference configuration ($\mathrm{d}V$) and current configuration ($\mathrm{d}v$) are written symbolically as

$$\begin{aligned}\mathrm{d}V &= \mathrm{d}\boldsymbol{X}_1 \cdot (\mathrm{d}\boldsymbol{X}_2 \times \mathrm{d}\boldsymbol{X}_3) \\ &= \epsilon_{ABC} \mathrm{d}X^A \mathrm{d}X^B \mathrm{d}X^C \\ &= \sqrt{G}\, \mathrm{d}X^1 \mathrm{d}X^2 \mathrm{d}X^3 \subset B_0,\end{aligned} \tag{3.36}$$

$$\begin{aligned} \mathrm{d}v &= \mathrm{d}\boldsymbol{x}_1 \cdot (\mathrm{d}\boldsymbol{x}_2 \times \mathrm{d}\boldsymbol{x}_3) \\ &= \epsilon_{abc}\mathrm{d}x^a\mathrm{d}x^b\mathrm{d}x^c \\ &= \sqrt{g}\,\mathrm{d}x^1\mathrm{d}x^2\mathrm{d}x^3 \subset B. \end{aligned} \tag{3.37}$$

It follows from (2.10) that

$$6\,\mathrm{d}X^A\mathrm{d}X^B\mathrm{d}X^C = \epsilon^{ABC}\mathrm{d}V. \tag{3.38}$$

Using linear transformation $\mathrm{d}x^a = F^a_{.A}\mathrm{d}X^A$ of (3.28),

$$\begin{aligned} \frac{\mathrm{d}v}{\mathrm{d}V} &= \frac{\epsilon_{abc}\mathrm{d}x^a\mathrm{d}x^b\mathrm{d}x^c}{\mathrm{d}V} \\ &= \frac{\epsilon_{abc}F^a_{.A}\mathrm{d}X^A F^b_{.B}\mathrm{d}X^B F^c_{.C}\mathrm{d}X^C}{\mathrm{d}V} \\ &= \epsilon_{abc}F^a_{.A}F^b_{.B}F^c_{.C}\frac{\mathrm{d}X^A\mathrm{d}X^B\mathrm{d}X^C}{\mathrm{d}V} \\ &= \epsilon_{abc}F^a_{.A}F^b_{.B}F^c_{.C}\frac{\epsilon^{ABC}}{6} \\ &= J. \end{aligned} \tag{3.39}$$

Thus the Jacobian determinant $J[\boldsymbol{F}(X,t), g(x), G(X)]$ is, from (2.40) and (2.41),

$$\begin{aligned} J &= 1/J^{-1} \\ &= \frac{1}{6}\epsilon^{ABC}\epsilon_{abc}\, F^a_{.A}F^b_{.B}F^c_{.C} \\ &= \frac{1}{6}\sqrt{g/G}\, e^{ABC}e_{abc}F^a_{.A}F^b_{.B}F^c_{.C} \\ &= \sqrt{g/G}\ \det \boldsymbol{F} \\ &= \sqrt{\det(g_{ab})/\det(G_{AB})}\ \det(F^a_{.A}) \\ &= \sqrt{\det(\partial_a \boldsymbol{x}\cdot\partial_b \boldsymbol{x})/\det(\partial_A \boldsymbol{X}\cdot\partial_B \boldsymbol{X})}\ \det(\partial_A x^a). \end{aligned} \tag{3.40}$$

Explicitly, the determinant of the mixed-variant, two-point deformation gradient tensor $\boldsymbol{F}$ entering J is

$$\begin{aligned} \det \boldsymbol{F} &= \det(F^a_{.A}) \\ &= \det(\partial_A x^a) \\ &= \det\begin{bmatrix} \partial x^1/\partial X^1 & \partial x^1/\partial X^2 & \partial x^1/\partial X^3 \\ \partial x^2/\partial X^1 & \partial x^2/\partial X^2 & \partial x^2/\partial X^3 \\ \partial x^3/\partial X^1 & \partial x^3/\partial X^2 & \partial x^3/\partial X^3 \end{bmatrix} \\ &= \det\begin{bmatrix} F^1_{.1} & F^1_{.2} & F^1_{.3} \\ F^2_{.1} & F^2_{.2} & F^2_{.3} \\ F^3_{.1} & F^3_{.2} & F^3_{.3} \end{bmatrix} \end{aligned}$$

$$
\begin{aligned}
&= F^1_{.1}(F^2_{.2}F^3_{.3} - F^3_{.2}F^2_{.3}) \\
&\quad - F^1_{.2}(F^2_{.1}F^3_{.3} - F^3_{.1}F^2_{.3}) + F^1_{.3}(F^2_{.1}F^3_{.2} - F^3_{.1}F^2_{.2}) \\
&= F^1_{.1}F^2_{.2}F^3_{.3} + F^1_{.2}F^2_{.3}F^3_{.1} + F^1_{.3}F^2_{.1}F^3_{.2} \\
&\quad - F^1_{.3}F^2_{.2}F^3_{.1} - F^1_{.2}F^2_{.1}F^3_{.3} - F^1_{.1}F^2_{.3}F^3_{.2}. \qquad (3.41)
\end{aligned}
$$

Similarly, inverse Jacobian determinant $J^{-1}[\boldsymbol{F}^{-1}(x,t), G(X), g(x)]$ is

$$
\begin{aligned}
J^{-1} &= 1/J \\
&= \frac{1}{6}\epsilon^{abc}\epsilon_{ABC}\, F^{-1A}_{\ \ .a}F^{-1B}_{\ \ .b}F^{-1C}_{\ \ .c} \\
&= \frac{1}{6}\sqrt{G/g}\; e^{abc}e_{ABC}F^{-1A}_{\ \ .a}F^{-1B}_{\ \ .b}F^{-1C}_{\ \ .c} \\
&= \sqrt{G/g}\ \det \boldsymbol{F}^{-1} \\
&= \sqrt{\det(G_{AB})/\det(g_{ab})}\ \det(F^{-1A}_{\ \ .a}) \\
&= \sqrt{\det(\partial_A \boldsymbol{X}\cdot\partial_B \boldsymbol{X})/\det(\partial_a \boldsymbol{x}\cdot\partial_b \boldsymbol{x})}\ \det(\partial_a X^A). \qquad (3.42)
\end{aligned}
$$

Explicitly, the determinant of the mixed-variant, two-point inverse deformation gradient tensor $\boldsymbol{F}^{-1}$ entering J^{-1} is

$$
\begin{aligned}
\det \boldsymbol{F}^{-1} &= \det(\boldsymbol{F}^{-1}) \\
&= (\det \boldsymbol{F})^{-1} \\
&= \det(F^{-1A}_{\ \ .a}) \\
&= \det(\partial_a X^A) \\
&= \det \begin{bmatrix} \partial X^1/\partial x^1 & \partial X^1/\partial x^2 & \partial X^1/\partial x^3 \\ \partial X^2/\partial x^1 & \partial X^2/\partial x^2 & \partial X^2/\partial x^3 \\ \partial X^3/\partial x^1 & \partial X^3/\partial x^2 & \partial X^3/\partial x^3 \end{bmatrix} \\
&= \det \begin{bmatrix} (F^{-1})^1_{.1} & (F^{-1})^1_{.2} & (F^{-1})^1_{.3} \\ (F^{-1})^2_{.1} & (F^{-1})^2_{.2} & (F^{-1})^2_{.3} \\ (F^{-1})^3_{.1} & (F^{-1})^3_{.2} & (F^{-1})^3_{.3} \end{bmatrix} \\
&= (F^{-1})^1_{.1}[(F^{-1})^2_{.2}(F^{-1})^3_{.3} - (F^{-1})^3_{.2}(F^{-1})^2_{.3}] \\
&\quad - (F^{-1})^1_{.2}[(F^{-1})^2_{.1}(F^{-1})^3_{.3} - (F^{-1})^3_{.1}(F^{-1})^2_{.3}] \\
&\quad + (F^{-1})^1_{.3}[(F^{-1})^2_{.1}(F^{-1})^3_{.2} - (F^{-1})^3_{.1}(F^{-1})^2_{.2}] \\
&= (F^{-1})^1_{.1}(F^{-1})^2_{.2}(F^{-1})^3_{.3} + (F^{-1})^1_{.2}(F^{-1})^2_{.3}(F^{-1})^3_{.1} \\
&\quad + (F^{-1})^1_{.3}(F^{-1})^2_{.1}(F^{-1})^3_{.2} - (F^{-1})^1_{.3}(F^{-1})^2_{.2}(F^{-1})^3_{.1} \\
&\quad - (F^{-1})^1_{.2}(F^{-1})^2_{.1}(F^{-1})^3_{.3} - (F^{-1})^1_{.1}(F^{-1})^2_{.3}(F^{-1})^3_{.2}. \qquad (3.43)
\end{aligned}
$$

The following identities are noted for determinants of generic non-singular square matrices $\boldsymbol{A}$ and $\boldsymbol{B}$:

$$
\det(\boldsymbol{AB}) = \det \boldsymbol{A} \det \boldsymbol{B}, \qquad \det(\boldsymbol{A}^{-1}) = (\det \boldsymbol{A})^{-1}. \qquad (3.44)
$$

When the motion of a body is restricted to rigid translation (or to no motion at all), then $F^a_{\bullet A} = g^a_A$ [the shifter of (2.47)], and from (2.54),

$$\boldsymbol{F} = g^a_A \boldsymbol{g}_a \otimes \boldsymbol{G}^A = \boldsymbol{g}_a \otimes \boldsymbol{g}^a = \boldsymbol{G}_A \otimes \boldsymbol{G}^A$$
$$\Rightarrow J = \sqrt{g/G}\det(g^a_A) = 1 \quad \text{(rigid translation).} \tag{3.45}$$

From identity (2.142), it follows that

$$\begin{aligned} \frac{\partial J}{\partial F^a_{\bullet A}} &= \frac{\partial(\sqrt{g/G}\ \det \boldsymbol{F})}{\partial F^a_{\bullet A}} \\ &= \sqrt{g/G}\ \frac{\partial \det \boldsymbol{F}}{\partial F^a_{\bullet A}} \\ &= \sqrt{g/G}\ \det \boldsymbol{F}\ F^{-1A}_{\ \ \bullet a} \\ &= JF^{-1A}_{\ \ \bullet a}. \end{aligned} \tag{3.46}$$

Similarly,

$$\begin{aligned} \frac{\partial J^{-1}}{\partial F^{-1A}_{\ \ \bullet a}} &= \frac{\partial(\sqrt{G/g}\ \det \boldsymbol{F}^{-1})}{\partial F^{-1A}_{\ \ \bullet a}} \\ &= \sqrt{G/g}\ \frac{\partial \det \boldsymbol{F}^{-1}}{\partial F^{-1A}_{\ \ \bullet a}} \\ &= \sqrt{G/g}\ \det \boldsymbol{F}^{-1}\ F^a_{\bullet A} \\ &= J^{-1}F^a_{\bullet A}. \end{aligned} \tag{3.47}$$

Therefore, inverses obey

$$F^{-1A}_{\ \ \bullet a} = \frac{\partial \ln J}{\partial F^a_{\bullet A}}, \tag{3.48}$$

$$F^a_{\bullet A} = \frac{\partial \ln(J^{-1})}{\partial F^{-1A}_{\ \ \bullet a}}. \tag{3.49}$$

Furthermore, from the chain rule,

$$\frac{\partial J^{-1}}{\partial F^a_{\bullet A}} = \frac{\partial J^{-1}}{\partial J}\frac{\partial J}{\partial F^a_{\bullet A}} = -J^{-1}F^{-1A}_{\ \ \bullet a}, \tag{3.50}$$

$$\frac{\partial J}{\partial F^{-1A}_{\ \ \bullet a}} = \frac{\partial J}{\partial J^{-1}}\frac{\partial J^{-1}}{\partial F^{-1A}_{\ \ \bullet a}} = -JF^a_{\bullet A}. \tag{3.51}$$

Application of (2.143) to the deformation gradient and its inverse leads to

$$\begin{aligned} \partial_A[\ln(\det \boldsymbol{F})] &= \frac{1}{\det \boldsymbol{F}}\partial_A \det \boldsymbol{F} \\ &= \frac{1}{\det \boldsymbol{F}}\frac{\partial \det \boldsymbol{F}}{\partial F^a_{\bullet B}}\partial_A F^a_{\bullet B} \\ &= F^{-1B}_{\ \ \bullet a}\partial_A F^a_{\bullet B}, \end{aligned} \tag{3.52}$$

$$\begin{aligned}\partial_a[\ln(\det \boldsymbol{F}^{-1})] &= \frac{1}{\det \boldsymbol{F}^{-1}}\partial_a \det \boldsymbol{F}^{-1} \\ &= \frac{1}{\det \boldsymbol{F}^{-1}}\frac{\partial \det \boldsymbol{F}^{-1}}{\partial F^{-1A}_{\ .b}}\partial_a F^{-1A}_{\ .b} \\ &= F^b_{.A}\partial_a F^{-1A}_{\ .b}.\end{aligned} \tag{3.53}$$

The following additional identities can be derived [7,31]:

$$JF^{-1A}_{\ .a} = \frac{1}{2}\epsilon_{abc}\epsilon^{ABC}F^b_{.B}F^c_{.C}, \tag{3.54}$$

$$J^{-1}F^a_{.A} = \frac{1}{2}\epsilon_{ABC}\epsilon^{abc}F^{-1B}_{\ .b}F^{-1C}_{\ .c}, \tag{3.55}$$

$$J\delta^A_B = \frac{1}{2}\epsilon_{acd}\epsilon^{ACD}F^c_{.C}F^d_{.D}F^a_{.B}, \tag{3.56}$$

$$J^{-1}\delta^a_b = \frac{1}{2}\epsilon_{ACD}\epsilon^{acd}F^{-1C}_{\ .c}F^{-1D}_{\ .d}F^{-1A}_{\ .b}. \tag{3.57}$$

Permutation tensors map between configurations via

$$\epsilon^{ABC} = J\epsilon^{abc}F^{-1A}_{\ .a}F^{-1B}_{\ .b}F^{-1C}_{\ .c}, \tag{3.58}$$

$$\epsilon_{abc} = J\epsilon_{ABC}F^{-1A}_{\ .a}F^{-1B}_{\ .b}F^{-1C}_{\ .c}, \tag{3.59}$$

$$\epsilon_{ABC} = J^{-1}\epsilon_{abc}F^a_{.A}F^b_{.B}F^c_{.C}, \tag{3.60}$$

$$\epsilon^{abc} = J^{-1}\epsilon^{ABC}F^a_{.A}F^b_{.B}F^c_{.C}. \tag{3.61}$$

One of Piola's identities is derived by taking the divergence of (3.47) [7]:

$$\begin{aligned}\left(\frac{\partial J^{-1}}{\partial F^{-1A}_{\ .a}}\right)_{:a} &= (J^{-1}F^a_{.A})_{:a} \\ &= (J^{-1}F^a_{.A})_{;a} - J^{-1}\overset{G}{\Gamma}{}^{..C}_{BA}F^a_{.C}F^{-1B}_{\ .a} \\ &= \partial_a(J^{-1}F^a_{.A}) + J^{-1}\overset{g}{\Gamma}{}^{..a}_{ab}F^b_{.A} - J^{-1}\overset{G}{\Gamma}{}^{..C}_{BA}\delta^B_C \\ &= \partial_a(J^{-1}F^a_{.A}) + J^{-1}\partial_a(\ln\sqrt{g})F^a_{.A} - J^{-1}\overset{G}{\Gamma}{}^{..B}_{BA} \\ &= \frac{1}{\sqrt{g}}\partial_a(\sqrt{g}J^{-1}F^a_{.A}) - J^{-1}\overset{G}{\Gamma}{}^{..B}_{BA} \\ &= J^{-1}\partial_B(F^a_{.A})F^{-1B}_{\ .a} + \frac{1}{\sqrt{g}}\partial_a(\sqrt{g}J^{-1})F^a_{.A} - J^{-1}\overset{G}{\Gamma}{}^{..B}_{BA} \\ &= J^{-1}F^{-1B}_{\ .a}\partial_B F^a_{.A} + \frac{1}{\sqrt{g}}\partial_a(\sqrt{G}\det \boldsymbol{F}^{-1})F^a_{.A} - J^{-1}\overset{G}{\Gamma}{}^{..B}_{BA} \\ &= J^{-1}F^{-1B}_{\ .a}\partial_B F^a_{.A} + \frac{1}{\sqrt{g}}\partial_a(\sqrt{G})\det \boldsymbol{F}^{-1}F^a_{.A} \\ &\quad + \frac{\sqrt{G}}{\sqrt{g}}\partial_a(\det \boldsymbol{F}^{-1})F^a_{.A} - J^{-1}\overset{G}{\Gamma}{}^{..B}_{BA}\end{aligned}$$

$$
\begin{aligned}
&= J^{-1}F^{-1B}_{\cdot a}\partial_B F^a_{\cdot A} + \frac{\sqrt{G}}{\sqrt{g}}\det \boldsymbol{F}^{-1}\frac{\partial_A(\sqrt{G})}{\sqrt{G}} \\
&\quad - \frac{\sqrt{G}}{\sqrt{g}}(\det \boldsymbol{F})^{-2}\partial_a(\det \boldsymbol{F})F^a_{\cdot A} - J^{-1}\overset{G}{\Gamma}{}^{\cdot\cdot B}_{BA} \\
&= J^{-1}F^{-1B}_{\cdot a}\partial_B F^a_{\cdot A} + J^{-1}\partial_A(\ln\sqrt{G}) \\
&\quad - \frac{\sqrt{G}}{\sqrt{g}}\frac{1}{(\det \boldsymbol{F})^2}\frac{\partial \det \boldsymbol{F}}{\partial F^b_{\cdot B}}\partial_a(F^b_{\cdot B})F^a_{\cdot A} - J^{-1}\overset{G}{\Gamma}{}^{\cdot\cdot B}_{BA} \\
&= J^{-1}F^{-1B}_{\cdot a}\partial_B F^a_{\cdot A} - \frac{\sqrt{G}}{\sqrt{g}}\frac{1}{(\det \boldsymbol{F})^2}\frac{\partial \det \boldsymbol{F}}{\partial F^b_{\cdot B}}\partial_A F^b_{\cdot B} \\
&\quad + J^{-1}\left(\overset{G}{\Gamma}{}^{\cdot\cdot B}_{BA} - \overset{G}{\Gamma}{}^{\cdot\cdot B}_{BA}\right) \\
&= J^{-1}F^{-1B}_{\cdot a}\partial_B F^a_{\cdot A} - \frac{1}{J\det \boldsymbol{F}}\frac{\partial \det \boldsymbol{F}}{\partial F^b_{\cdot B}}\partial_A F^b_{\cdot B} \\
&= J^{-1}F^{-1B}_{\cdot a}\partial_B F^a_{\cdot A} - \frac{1}{J^2}\frac{\partial J}{\partial F^b_{\cdot B}}\partial_A F^b_{\cdot B} \\
&= J^{-1}F^{-1B}_{\cdot a}\partial_B F^a_{\cdot A} - \frac{1}{J^2}(JF^{-1B}_{\cdot b})\partial_A F^b_{\cdot B} \\
&= J^{-1}\left(F^{-1B}_{\cdot a}\partial_B F^a_{\cdot A} - F^{-1B}_{\cdot b}\partial_A F^b_{\cdot B}\right) \\
&= J^{-1}F^{-1B}_{\cdot a}\left[\partial_B F^a_{\cdot A} - \partial_A F^a_{\cdot B}\right] \\
&= J^{-1}F^{-1B}_{\cdot a}\left[\partial_B(\partial_A x^a) - \partial_A(\partial_B x^a)\right] \\
&= 0.
\end{aligned}
\tag{3.62}
$$

The same steps performed on (3.46) with reference and spatial coordinates interchanged produce Piola's identity

$$
\begin{aligned}
\left(\frac{\partial J}{\partial F^a_{\cdot A}}\right)_{:A} &= (JF^{-1A}_{\cdot a})_{:A} \\
&= JF^b_{\cdot A}\left[\partial_b F^{-1A}_{\cdot a} - \partial_a F^{-1A}_{\cdot b}\right] \\
&= JF^b_{\cdot A}\left[\partial_b(\partial_a X^A) - \partial_a(\partial_b X^A)\right] \\
&= 0.
\end{aligned}
\tag{3.63}
$$

Setting $G_{AB} = \delta_{AB}$ and $g_{ab} = \delta_{ab}$, connection coefficients vanish and total covariant differentiation reduces to partial coordinate differentation, such that (3.62) and (3.63) yield the simpler identities

$$
\partial_a(F^a_{\cdot A}\det \boldsymbol{F}^{-1}) = 0, \qquad \partial_A(F^{-1A}_{\cdot a}\det \boldsymbol{F}) = 0. \tag{3.64}
$$

The referential gradient of the Jacobian determinant obeys, from the chain rule, (2.145) and (3.46),

$$\begin{aligned}
\partial_A J &= \frac{\partial J}{\partial X^A} \\
&= \frac{\partial J}{\partial F^a_{.B}}\partial_A F^a_{.B} + \frac{\partial J}{\partial g}\partial_A g + \frac{\partial J}{\partial G}\partial_A G \\
&= J F^{-1B}_{.a}\partial_A F^a_{.B} + 2g \overset{g}{\Gamma}{}^{..b}_{ab} F^a_{.A}\frac{\partial J}{\partial g} + 2G \overset{G}{\Gamma}{}^{..B}_{AB}\frac{\partial J}{\partial G} \\
&= J F^{-1B}_{.a}\partial_A F^a_{.B} + J \overset{g}{\Gamma}{}^{..b}_{ab} F^a_{.A} - J \overset{G}{\Gamma}{}^{..B}_{AB} \\
&= J F^{-1B}_{.a}(\partial_A F^a_{.B} + \overset{g}{\Gamma}{}^{..a}_{cb} F^c_{.A} F^b_{.B} - \overset{G}{\Gamma}{}^{..C}_{AB} F^a_{.C}) \\
&= J F^{-1B}_{.a} F^a_{.B:A} \\
&= \frac{\partial J}{\partial F^a_{.B}} F^a_{.B:A}.
\end{aligned} \tag{3.65}$$

The spatial gradient of the Jacobian determinant then follows as

$$\partial_a J = \frac{\partial J}{\partial x^a} = F^{-1A}_{.a}\partial_A J = J F^{-1A}_{.b} F^b_{.A:a} = \frac{\partial J}{\partial F^b_{.A}} F^b_{.A:a}. \tag{3.66}$$

The referential gradient of the inverse of the Jacobian determinant obeys

$$\begin{aligned}
\partial_A (J^{-1}) &= -\frac{1}{J^2}\partial_A J \\
&= -J^{-1} F^{-1B}_{.a} F^a_{.B:A} \\
&= J^{-1} F^a_{.B} F^{-1B}_{.a:A} \\
&= \frac{\partial J^{-1}}{\partial F^{-1B}_{.a}} F^{-1B}_{.a:A} \\
&= F^a_{.A}\partial_a (J^{-1}),
\end{aligned} \tag{3.67}$$

where (3.47) and the following identity have been applied:

$$0 = (\delta^B_C)_{:A} = (F^{-1B}_{.a} F^a_{.C})_{:A} = F^{-1B}_{.a:A} F^a_{.C} + F^a_{.C:A} F^{-1B}_{.a}. \tag{3.68}$$

From (3.66) and (3.67), Piola's identities can be derived succinctly as

$$\begin{aligned}
(J^{-1} F^a_{.A})_{:a} &= J^{-1} F^a_{.A:a} + F^a_{.A}\partial_a (J^{-1}) \\
&= J^{-1}(F^a_{.A:a} - F^a_{.A} F^{-1B}_{.b} F^b_{.B:a}) \\
&= J^{-1}(F^a_{.A:a} - F^{-1B}_{.b} F^b_{.B:A}) \\
&= J^{-1}(F^a_{.A:a} - F^{-1B}_{.b} F^b_{.A:B}) \\
&= J^{-1}(F^a_{.A:a} - F^a_{.A:a}) \\
&= 0,
\end{aligned} \tag{3.69}$$

$$
\begin{aligned}
(JF^{-1A}_{\cdot a})_{:A} &= JF^{-1A}_{\cdot a:A} + F^{-1A}_{\cdot a}\partial_a J \\
&= J(F^{-1A}_{\cdot a:A} + F^{-1A}_{\cdot a}F^{b}_{\cdot B:A}F^{-1B}_{\cdot b}) \\
&= J(F^{-1A}_{\cdot a:A} - F^{-1A}_{\cdot a}F^{b}_{\cdot B}F^{-1B}_{\cdot b:A}) \\
&= J(F^{-1A}_{\cdot a:A} - F^{b}_{\cdot B}F^{-1B}_{\cdot b:a}) \\
&= J(F^{-1A}_{\cdot a:A} - F^{b}_{\cdot B}F^{-1B}_{\cdot a:b}) \\
&= J(F^{-1A}_{\cdot a:A} - F^{-1A}_{\cdot a:A}) \\
&= 0. \qquad (3.70)
\end{aligned}
$$

Let vector field $\boldsymbol{A}(X) = A^A \boldsymbol{G}_A$ be the Piola transform of $\boldsymbol{a}(x) = a^a \boldsymbol{g}_a$:

$$A^A = JF^{-1A}_{\cdot a}a^a. \qquad (3.71)$$

Taking the divergence of (3.71) and applying the product rule for covariant differentiation along with identity (3.63) results in

$$
\begin{aligned}
A^A_{;A} &= A^A_{:A} \\
&= (JF^{-1A}_{\cdot a})_{:A}a^a + JF^{-1A}_{\cdot a}a^a_{:A} \\
&= JF^{-1A}_{\cdot a}a^a_{:A} \\
&= Ja^a_{:a} \\
&= Ja^a_{;a}. \qquad (3.72)
\end{aligned}
$$

The physical requirement that referential and spatial volume elements remain positive and bounded leads to the constraints

$$0 < \mathrm{d}v < \infty \Rightarrow 0 < J < \infty, \qquad 0 < \mathrm{d}V < \infty \Rightarrow 0 < J^{-1} < \infty. \qquad (3.73)$$

Since determinants of metric tensors are positive over any volume, it follows that $\det \boldsymbol{F} > 0$ and $\det \boldsymbol{F}^{-1} > 0$ as imposed already in (3.6) and (3.7). If, instead of (3.39), volume is defined as positive after inversion, then $\mathrm{d}v = |J|\mathrm{d}V$, regardless of the algebraic sign of $\det \boldsymbol{F}$. Because J is regarded as a continuous function of time t, attainment of a local state with $J(X,t) < 0$ from a reference state wherein $J(X,0) = 1$ would require that $J = 0$ at some instant during the deformation process, which is ruled out by the invertibility property of $\boldsymbol{F}$. Therefore, conditions (3.73) must hold regardless of whether volume is defined as positive after inversion.

3.1.3 *Rotation, stretch, and strain*

According to the polar decomposition theorem, the non-singular deformation gradient $\boldsymbol{F}$ can be decomposed into a product of an orthogonal tensor $\boldsymbol{R}$ and a stretch tensor $\boldsymbol{U}$ or $\boldsymbol{V}$:

$$\boldsymbol{F} = \boldsymbol{R}\boldsymbol{U} = \boldsymbol{V}\boldsymbol{R}, \qquad F^a_{\cdot A} = R^a_{\cdot B}U^B_{\cdot A} = V^a_{\cdot b}R^b_{\cdot A}. \qquad (3.74)$$

Tensor $\boldsymbol{R}(X,t)$ is a two-point tensor obeying the orthogonality conditions

$$\boldsymbol{R}^{-1} = \boldsymbol{R}^{\mathrm{T}}, \quad R^{a}_{.A} R^{\mathrm{T}\,A}_{\;.b} = \delta^{a}_{b}, \quad R^{a}_{.B} R^{\mathrm{T}\,A}_{\;.a} = \delta^{A}_{B}; \tag{3.75}$$

and because $\det \boldsymbol{F} > 0$, tensor $\boldsymbol{R}$ is proper orthogonal, *i.e.*, a rotation tensor:

$$\frac{1}{6}\epsilon^{ABC}\epsilon_{abc}\, R^{a}_{.A} R^{b}_{.B} R^{c}_{.C} = \sqrt{g/G}\, \det \boldsymbol{R} = 1. \tag{3.76}$$

From (3.75),

$$\boldsymbol{R}^{\mathrm{T}}\boldsymbol{R} = \boldsymbol{1} \Leftrightarrow R^{a}_{.A} R^{b}_{.B} g_{ab} = G_{AB}. \tag{3.77}$$

Right stretch tensor $\boldsymbol{U}$ and left stretch tensor $\boldsymbol{V}$ are symmetric and positive definite:

$$G_{AB} U^{B}_{.C} = G_{CB} U^{B}_{.A}, \qquad g_{ab} V^{b}_{.c} = g_{cb} V^{b}_{.a}; \tag{3.78}$$

$$\det \boldsymbol{U} > 0, \qquad \det \boldsymbol{V} > 0. \tag{3.79}$$

From (3.40), (3.44), (3.74), and (3.76), it follows that the Jacobian determinant is fully determined by the determinant of either stretch tensor:

$$J = \sqrt{g/G}\, \det \boldsymbol{R} \det \boldsymbol{U} = \sqrt{g/G}\, \det \boldsymbol{V} \det \boldsymbol{R} = \det \boldsymbol{U} = \det \boldsymbol{V}. \tag{3.80}$$

In coordinates, transposes of two-point tensors $\boldsymbol{F}$ and $\boldsymbol{R}$ are

$$\begin{aligned} \boldsymbol{F}^{\mathrm{T}} &= F^{\mathrm{T}\,A}_{\;.a}\, \boldsymbol{G}_A \otimes \boldsymbol{g}^a \\ &= F^{b}_{.B} g_{ab} G^{AB}\, \boldsymbol{G}_A \otimes \boldsymbol{g}^a \\ &= \partial_B x^b g_{ab} G^{AB}\, \boldsymbol{G}_A \otimes \boldsymbol{g}^a, \end{aligned} \tag{3.81}$$

$$\begin{aligned} \boldsymbol{R}^{\mathrm{T}} &= R^{\mathrm{T}\,A}_{\;.a}\, \boldsymbol{G}_A \otimes \boldsymbol{g}^a \\ &= R^{b}_{.B} g_{ab} G^{AB}\, \boldsymbol{G}_A \otimes \boldsymbol{g}^a \\ &= R^{-1\,A}_{\;.a}\, \boldsymbol{G}_A \otimes \boldsymbol{g}^a. \end{aligned} \tag{3.82}$$

The transpose of the inverse of the deformation gradient $\boldsymbol{F}^{-1}$ is

$$\begin{aligned} \boldsymbol{F}^{-\mathrm{T}} &= (\boldsymbol{F}^{-1})^{\mathrm{T}} \\ &= (\boldsymbol{F}^{\mathrm{T}})^{-1} \\ &= F^{-\mathrm{T}\,a}_{\;.A}\, \boldsymbol{g}_a \otimes \boldsymbol{G}^A \\ &= F^{-1\,B}_{\;.b} G_{AB} g^{ab}\, \boldsymbol{g}_a \otimes \boldsymbol{G}^A \\ &= \partial_b X^B G_{AB} g^{ab}\, \boldsymbol{g}_a \otimes \boldsymbol{G}^A. \end{aligned} \tag{3.83}$$

Noting that

$$\det \boldsymbol{F}^{\mathrm{T}} = \det(F^{b}_{.B}) \det(g_{ab}) \det(G^{AB}) = (g/G) \det \boldsymbol{F}, \tag{3.84}$$

Jacobian determinant J obeys

$$J = \sqrt{g/G}\det \boldsymbol{F} = \sqrt{G/g}[(g/G)\det \boldsymbol{F}] = \sqrt{G/g}\det \boldsymbol{F}^{\mathrm{T}}. \tag{3.85}$$

Symmetric deformation tensors are introduced as

$$\boldsymbol{C} = \boldsymbol{F}^{\mathrm{T}}\boldsymbol{F} = \boldsymbol{U}^2, \qquad C_{AB} = F^a_{.A} g_{ab} F^b_{.B}; \tag{3.86}$$

$$\boldsymbol{B} = \boldsymbol{F}\boldsymbol{F}^{\mathrm{T}} = \boldsymbol{V}^2, \qquad B^{ab} = F^a_{.A} G^{AB} F^b_{.B}. \tag{3.87}$$

These positive definite tensors are related by

$$\boldsymbol{B} = \boldsymbol{F}\boldsymbol{F}^{\mathrm{T}} = \boldsymbol{R}\boldsymbol{U}\boldsymbol{U}^{\mathrm{T}}\boldsymbol{R}^{\mathrm{T}} = \boldsymbol{R}\boldsymbol{C}\boldsymbol{R}^{-1}. \tag{3.88}$$

Deformation tensor $\boldsymbol{C}(X,t)$ will be used frequently later in this text and is given further attention. From linear transformation $\mathrm{d}x^a = F^a_{.A}\mathrm{d}X^A$ of (3.28), $\boldsymbol{C}$ can be used to determine the length of a deformed line element:

$$\begin{aligned} |\mathrm{d}\boldsymbol{x}|^2 &= \mathrm{d}\boldsymbol{x}\cdot\mathrm{d}\boldsymbol{x} \\ &= \mathrm{d}x^a\,\mathrm{d}x_a \\ &= F^a_{.A}\,\mathrm{d}X^A\,g_{ab}\,F^b_{.B}\,\mathrm{d}X^B \\ &= C_{AB}\,\mathrm{d}X^A\,\mathrm{d}X^B \\ &= \langle \boldsymbol{C}\,\mathrm{d}\boldsymbol{X}, \mathrm{d}\boldsymbol{X}\rangle. \end{aligned} \tag{3.89}$$

The following identity is also noted:

$$\boldsymbol{C} = \boldsymbol{U}^2 \Leftrightarrow C_{AB} = U^C_{.A} U^D_{.B} G_{CD}. \tag{3.90}$$

Furthermore, from (3.12), $\boldsymbol{C}$ obeys

$$C_{AB} = F^a_{.A} g_{ab} F^b_{.B} = g_{ab} F^a_{.B} F^b_{.A} = -g_{ab}\frac{\partial F^a_{.A}}{\partial F^{-1B}_{\ \ .b}} = -g_{ab}\frac{\partial(\partial_A x^a)}{\partial(\partial_b X^B)}. \tag{3.91}$$

The covariant right Cauchy-Green strain tensor $\boldsymbol{E}(X,t)$ is also encountered frequently:

$$\boldsymbol{E} = \frac{1}{2}(\boldsymbol{F}^{\mathrm{T}}\boldsymbol{F} - \boldsymbol{G}), \qquad E_{AB} = \frac{1}{2}(F^a_{.A} g_{ab} F^b_{.B} - G_{AB}). \tag{3.92}$$

Notice that $\boldsymbol{E}$ provides the difference in squared lengths of deformed and referential line elements:

$$|\mathrm{d}\boldsymbol{x}|^2 - |\mathrm{d}\boldsymbol{X}|^2 = \mathrm{d}\boldsymbol{x}\cdot\mathrm{d}\boldsymbol{x} - \mathrm{d}\boldsymbol{X}\cdot\mathrm{d}\boldsymbol{X} = 2\langle \boldsymbol{E}\,\mathrm{d}\boldsymbol{X}, \mathrm{d}\boldsymbol{X}\rangle. \tag{3.93}$$

The mixed-variant version of $\boldsymbol{E}$ is

$$\boldsymbol{E} = \frac{1}{2}(\boldsymbol{C} - \boldsymbol{1}), \qquad E^A_{.B} = \frac{1}{2}(F^{.A}_{a} F^a_{.B} - \delta^A_B) = E_{BC} G^{CA}. \tag{3.94}$$

The following useful identities are noted:

$$\frac{\partial C_{AB}}{\partial g_{ab}} = F^a_{.A} F^b_{.B}, \quad \frac{\partial C_{AB}}{\partial F^a_{.C}} = 2 g_{ab} \delta^C_{(A} F^b_{.B)}, \quad \frac{\partial E_{AB}}{\partial C_{CD}} = \frac{1}{2} \delta^{(C}_{(A} \delta^{D)}_{B)}. \tag{3.95}$$

Noting that the determinant of the mixed-variant version of $\boldsymbol{C}$ is

$$\det \boldsymbol{C} = \det(C^A_{.B}) = \det(g_{ab} G^{AC} F^a_{.C} F^b_{.B}) = (g/G)(\det \boldsymbol{F})^2 = J^2, \tag{3.96}$$

and using (2.142), the following identity is derived, holding G_{AB} fixed:

$$\begin{aligned} \frac{\partial J}{\partial E_{AB}} &= 2 \frac{\partial J}{\partial C_{AB}} \\ &= \frac{1}{J} \frac{\partial (J^2)}{\partial C_{AB}} \\ &= \frac{1}{J} G^{AC} \frac{\partial (J^2)}{\partial C^C_{.B}} \\ &= \frac{1}{J} G^{AC} \frac{\partial \det \boldsymbol{C}}{\partial C^C_{.B}} \\ &= \frac{1}{J} G^{AC} (\det \boldsymbol{C}) C^{-1B}_{.C} \\ &= J C^{-1BA} \\ &= J C^{-1AB} \\ &= J F^{-1A}_{.a} g^{ab} F^{-1B}_{.b} \\ &= J\, \partial_a X^A \, g^{ab} \, \partial_b X^B. \end{aligned} \tag{3.97}$$

Similarly,

$$\begin{aligned} \frac{\partial \ln J}{\partial C_{AB}} &= \frac{1}{J} \frac{\partial J}{\partial C_{AB}} \\ &= \frac{1}{2J^2} \frac{\partial (J^2)}{\partial C_{AB}} \\ &= \frac{1}{2J^2} \frac{\partial \det \boldsymbol{C}}{\partial C_{AB}} \\ &= \frac{1}{2J^2} (\det \boldsymbol{C}) C^{-1BA} \\ &= \frac{1}{2} C^{-1AB} \\ &= \frac{1}{2} \partial_a X^A \, g^{ab} \, \partial_b X^B. \end{aligned} \tag{3.98}$$

3.1.4 *Displacement*

The displacement field $\boldsymbol{u}$ can be introduced in Euclidean space:

$$\boldsymbol{u}(X,t) = u^A(X,t)\boldsymbol{G}_A(X), \qquad \boldsymbol{u}(x,t) = u^a(x,t)\boldsymbol{g}_a(x). \tag{3.99}$$

Referential and spatial components of displacement are related via the shifter:

$$u^a(x,t) = g^a_A(x,X)U^A(X,t), \qquad u^A(X,t) = g^A_a(X,x)u^a(x,t). \tag{3.100}$$

Displacement gradients are related to deformation gradients as follows [7, 15]:

$$F^a_{\bullet A} = g^a_B(u^B_{;A} + \delta^B_A), \qquad F^{-1A}_{\bullet a} = g^A_b(\delta^b_a - u^b_{;a}). \tag{3.101}$$

It follows from (2.193) and (3.28) that to first order, differential line elements in reference and current configurations are related by

$$\begin{aligned} \mathrm{d}x^a - g^a_A \mathrm{d}X^A &= (F^a_{\bullet A} - g^a_A)\mathrm{d}X^A \\ &= g^a_B u^B_{;A}\mathrm{d}X^A \\ &= (g^a_B u^B)_{;A}\mathrm{d}X^A \\ &= u^a_{;A}\mathrm{d}X^A. \end{aligned} \tag{3.102}$$

The Jacobian determinant and its inverse can be expressed as

$$\begin{aligned} J &= \sqrt{g/G}\det(F^a_{\bullet B}) \\ &= \sqrt{g/G}\det(g^a_A)\det(u^A_{;B} + \delta^A_B) \\ &= \det(u^A_{;B} + \delta^A_B), \end{aligned} \tag{3.103}$$

$$\begin{aligned} J^{-1} &= \sqrt{G/g}\det(F^{-1A}_{\bullet b}) \\ &= \sqrt{G/g}\det(g^A_a)\det(\delta^a_b - u^a_{;b}) \\ &= \det(\delta^a_b - u^a_{;b}). \end{aligned} \tag{3.104}$$

The deformation tensor $\boldsymbol{C}$ of (3.86) is, in terms of displacement,

$$\begin{aligned} C_{AB} &= F_{aA}F^a_{\bullet B} \\ &= g_{aD}(u^D_{;A} + \delta^D_A)g^a_C(u^C_{;B} + \delta^C_B) \\ &= G_{CD}(u^D_{;A} + \delta^D_A)(u^C_{;B} + \delta^C_B) \\ &= (u^C_{;A} + \delta^C_A)(u_{C;B} + G_{BC}) \\ &= u^C_{;A}G_{BC} + u_{C;B}\delta^C_A + u^C_{;A}u_{C;B} + G_{BC}\delta^C_A \\ &= u_{A;B} + u_{B;A} + u^C_{;A}u_{C;B} + G_{AB}. \end{aligned} \tag{3.105}$$

The strain tensor $\boldsymbol{E}$ of (3.92) is then

$$E_{AB} = \frac{1}{2}(C_{AB} - G_{AB}) = u_{(A;B)} + \frac{1}{2}u^C_{;A}u_{C;B}. \tag{3.106}$$

3.1.5 *Displacement potential*

Let deformation gradient $\boldsymbol{F}$ be restricted to a pure stretch field:

$$\begin{aligned}&\boldsymbol{F}(X,t)=\boldsymbol{U}(X,t),\\&F^a_{.A}=\partial_A x^a=g^a_B U^B_{.A}=g^{aB}U_{BA}=g^{aB}U_{AB}\quad\text{(stretch)}.\end{aligned}\tag{3.107}$$

Assign basis vectors $\boldsymbol{g}_a$ and $\boldsymbol{G}_A$ in such a way that shifter components g^a_B and g^{aB} are constant over the domain. Define shifted spatial coordinates:

$$\begin{aligned}x^B(X,t)&=g^B_a x^a(X,t)=g^{aB}x_a(X,t),\\x^a(X,t)&=g^a_B x^B(X,t)=g^{aB}x_B(X,t).\end{aligned}\tag{3.108}$$

The deformation gradient becomes

$$F^a_{.A}=\partial_A x^a=\partial_A(g^{aB}x_B)=g^{aB}\partial_A x_B.\tag{3.109}$$

Comparing (3.107) and (3.109), it follows that

$$U_{AB}=\partial_A x_B=\partial_B x_A=U_{BA}.\tag{3.110}$$

These are necessary and sufficient conditions for the existence of a local displacement potential $\vartheta(X,t)$:

$$x_A=\partial_A\vartheta=\frac{\partial\vartheta}{\partial X^A}.\tag{3.111}$$

When domain B_0 is simply connected, then scalar field ϑ is single valued for all $X\in B_0$. The stretch field obeys

$$U_{AB}=\partial_A\partial_B\vartheta=\frac{\partial^2\vartheta}{\partial X^A\partial X^B}=\frac{\partial^2\vartheta}{\partial X^B\partial X^A}=\partial_B\partial_A\vartheta=U_{BA},\tag{3.112}$$

and since $\boldsymbol{U}$ is positive definite,

$$\det\left[\frac{\partial^2\vartheta}{\partial X^A\partial X^B}\right]>0,\tag{3.113}$$

implying that $\vartheta(X,t)$ is a convex function of X^A at fixed t. A simple example of a displacement potential is $\vartheta=(c/2)\boldsymbol{X}\cdot\boldsymbol{X}$, with $c>0$ a constant, which produces a volumetric deformation $x_A=cX_A$ and a spherically symmetric deformation gradient of the form $\boldsymbol{F}=\boldsymbol{U}=c\boldsymbol{1}$.

3.1.6 *Compatibility conditions*

In what follows in Section 3.1.6, Cartesian basis vectors $\boldsymbol{g}_a = \boldsymbol{e}_a$ are selected in the current configuration such that the deformation gradient

$$\boldsymbol{F}(X,t) = F^a_{.A}(X,t)\ \boldsymbol{e}_a \otimes \boldsymbol{G}^A(X). \tag{3.114}$$

By definition, Cartesian basis vectors obey

$$\langle \boldsymbol{e}^a, \boldsymbol{e}_b \rangle = \delta^a_b, \qquad \boldsymbol{e}^a \cdot \boldsymbol{e}^b = \delta^{ab}, \qquad \boldsymbol{e}_a \cdot \boldsymbol{e}_b = \delta_{ab}. \tag{3.115}$$

Convected basis vectors and their reciprocals are defined as

$$\boldsymbol{G}'^A(x,t) = F^{-1A}_{\ \ .a}(x,t)\boldsymbol{e}^a, \qquad \boldsymbol{G}'_A(X,t) = F^a_{.A}(X,t)\boldsymbol{e}_a. \tag{3.116}$$

It follows that

$$\langle \boldsymbol{G}'^A, \boldsymbol{G}'_B \rangle = F^{-1A}_{\ \ .a} F^b_{.B} \langle \boldsymbol{e}^a, \boldsymbol{e}_b \rangle = F^{-1A}_{\ \ .a} F^b_{.B} \delta^a_b = \delta^A_B. \tag{3.117}$$

In convected coordinate form, (3.114) becomes

$$\begin{aligned} \boldsymbol{F}(X,t) &= [F^a_{.A}(X,t)\ \boldsymbol{e}_a] \otimes \boldsymbol{G}^A(X) \\ &= \boldsymbol{G}'_A(X,t) \otimes \boldsymbol{G}^A(X) \\ &= \delta^A_B \boldsymbol{G}'_A(X,t) \otimes \boldsymbol{G}^B(X). \end{aligned} \tag{3.118}$$

The metric tensor formed from convected basis vectors is the deformation tensor of (3.86) when Cartesian spatial coordinates are used:

$$\boldsymbol{G}'_A \cdot \boldsymbol{G}'_B = F^a_{.A} F^b_{.B} \boldsymbol{e}_a \cdot \boldsymbol{e}_b = F^a_{.A} \delta_{ab} F^b_{.B} = C_{AB}. \tag{3.119}$$

Referential gradients of convected basis vectors are

$$\partial_B \boldsymbol{G}'_A = \partial_B F^a_{.A} \boldsymbol{e}_a = F^{-1D}_{\ \ .a} \partial_B F^a_{.A} \boldsymbol{G}'_D = \overset{C}{\Gamma}{}^{..D}_{BA} \boldsymbol{G}'_D, \tag{3.120}$$

where the following connection coefficients are defined:

$$\overset{C}{\Gamma}{}^{..D}_{BA} = F^{-1D}_{\ \ .a} \partial_B F^a_{.A} = -F^a_{.A} \partial_B F^{-1D}_{\ \ .a}. \tag{3.121}$$

The final equality is obtained from $\partial_B(F^a_{.A} F^{-1D}_{\ \ .a}) = \partial_B \delta^D_A = 0$. The torsion of this connection vanishes identically as a consequence of the compatibility conditions (*i.e.*, local integrability conditions) for $\boldsymbol{F}$:

$$\begin{aligned} \partial_B F^a_{.A} &= \frac{\partial^2 x^a}{\partial X^B \partial X^A} = \frac{\partial^2 x^a}{\partial X^A \partial X^B} = \partial_A F^a_{.B} \\ &\Rightarrow \partial_{[B} F^a_{.A]} = \partial_{[B} \partial_{A]} x^a = 0; \end{aligned} \tag{3.122}$$

$$\overset{C}{T}{}^{..D}_{BA} = \overset{C}{\Gamma}{}^{..D}_{[BA]} = F^{-1D}_{\ \ .a} \partial_{[B} F^a_{.A]} = 0. \tag{3.123}$$

The coefficients of (3.121) are equivalent to those of the Levi-Civita connection associated with the metric C_{AB}:

$$\begin{aligned}
2\overset{C}{\Gamma}{}^{..A}_{BC} &= 2\overset{C}{\Gamma}{}^{..A}_{(BC)} \\
&= 2F^{-1A}_{\ .a}\partial_{(B}F^{a}_{.C)} \\
&= 2F^{-1A}_{\ .a}\partial_{(B}F^{a}_{.C)} \\
&\quad + 2F^{-1A}_{\ .a}\delta^{ab}F^{-1D}_{\ .b}\delta_{cd}[F^{c}_{.B}\partial_{[C}F^{d}_{.D]} + F^{c}_{.C}\partial_{[B}F^{d}_{.D]}] \\
&= F^{-1A}_{\ .a}\delta^{ab}F^{-1D}_{\ .b}\delta_{cd}[F^{d}_{.D}\partial_C F^{c}_{.B} + F^{d}_{.D}\partial_B F^{c}_{.C} + F^{c}_{.B}\partial_C F^{d}_{.D} \\
&\quad - F^{c}_{.B}\partial_D F^{d}_{.C} + F^{c}_{.C}\partial_B F^{d}_{.D} - F^{c}_{.C}\partial_D F^{d}_{.B}] \\
&= F^{-1A}_{\ .a}\delta^{ab}F^{-1D}_{\ .b}\delta_{cd}[(F^{d}_{.D}\partial_C F^{c}_{.B} + F^{c}_{.B}\partial_C F^{d}_{.D}) \\
&\quad + (F^{d}_{.D}\partial_B F^{c}_{.C} + F^{c}_{.C}\partial_B F^{d}_{.D}) \\
&\quad - (F^{d}_{.C}\partial_D F^{c}_{.B} + F^{c}_{.B}\partial_D F^{d}_{.C})] \\
&= C^{-1AD}[\partial_C C_{BD} + \partial_B C_{CD} - \partial_D C_{BC}].
\end{aligned} \tag{3.124}$$

Differentiating the connection coefficients of (3.121),

$$\begin{aligned}
\partial_E(F^{a}_{.D}\overset{C}{\Gamma}{}^{..D}_{BA}) &= \partial_E(F^{a}_{.D}F^{-1D}_{\ .b}\partial_B F^{b}_{.A}) \\
&= \partial_E(\partial_B F^{a}_{.A}) \\
&= \partial_E\partial_B\partial_A x^a.
\end{aligned} \tag{3.125}$$

Applying the product rule to the left of (3.125),

$$\begin{aligned}
\partial_E(F^{a}_{.D}\overset{C}{\Gamma}{}^{..D}_{BA}) &= \partial_E F^{a}_{.D}\overset{C}{\Gamma}{}^{..D}_{BA} + F^{a}_{.D}\partial_E\overset{C}{\Gamma}{}^{..D}_{BA} \\
&= F^{a}_{.D}\left(\overset{C}{\Gamma}{}^{..D}_{EF}\overset{C}{\Gamma}{}^{..F}_{BA} + \partial_E\overset{C}{\Gamma}{}^{..D}_{BA}\right).
\end{aligned} \tag{3.126}$$

Skew parts of (3.125) and (3.126) provide

$$\begin{aligned}
F^{a}_{.D}\left(\partial_{[E}\overset{C}{\Gamma}{}^{..D}_{B]A} + \overset{C}{\Gamma}{}^{..D}_{[E|F|}\overset{C}{\Gamma}{}^{..F}_{B]A}\right) &= \frac{1}{2}F^{a}_{.D}\overset{C}{R}{}^{...D}_{EBA} \\
&= \partial_{[E}\partial_{B]}\partial_A x^a \\
&= 0.
\end{aligned} \tag{3.127}$$

Since $\boldsymbol{F}$ is non-singular by definition, compatibility conditions for $\boldsymbol{C}$ correspond to vanishing of the Riemann-Christoffel curvature tensor formed from the Levi-Civita connection obtained from metric components C_{AB}, *i.e.*,

$$\overset{C}{R}{}^{...D}_{EBA} = 0. \tag{3.128}$$

Compatibility conditions for the deformation can thus be summarized as

$$F^{a}_{.A} = \partial_A x^a \Rightarrow \partial_{[B}F^{a}_{.A]} = 0 \Rightarrow \overset{C}{R}{}^{...D}_{EBA} = 0. \tag{3.129}$$

Considering the reverse argument, given a twice differentiable, symmetric, positive definite tensor field $\boldsymbol{C}(X)$ whose Riemann-Christoffel curvature tensor vanishes at a point X, a compatible mapping $\varphi(X)$ exists that yields $C_{AB} = (\partial_A \varphi^a) g_{ab} (\partial_B \varphi^a)$ in the local neighborhood of $X \in B_0$ [24]. However, mapping φ need not continuously cover all of B_0 even if the curvature tensor of $C_{AB}(X)$ vanishes at each point X. Furthermore, vanishing of the curvature tensor of $\boldsymbol{C}$, while necessary, is not sufficient to ensure integrability of an arbitrary twice differentiable and invertible tensor field $\boldsymbol{F}$ that is not symmetric, since rotation $\boldsymbol{R}$ associated with $\boldsymbol{F}$ may have non-vanishing contortion (see Section 4.1.6), leading to incompatibility of $\boldsymbol{F}$.

3.1.7 *Nanson's formula*

Oriented area elements in reference and current configurations can be written respectively as

$$N_A \mathrm{d}S = \epsilon_{ABC} \mathrm{d}X^B \mathrm{d}X^C, \qquad n_a \mathrm{d}s = \epsilon_{abc} \mathrm{d}x^b \mathrm{d}x^c. \tag{3.130}$$

It is understood that $\mathrm{d}X^B$ and $\mathrm{d}X^C$ represent components of different line segments [hence $\mathrm{d}X^B \mathrm{d}X^C \neq \mathrm{d}X^{(B} \mathrm{d}X^{C)}$], and similarly for $\mathrm{d}x^b$ and $\mathrm{d}x^c$. In configuration B_0, $\boldsymbol{N}(X)$ is a unit normal covariant vector to scalar differential area element $\mathrm{d}S$. In configuration B, $\boldsymbol{n}(x,t)$ is a unit normal covariant vector to scalar differential area element $\mathrm{d}s$. Oriented area elements map between configurations via Nanson's formula, which can be derived directly using (3.28) and (3.59):

$$\begin{aligned} n_a \mathrm{d}s &= (\epsilon_{abc})(\mathrm{d}x^b \mathrm{d}x^c) \\ &= (J \epsilon_{AEF} F^{-1A}_{.a} F^{-1E}_{.b} F^{-1F}_{.c})(F^b_{.B} \mathrm{d}X^B F^c_{.C} \mathrm{d}X^C) \\ &= J F^{-1A}_{.a} \epsilon_{ABC} \mathrm{d}X^B \mathrm{d}X^C \\ &= J F^{-1A}_{.a} N_A \mathrm{d}S. \end{aligned} \tag{3.131}$$

Squaring (3.131), noting that $N_A N^A = n_a n^a = 1$, dividing by $\mathrm{d}S^2$, and taking the square root leads to the ratio

$$\frac{\mathrm{d}s}{\mathrm{d}S} = (J^2 F^{-1A}_{.a} g^{ab} F^{-1B}_{.b} N_A N_B)^{1/2} = J (C^{-1AB} N_A N_B)^{1/2}. \tag{3.132}$$

3.2 Integral Theorems

General forms of two integral theorems used frequently in continuum mechanics are listed. Gauss's theorem relates volume and surface integrals. Stokes's theorem relates surface and line integrals.

3.2.1 Gauss's theorem

Let $\boldsymbol{A}$ be a scalar, vector, or tensor field of arbitrary order with continuous first derivatives with respect to local coordinates, and let ∇ be the covariant derivative or gradient operator with respect to the Levi-Civita connection of these coordinates. Let V denote a simply connected volume enclosed by a single continuous surface S with unique unit outward normal covariant vector $\boldsymbol{N}$. The generalized Gauss's theorem is written [22]

$$\int_V \nabla \star \boldsymbol{A}\, \mathrm{d}V = \oint_S \boldsymbol{N} \star \boldsymbol{A}\, \mathrm{d}S, \tag{3.133}$$

where $\star$ denotes a generic operator obeying the distributive property. Standard examples include the scalar products $\cdot$ and $\langle\ ,\ \rangle$, the vector cross product $\times$, and the outer product $\otimes$. Gauss's theorem can also be applied over the union of disconnected surfaces completely enclosing a volume that is not simply connected.

As a first example, let $\boldsymbol{A} \to \boldsymbol{V}(X) = V^A \boldsymbol{G}_A$ be a differentiable vector field in the reference configuration, and let $\star \to \langle\ ,\ \rangle$, *i.e.*, the scalar product operator. The divergence theorem for a vector field is then obtained from (3.133):

$$\int_V \langle \overset{G}{\nabla}, \boldsymbol{V} \rangle\, \mathrm{d}V = \oint_S \langle \boldsymbol{N}, \boldsymbol{V} \rangle\, \mathrm{d}S, \tag{3.134}$$

or in coordinates,

$$\int_V V^A_{;A}\, \mathrm{d}V = \oint_S V^A N_A\, \mathrm{d}S. \tag{3.135}$$

As a second example, let $\boldsymbol{A} \to \boldsymbol{x}(X,t)$ be the spatial position, and let $\star \to \otimes$:

$$\int_V \overset{G}{\nabla} \otimes \boldsymbol{x}\, \mathrm{d}V = \oint_S \boldsymbol{N} \otimes \boldsymbol{x}\, \mathrm{d}S, \tag{3.136}$$

or in coordinates,

$$\int_V x^a_{;A}\, \mathrm{d}V = \oint_S x^a N_A\, \mathrm{d}S. \tag{3.137}$$

Because (3.137) involves integration of components of vector fields on each configuration, basis vectors in configurations B_0 and B must be held fixed during the integration. Hence, in this second example only, $\partial_B \boldsymbol{G}_A = 0$ and $(\cdot)_{;A} = \partial_A(\cdot)$, and similarly for the current configuration. It follows that (3.137) describes the referential volume integral of the deformation gradient $\boldsymbol{F}(X,t)$, *i.e.*,

$$\oint_S x^a N_A\, \mathrm{d}S = \int_V x^a_{;A}\, \mathrm{d}V = \int_V \partial_A x^a\, \mathrm{d}V = \int_V F^a_{.A}\, \mathrm{d}V. \tag{3.138}$$

As a third example of Gauss's theorem, integrated spatial volume $v(t)$ of a simply connected body can be determined from motion of its referential boundary S via [6]

$$\begin{aligned}
v &= \int_v \mathrm{d}v \\
&= \int_V J\mathrm{d}V \\
&= \frac{1}{6}\int_V \epsilon_{abc}\epsilon^{ABC} F^a_{.A} F^b_{.B} F^c_{.C}\mathrm{d}V \\
&= \frac{1}{6}\int_V \epsilon_{abc}\epsilon^{ABC} \partial_A x^a \partial_B x^b \partial_C x^c \mathrm{d}V \\
&= \frac{1}{6}\int_V \epsilon_{abc}\epsilon^{ABC} x^a_{:A} \partial_B x^b \partial_C x^c \mathrm{d}V \\
&= \frac{1}{6}\int_V (\epsilon_{abc}\epsilon^{ABC} x^a \partial_B x^b \partial_C x^c)_{:A}\mathrm{d}V \\
&= \frac{1}{6}\int_V (\epsilon_{abc}\epsilon^{ABC} x^a \partial_B x^b \partial_C x^c)_{;A}\mathrm{d}V \\
&= \frac{1}{6}\oint_S (\epsilon_{abc}\epsilon^{ABC} \partial_B x^b \partial_C x^c) x^a N_A \mathrm{d}S \\
&= \frac{1}{3}\oint_S JF^{-1A}_{.a} x^a N_A \mathrm{d}S \\
&= \frac{1}{3}\oint_S \langle J\boldsymbol{F}^{-1}\boldsymbol{x}, \boldsymbol{N}\rangle \mathrm{d}S,
\end{aligned} \tag{3.139}$$

where (3.54) has been used along with integrability conditions $F^a_{.[A:B]} = 0$. Setting $\boldsymbol{F}^{-1} = \mathbf{1}$ and $J = 1$, the referential volume can be recovered from (3.139):

$$\begin{aligned}
V &= \frac{1}{3}\int_V \delta^A_A \mathrm{d}V \\
&= \frac{1}{3}\int_V X^A_{;A}\mathrm{d}V \\
&= \frac{1}{3}\oint_S X^A N_A \mathrm{d}S \\
&= \frac{1}{3}\oint_S \langle \boldsymbol{X}, \boldsymbol{N}\rangle \mathrm{d}S.
\end{aligned} \tag{3.140}$$

Analogously, integrated referential volume V can be determined from the inverse motion of spatial boundary s as

$$\begin{aligned}
V &= \int_V \mathrm{d}V \\
&= \int_v J^{-1}\mathrm{d}v \\
&= \frac{1}{6}\int_v \epsilon^{abc}\epsilon_{ABC}F^{-1A}_{.a}F^{-1B}_{.b}F^{-1C}_{.c}\mathrm{d}v \\
&= \frac{1}{6}\int_v \epsilon^{abc}\epsilon_{ABC}\partial_a X^A \partial_b X^B \partial_c X^C \mathrm{d}v \\
&= \frac{1}{6}\int_v \epsilon^{abc}\epsilon_{ABC}X^A_{:a}\partial_b X^B \partial_c X^C \mathrm{d}v \\
&= \frac{1}{6}\int_v (\epsilon^{abc}\epsilon_{ABC}X^A \partial_b X^B \partial_c X^C)_{:a}\mathrm{d}v \\
&= \frac{1}{6}\int_v (\epsilon^{abc}\epsilon_{ABC}X^A \partial_b X^B \partial_c X^C)_{;a}\mathrm{d}v \\
&= \frac{1}{6}\oint_s (\epsilon^{abc}\epsilon_{ABC}\partial_b X^B \partial_c X^C) X^A n_a \mathrm{d}s \\
&= \frac{1}{3}\oint_s J^{-1}F^a_{.A}X^A n_a \mathrm{d}s \\
&= \frac{1}{3}\oint_s \langle J^{-1}\boldsymbol{F}\boldsymbol{X}, \boldsymbol{n}\rangle \mathrm{d}s,
\end{aligned} \tag{3.141}$$

where (3.55) has been used along with integrability conditions $F^{-1A}_{.[a:b]} = 0$. Setting $\boldsymbol{F} = \mathbf{1}$ and $J^{-1} = 1$, the spatial volume can be recovered from (3.141):

$$v = \frac{1}{3}\int_v \delta^a_a \mathrm{d}v = \frac{1}{3}\int_v x^a_{;a}\mathrm{d}v = \frac{1}{3}\oint_s x^a n_a \mathrm{d}s = \frac{1}{3}\oint_s \langle \boldsymbol{x}, \boldsymbol{n}\rangle \mathrm{d}s. \tag{3.142}$$

3.2.2 *Stokes's theorem*

Let $\boldsymbol{A}$ be a scalar, vector, or tensor field of arbitrary order with continuous first derivatives with respect to local coordinates, and let ∇ be the covariant derivative or gradient operator with respect to the Levi-Civita connection of these coordinates. Let C denote a continuous curve with coordinates $\boldsymbol{X}$ enclosing simply connected surface S, the latter having unique unit outward normal covariant vector $\boldsymbol{N}$. The generalized Stokes's theorem is written [22]

$$\int_S (\boldsymbol{N} \times \nabla) \star \boldsymbol{A}\, \mathrm{d}S = \oint_C \mathrm{d}\boldsymbol{X} \star \boldsymbol{A}, \tag{3.143}$$

where $\star$ denotes a generic operator as discussed in Section 3.2.1. When a surface is not simply connected, Stokes's theorem can be applied by line integration over a collection of bounding curves if such curves exist.

As a first example, let $\boldsymbol{A} \to \boldsymbol{\alpha}(X) = \alpha_A \boldsymbol{G}^A$ be a differentiable covector field in the reference configuration, and let $\star \to \langle\, ,\, \rangle$. Stokes's theorem is then applied using (3.143) as

$$\int_S \langle \boldsymbol{N} \times \overset{G}{\nabla}, \boldsymbol{\alpha} \rangle \, \mathrm{d}S = \oint_C \langle \mathrm{d}\boldsymbol{X}, \boldsymbol{\alpha} \rangle, \tag{3.144}$$

or in coordinates,

$$\begin{aligned} \int_S \epsilon^{ABC} \alpha_{A;C} \, N_B \, \mathrm{d}S &= \int_S \epsilon^{ABC} \partial_C \alpha_A \, N_B \, \mathrm{d}S \\ &= \int_S \epsilon^{ABC} \partial_A \alpha_B \, N_C \, \mathrm{d}S \\ &= \oint_C \alpha_A \mathrm{d}X^A. \end{aligned} \tag{3.145}$$

Here the covariant derivative can be replaced by a partial derivative because of the symmetry of the covariant coefficients of the Levi-Civita connection.

As a second example application of Stokes's theorem, let $\boldsymbol{A} \to \boldsymbol{F}(X,t) = F^a_{.A} \boldsymbol{g}_a \otimes \boldsymbol{G}^A$ be the deformation gradient, and assign constant spatial basis vectors $\boldsymbol{g}_a$ over the domain in current configuration B with corresponding vanishing Christoffel symbols. It follows that

$$\begin{aligned} \int_S \epsilon^{ABC} F^a_{.A;C} \, N_B \, \mathrm{d}S &= \int_S \epsilon^{ABC} \partial_A F^a_{.B} \, N_C \, \mathrm{d}S \\ &= \int_S \epsilon^{ABC} \partial_A \partial_B x^a \, N_C \, \mathrm{d}S \\ &= \int_S \epsilon^{[AB]C} \partial_{(A} \partial_{B)} x^a \, N_C \, \mathrm{d}S \\ &= \oint_C F^a_{.A} \mathrm{d}X^A \\ &= 0. \end{aligned} \tag{3.146}$$

Analogously, for the inverse deformation gradient and constant reference basis vectors $\boldsymbol{G}_A$, integrating over an appropriate domain in the current configuration,

$$\begin{aligned}\int_s \epsilon^{abc} F^{-1A}_{\bullet a;c}\, n_b\, \mathrm{d}s &= \int_s \epsilon^{abc} \partial_a F^{-1A}_{\bullet b}\, n_c\, \mathrm{d}s \\ &= \int_s \epsilon^{abc} \partial_a \partial_b X^A\, n_c\, \mathrm{d}s \\ &= \int_s \epsilon^{[ab]c} \partial_{(a} \partial_{b)} X^A\, n_c\, \mathrm{d}s \\ &= \oint_c F^{-1A}_{\bullet a} \mathrm{d}x^a \\ &= 0. \end{aligned} \tag{3.147}$$

Letting C and c represent all possible closed curves in a body with the requisite properties, relationships (3.146) and (3.147) can then be interpreted as global compatibility conditions for the deformation gradient and its inverse, respectively, over areas S and s in configurations B_0 and B.

3.3 Velocities and Time Derivatives

Velocity fields are defined. The material time derivative is introduced, and relevant examples are given. The Lie derivative is defined.

3.3.1 *Velocity fields*

The spatial velocity $\boldsymbol{v}(x, t)$ measures the time rate of change of position of a material particle. Using (2.1), the spatial velocity can be written

$$\boldsymbol{v} = \frac{\partial \boldsymbol{x}(X, t)}{\partial t} = \frac{\partial \varphi^a}{\partial t}\bigg|_X \boldsymbol{g}_a(x). \tag{3.148}$$

Components of the spatial velocity $\boldsymbol{v}(x, t)$ are related to those of the material velocity $\boldsymbol{V}(X, t)$ as follows:

$$V^a(X, t) = v^a[\varphi(X, t), t]. \tag{3.149}$$

Here and henceforward, it is assumed that spatial coordinates $x^a(X, t)$ are at least twice differentiable with respect to time t.

3.3.2 *Material time derivatives*

The material time derivative measures the time rate of change of a given quantity associated with a fixed material particle X. In contrast, the spatial partial time derivative measures the time rate of change of a given quantity associated with a fixed spatial point x. First consider a generic scalar field $f(x, t)$. The material time derivative of f is

$$\begin{aligned}
\dot{f}(x,t) &= \frac{\mathrm{d}f}{\mathrm{d}t} \\
&= \left.\frac{\partial f}{\partial t}\right|_X \\
&= \left.\frac{\partial f}{\partial t}\right|_x + \left.\frac{\partial f}{\partial x^a}\right|_t \left.\frac{\partial x^a}{\partial t}\right|_X \\
&= \left.\frac{\partial f}{\partial t}\right|_x + (\partial_a f)\, v^a \\
&= \left.\frac{\partial f}{\partial t}\right|_x + f_{;a} v^a,
\end{aligned} \tag{3.150}$$

recalling from (2.167) the equivalence of covariant and partial differentiation of scalar fields. Notation $\mathrm{d}(\cdot)/\mathrm{d}t = \dot{(\cdot)}$ is used interchangeably for the material time derivative operation. The material time derivative is extended to spatial vector fields and tensor fields of higher order as follows:

$$\begin{aligned}
\dot{f}^a(x,t) &= \left.\frac{\partial f^a}{\partial t}\right|_x + f^a_{;b} v^b, \\
\dot{f}^{a\ldots c}_{d\ldots f}(x,t) &= \left.\frac{\partial f^{a\ldots c}_{d\ldots f}}{\partial t}\right|_x + f^{a\ldots c}_{d\ldots f\,;\,k} v^k.
\end{aligned} \tag{3.151}$$

For a vector or tensor field with all indices referred to reference coordinates, the material time derivative is simply

$$\begin{aligned}
\dot{f}^A(X,t) &= \left.\frac{\partial f^A}{\partial t}\right|_X, \\
\dot{f}^{A\ldots C}_{D\ldots F}(X,t) &= \left.\frac{\partial f^{A\ldots C}_{D\ldots F}}{\partial t}\right|_X.
\end{aligned} \tag{3.152}$$

For a field $f(X, t)$ with one or more indices referred to spatial coordinates, the material time derivative is [31]

$$\begin{aligned}
\dot{f}^{A\ldots C\;a\ldots c}_{D\ldots F\;d\ldots f}(X,t) = {} & \frac{\mathrm{D} f^{A\ldots C\;a\ldots c}_{D\ldots F\;d\ldots f}}{\mathrm{D}t} \\
& + \Big(\overset{g}{\Gamma}{}^{\cdot\cdot a}_{bk} f^{A\ldots C\;k\ldots c}_{D\ldots F\;d\ldots f} + \cdots \\
& \qquad - \overset{g}{\Gamma}{}^{\cdot\cdot k}_{bd} f^{A\ldots C\;a\ldots c}_{D\ldots F\;k\ldots f} - \cdots \Big) V^b,
\end{aligned} \tag{3.153}$$

where $\mathrm{D}(\cdot)/\mathrm{D}t$ denotes the partial time derivative with X and $\boldsymbol{g}(x)$ held fixed. As will be clear from subsequent examples, terms involving spatial connection coefficients account for partial time derivatives of spatial basis vectors at fixed X.

Several important examples of material time derivatives are derived next. The spatial acceleration is an example of (3.151):

$$\begin{aligned}\boldsymbol{a}(x,t) &= a^a \boldsymbol{g}_a \\ &= \dot{\boldsymbol{v}} \\ &= \frac{\partial \boldsymbol{v}}{\partial t}\bigg|_x + (\partial_b \boldsymbol{v} \otimes \boldsymbol{g}^b)\boldsymbol{v} \\ &= \left(\frac{\partial v^a}{\partial t}\bigg|_x + v^a_{;b} v^b \right) \boldsymbol{g}_a. \end{aligned} \tag{3.154}$$

The material acceleration is an example of (3.153):

$$\begin{aligned}\boldsymbol{A}(X,t) &= A^a \boldsymbol{g}_a \\ &= \frac{\partial \boldsymbol{V}}{\partial t}\bigg|_X \\ &= \frac{\partial (V^a \boldsymbol{g}_a)}{\partial t}\bigg|_X \\ &= \frac{\partial V^a}{\partial t}\bigg|_X \boldsymbol{g}_a + V^a \frac{\partial \boldsymbol{g}_a(x)}{\partial t}\bigg|_X \\ &= \frac{\partial V^a}{\partial t}\bigg|_X \boldsymbol{g}_a + V^c \partial_b \boldsymbol{g}_c \frac{\partial x^b}{\partial t}\bigg|_X \\ &= \left(\frac{\partial V^a}{\partial t}\bigg|_X + \overset{g}{\Gamma}{}^{\cdot\cdot a}_{bc} V^c V^b \right) \boldsymbol{g}_a \\ &= \left(\frac{\mathrm{D} V^a}{\mathrm{D} t} + \overset{g}{\Gamma}{}^{\cdot\cdot a}_{bc} V^b V^c \right) \boldsymbol{g}_a. \end{aligned} \tag{3.155}$$

According to definitions and notation used here, material time derivatives of spatial basis vectors vanish:

$$\begin{aligned}\dot{\boldsymbol{g}}_a[x(X,t)] &= \frac{\partial \boldsymbol{g}_a(x)}{\partial t}\bigg|_x + \boldsymbol{g}_{a;b} v^b \\ &= \boldsymbol{g}_{a;b} v^b \\ &= V^b \partial_b \boldsymbol{g}_a - \overset{g}{\Gamma}{}^{\cdot\cdot c}_{ba} \boldsymbol{g}_c V^b \\ &= \frac{\partial x^b}{\partial t}\bigg|_X \frac{\partial \boldsymbol{g}_a}{\partial x^b} - \overset{g}{\Gamma}{}^{\cdot\cdot c}_{ba} \boldsymbol{g}_c V^b \\ &= \frac{\partial \boldsymbol{g}_a(x)}{\partial t}\bigg|_X - \overset{g}{\Gamma}{}^{\cdot\cdot c}_{ba} \boldsymbol{g}_c V^b \\ &= 0. \end{aligned} \tag{3.156}$$

Furthermore, for referential basis vectors, the following identity is trivial:

$$\dot{\boldsymbol{G}}_A(X) = \frac{\partial \boldsymbol{G}_A(X)}{\partial t}\bigg|_X = 0. \tag{3.157}$$

The following notation is often used for velocities:

$$v^a[\varphi(X,t),t] = V^a(X,t) = \dot{x}^a(X,t), \tag{3.158}$$

and similarly for accelerations:

$$a^a[\varphi(X,t),t] = A^a(X,t) = \ddot{x}^a(X,t). \tag{3.159}$$

The material time derivative of the deformation gradient $\boldsymbol{F}(X,t)$ is another example of (3.153):

$$\begin{aligned}
\dot{F}^a_{.A} &= \frac{\mathrm{d}}{\mathrm{d}t}(\partial_A x^a) \\
&= \frac{\partial F^a_{.A}}{\partial t}\bigg|_x + F^a_{.A;b}v^b \\
&= \frac{\partial F^a_{.A}}{\partial t}\bigg|_x + \left(\partial_b F^a_{.A} + \overset{g}{\Gamma}{}^{..a}_{bc} F^c_{.A}\right) v^b \\
&= \frac{\mathrm{D}F^a_{.A}}{\mathrm{D}t} + \overset{g}{\Gamma}{}^{..a}_{bc} F^c_{.A} V^b \\
&= \partial_A V^a + \overset{g}{\Gamma}{}^{..a}_{bc} F^c_{.A} V^b \\
&= \left(\partial_c v^a + \overset{g}{\Gamma}{}^{..a}_{cb} v^b\right) F^c_{.A} \\
&= v^a_{;c} F^c_{.A} \\
&= L^a_{.c} F^c_{.A},
\end{aligned} \tag{3.160}$$

where the symmetry of Levi-Civita coefficients $\overset{g}{\Gamma}{}^{..a}_{bc} = \overset{g}{\Gamma}{}^{..a}_{cb}$ has been applied. The second equality in (3.160) can be derived directly as follows:

$$\begin{aligned}
\dot{\boldsymbol{F}} &= \frac{\partial \boldsymbol{F}}{\partial t}\bigg|_X \\
&= \frac{\partial}{\partial t}\bigg|_X \{F^a_{.A}(X,t)\, \boldsymbol{g}_a[x(X,t)] \otimes \boldsymbol{G}^A(X)\} \\
&= \frac{\partial F^a_{.A}}{\partial t}\bigg|_X \boldsymbol{g}_a \otimes \boldsymbol{G}^A + F^a_{.A} \frac{\partial \boldsymbol{g}_a}{\partial t}\bigg|_X \otimes \boldsymbol{G}^A \\
&= \left(\frac{\partial F^a_{.A}}{\partial t}\bigg|_x + \frac{\partial F^a_{.A}}{\partial x^b}\bigg|_t \frac{\partial x^b}{\partial t}\bigg|_X\right) \boldsymbol{g}_a \otimes \boldsymbol{G}^A + F^a_{.A} v^b \partial_b \boldsymbol{g}_a \otimes \boldsymbol{G}^A \\
&= \left(\frac{\partial F^a_{.A}}{\partial t}\bigg|_x + \frac{\partial F^a_{.A}}{\partial x^b}\bigg|_t v^b\right) \boldsymbol{g}_a \otimes \boldsymbol{G}^A + F^a_{.A} v^b \overset{g}{\Gamma}{}^{..c}_{ba} \boldsymbol{g}_c \otimes \boldsymbol{G}^A \\
&= \left(\frac{\partial F^a_{.A}}{\partial t}\bigg|_x + v^b \partial_b F^a_{.A} + v^b F^c_{.A} \overset{g}{\Gamma}{}^{..a}_{bc}\right) \boldsymbol{g}_a \otimes \boldsymbol{G}^A \\
&= \left(\frac{\partial F^a_{.A}}{\partial t}\bigg|_x + v^b F^a_{.A;b}\right) \boldsymbol{g}_a \otimes \boldsymbol{G}^A.
\end{aligned} \tag{3.161}$$

Spatial velocity gradient $\boldsymbol{L}(x,t)$ is defined as

$$\boldsymbol{L} = \partial_b \boldsymbol{v} \otimes \boldsymbol{g}^b = v^a_{;b}\boldsymbol{g}_a \otimes \boldsymbol{g}^b = \dot{\boldsymbol{F}}\boldsymbol{F}^{-1},$$
$$L^a_{.b} = v^a_{;b} = (\dot{x}^a)_{;b} = \dot{F}^a_{.A}F^{-1A}_{.b}. \tag{3.162}$$

The material time derivative of the inverse deformation gradient is computed as follows using the product rule:

$$0 = \frac{\mathrm{d}}{\mathrm{d}t}(\delta^a_b) = \frac{\mathrm{d}}{\mathrm{d}t}(F^a_{.A}F^{-1A}_{.b}) = F^a_{.A}\frac{\mathrm{d}}{\mathrm{d}t}(F^{-1A}_{.b}) + F^{-1A}_{.b}\frac{\mathrm{d}}{\mathrm{d}t}F^a_{.A}$$
$$\Rightarrow \frac{\mathrm{d}}{\mathrm{d}t}(F^{-1A}_{.a}) = \frac{\mathrm{d}}{\mathrm{d}t}(\partial_a X^A) = -F^{-1A}_{.b}L^b_{.a} = -\partial_b X^A(\dot{x}^b)_{;a}. \tag{3.163}$$

From (3.28), the velocity gradient quantifies the time rate of change of a differential line element:

$$\mathrm{d}v^a = \mathrm{d}\dot{x}^a = \dot{F}^a_{.A}\mathrm{d}X^A = \dot{F}^a_{.A}F^{-1A}_{.b}\mathrm{d}x^b = L^a_{.b}\mathrm{d}x^b. \tag{3.164}$$

The covariant version of the velocity gradient can be divided into its symmetric part, the deformation rate tensor $\boldsymbol{D}(x,t)$, and its anti-symmetric part, the spin tensor $\boldsymbol{W}(x,t)$:

$$L_{ab} = g_{ac}L^c_{.b} = D_{ab} + W_{ab},$$
$$D_{ab} = L_{(ab)} = v_{(a;b)},$$
$$W_{ab} = L_{[ab]} = v_{[a;b]} = \partial_{[b}v_{a]}; \tag{3.165}$$

where the final identity for the spin tensor again follows from the symmetry of the Levi-Civita connection coefficients. The following relationships are noted among deformation rates and strain rates:

$$\begin{aligned}
\dot{C}_{AB} &= 2\dot{E}_{AB} \\
&= \dot{F}^a_{.A}g_{ab}F^b_{.B} + \dot{F}^b_{.B}g_{ab}F^a_{.A} \\
&= 2\dot{F}_{a(B}F^a_{.A)} \\
&= 2F^a_{.(A}\dot{F}_{a|C|}\delta^C_{B)} \\
&= 2F^a_{.(A}\dot{F}_{a|C|}F^{-1C}_{.b}F^b_{.B)} \\
&= 2F^a_{.(A}L_{ab}F^b_{.B)} \\
&= 2F^a_{.A}L_{(ab)}F^b_{.B} \\
&= 2F^a_{.A}D_{ab}F^b_{.B}.
\end{aligned} \tag{3.166}$$

Therefore,

$$\dot{\boldsymbol{E}} = \boldsymbol{F}^{\mathrm{T}}\boldsymbol{D}\boldsymbol{F}, \qquad \dot{E}_{AB} = F^a_{.A}D_{ab}F^b_{.B}; \tag{3.167}$$

$$\boldsymbol{D} = \boldsymbol{F}^{-\mathrm{T}}\dot{\boldsymbol{E}}\boldsymbol{F}^{-1}, \qquad D_{ab} = F^{-1A}_{\ .a}\dot{E}_{AB}F^{-1B}_{\ .b}. \tag{3.168}$$

Consistent with (3.156) and (3.157), material time derivatives of metric tensors with components G_{AB} and g_{ab} vanish identically:

$$\dot{G}_{AB}(X) = \frac{\partial G_{AB}}{\partial t}\bigg|_X = 0, \qquad \dot{g}_{ab}(x) = \frac{\partial g_{ab}}{\partial t}\bigg|_x + g_{ab;c}v^c = 0. \tag{3.169}$$

The second of (3.169) follows from (2.160). Similarly, from (3.156) and (3.157), material time derivatives of shifters vanish, *e.g.*,

$$\dot{g}^a_A = \frac{\mathrm{d}}{\mathrm{d}t}\langle \boldsymbol{g}^a, \boldsymbol{G}_A\rangle = \langle \dot{\boldsymbol{g}}^a, \boldsymbol{G}_A\rangle + \langle \boldsymbol{g}^a, \dot{\boldsymbol{G}}_A\rangle = 0. \tag{3.170}$$

It follows that material time differentiation commutes with raising, lowering, and shifting of indices.

Note that when the deformation consists only of rotation (*i.e.*, $\boldsymbol{F} \to \boldsymbol{R}$), then the associated velocity gradient $\dot{\boldsymbol{R}}\boldsymbol{R}^{-1}$ is skew:

$$\begin{aligned}
0 &= \frac{\mathrm{d}}{\mathrm{d}t}(g_{ab}) \\
&= \frac{\mathrm{d}}{\mathrm{d}t}(R_{aA}R^{-1A}_{\ .b}) \\
&= \frac{\mathrm{d}}{\mathrm{d}t}(R_{aA}R_b^{\cdot A}) \\
&= \dot{R}_{aA}R_b^{\cdot A} + \dot{R}_b^{\cdot A}R_{aA} \\
&\Rightarrow \dot{R}_{aA}R_b^{\cdot A} = \dot{R}_{aA}R^{-1A}_{\ .b} = -\dot{R}_{bA}R^{-1A}_{\ .a} = \dot{R}_{[a|A|}R^{-1A}_{\ .b]}.
\end{aligned} \tag{3.171}$$

From (3.46) and (3.160), the material time derivative of the Jacobian determinant is

$$\begin{aligned}
\dot{J} &= \frac{\mathrm{d}}{\mathrm{d}t}\{J[F^a_{.A}(X,t), g(x), G(X)]\} \\
&= \frac{\partial J}{\partial F^a_{.A}}\dot{F}^a_{.A} \\
&= JF^{-1A}_{\ .a}\dot{F}^a_{.A} \\
&= JL^a_{.a} \\
&= JD^a_{.a} \\
&= Jv^a_{;a}.
\end{aligned} \tag{3.172}$$

The following identities are noted, recalling that $\mathrm{d}v(x) \subset B$ and $\mathrm{d}V(X) \subset$

B_0 are differential volume elements:

$$\begin{aligned}\frac{\mathrm{d}}{\mathrm{d}t}\ln \mathrm{d}v &= \frac{\dot{\mathrm{d}v}}{\mathrm{d}v}\\ &= \frac{\dot{J}\mathrm{d}V}{J\mathrm{d}V}\\ &= \frac{\mathrm{d}}{\mathrm{d}t}\ln J\\ &= \dot{J}J^{-1}\\ &= L^{a}_{.a}\\ &= D^{a}_{.a}\\ &= v^{a}_{;a}\\ &= (\dot{x}^a)_{;a}.\end{aligned} \tag{3.173}$$

The material time derivative of an oriented area element is, from (3.131), (3.163), and (3.172)

$$\begin{aligned}\frac{\mathrm{d}}{\mathrm{d}t}(n_a\mathrm{d}s) &= \frac{\mathrm{d}}{\mathrm{d}t}(JF^{-1A}_{.a}N_A\mathrm{d}S)\\ &= \frac{\mathrm{d}}{\mathrm{d}t}(JF^{-1A}_{.a})N_A\mathrm{d}S\\ &= (\dot{J}F^{-1A}_{.a} + J\dot{F}^{-1A}_{.a})N_A\mathrm{d}S\\ &= J(L^{b}_{.b}F^{-1A}_{.a} - L^{b}_{.a}F^{-1A}_{.b})N_A\mathrm{d}S\\ &= (L^{b}_{.b}n_a - L^{b}_{.a}n_b)\mathrm{d}s\\ &= (v^{b}_{;b}n_a - v^{b}_{;a}n_b)\mathrm{d}s.\end{aligned} \tag{3.174}$$

Material time differentiation commutes with partial differentiation with respect to reference coordinates. For example, let $f[x(X,t),t]$ denote a differentiable scalar field. Then

$$\frac{\mathrm{d}}{\mathrm{d}t}\left(\frac{\partial f}{\partial X^A}\right) = \frac{\partial}{\partial X^A}\left(\frac{\partial f}{\partial t}\bigg|_X\right) = \partial_A \dot{f}. \tag{3.175}$$

Similarly, partial time differentiation at fixed x^a commutes with partial differentiation with respect to spatial coordinates:

$$\frac{\partial}{\partial t}\bigg|_x\left(\frac{\partial f}{\partial x^a}\right) = \frac{\partial}{\partial x^a}\left(\frac{\partial f}{\partial t}\bigg|_x\right) = \partial_a\left(\frac{\partial f}{\partial t}\bigg|_x\right). \tag{3.176}$$

However, material time differentiation does not generally commute with partial differentiation with respect to spatial coordinates:

$$\begin{aligned}\frac{\mathrm{d}}{\mathrm{d}t}\left(\frac{\partial f}{\partial x^a}\right) &= \frac{\mathrm{d}}{\mathrm{d}t}(\partial_a f)\\ &= \frac{\mathrm{d}}{\mathrm{d}t}(F^{-1A}_{\ \ \ \bullet a}\partial_A f)\\ &= F^{-1A}_{\ \ \ \bullet a}\partial_A \dot{f} + \dot{F}^{-1A}_{\ \ \ \bullet a}\partial_A f\\ &= \partial_a \dot{f} - F^{-1A}_{\ \ \ \bullet b}L^{b}_{\bullet a}\partial_A f\\ &= \partial_a \dot{f} - L^{b}_{\bullet a}\partial_b f. \end{aligned} \tag{3.177}$$

Analogously, partial spatial time differentiation does not always commute with partial differentation with respect to reference coordinates:

$$\begin{aligned}\left.\frac{\partial}{\partial t}\right|_x\left(\frac{\partial f}{\partial X^A}\right) &= \left.\frac{\partial}{\partial t}\right|_x(\partial_A f)\\ &= \left.\frac{\partial}{\partial t}\right|_x(F^{a}_{\bullet A}\partial_a f)\\ &= F^{a}_{\bullet A}\partial_a\left(\left.\frac{\partial f}{\partial t}\right|_x\right) + \left.\frac{\partial F^{a}_{\bullet A}}{\partial t}\right|_x\partial_a f\\ &= \partial_A\left(\left.\frac{\partial f}{\partial t}\right|_x\right) + F^{-1B}_{\ \ \ \bullet a}\left.\frac{\partial F^{a}_{\bullet A}}{\partial t}\right|_x\partial_B f. \end{aligned} \tag{3.178}$$

Rules listed in (3.175) and (3.176) also apply when partial coordinate differentiation is replaced with covariant differentiation with respect to the Levi-Civita connection of the same coordinate system. For example, letting $\boldsymbol{f} = f^A \boldsymbol{G}_A$ be a differentiable referential vector field,

$$\begin{aligned}\frac{\mathrm{d}}{\mathrm{d}t}(f^{A}_{;B}) &= \frac{\mathrm{d}}{\mathrm{d}t}(\partial_B f^A + \overset{G}{\Gamma}{}^{\ \cdot\cdot A}_{BC} f^C)\\ &= \partial_B \dot{f}^A + \overset{G}{\Gamma}{}^{\ \cdot\cdot A}_{BC}\dot{f}^C\\ &= (\dot{f}^A)_{;B}. \end{aligned} \tag{3.179}$$

Similarly, letting $\boldsymbol{f} = f^a \boldsymbol{g}_a$ be a differentiable spatial vector field,

$$\begin{aligned}\left.\frac{\partial}{\partial t}\right|_x(f^{a}_{;b}) &= \left.\frac{\partial}{\partial t}\right|_x(\partial_b f^a + \overset{g}{\Gamma}{}^{\ \cdot\cdot a}_{bc} f^c)\\ &= \partial_b\left(\left.\frac{\partial f^a}{\partial t}\right|_x\right) + \overset{g}{\Gamma}{}^{\ \cdot\cdot a}_{bc}\left.\frac{\partial f^c}{\partial t}\right|_x\\ &= \left(\left.\frac{\partial f^a}{\partial t}\right|_x\right)_{;b}. \end{aligned} \tag{3.180}$$

The following identities have been used in (3.179) and (3.180):

$$\frac{\mathrm{d}}{\mathrm{d}t}\left[\overset{G}{\Gamma}{}^{\ \cdot\cdot A}_{BC}(X)\right] = \left.\frac{\partial}{\partial t}\right|_X\left[\overset{G}{\Gamma}{}^{\ \cdot\cdot A}_{BC}(X)\right] = 0, \qquad \left.\frac{\partial}{\partial t}\right|_x\left[\overset{g}{\Gamma}{}^{\ \cdot\cdot a}_{bc}(x)\right] = 0. \tag{3.181}$$

Extending (3.180) to a tensor of arbitrary order,

$$\frac{\partial}{\partial t}\bigg|_x (f^{A\ldots C}_{D\ldots F}{}^{a\ldots c}_{d\ldots f;g}) = \left(\frac{\partial f^{A\ldots C}_{D\ldots F}{}^{a\ldots c}_{d\ldots f}}{\partial t}\bigg|_x\right)_{;g}, \tag{3.182}$$

the following identity is obtained relating the material time derivative and the spatial covariant derivative:

$$\begin{aligned}
\frac{\mathrm{d}}{\mathrm{d}t}(f^{A\ldots C}_{D\ldots F}{}^{a\ldots c}_{d\ldots f;g}) &= \frac{\partial}{\partial t}\bigg|_x (f^{A\ldots C}_{D\ldots F}{}^{a\ldots c}_{d\ldots f;g}) + f^{A\ldots C}_{D\ldots F}{}^{a\ldots c}_{d\ldots f;gh}v^h \\
&= \left(\frac{\partial f^{A\ldots C}_{D\ldots F}{}^{a\ldots c}_{d\ldots f}}{\partial t}\bigg|_x\right)_{;g} + f^{A\ldots C}_{D\ldots F}{}^{a\ldots c}_{d\ldots f;hg}v^h \\
&= \frac{\partial}{\partial t}\bigg|_x (f^{A\ldots C}_{D\ldots F}{}^{a\ldots c}_{d\ldots f;g}) + (f^{A\ldots C}_{D\ldots F}{}^{a\ldots c}_{d\ldots f;h}v^h)_{;g} \\
&\quad - f^{A\ldots C}_{D\ldots F}{}^{a\ldots c}_{d\ldots f;h}v^h_{;g} \\
&= (\dot{f}^{A\ldots C}_{D\ldots F}{}^{a\ldots c}_{d\ldots f})_{;g} - f^{A\ldots C}_{D\ldots F}{}^{a\ldots c}_{d\ldots f;h}L^h_{.g}.
\end{aligned} \tag{3.183}$$

For example, applying this identity to the partial covariant derivative of the deformation gradient gives

$$\dot{F}^a_{.A;b} = \frac{\mathrm{d}}{\mathrm{d}t}(F^a_{.A;b}) + F^a_{.A;c}L^c_{.b}. \tag{3.184}$$

The second gradient of the spatial velocity is then computed as follows in terms of time and space derivatives of the deformation gradient and its inverse:

$$\begin{aligned}
v^a_{;bc} &= (\dot{x}^a)_{;bc} \\
&= L^a_{b;c} \\
&= (\dot{F}^a_{.A}F^{-1A}_{\ .b})_{;c} \\
&= \dot{F}^a_{.A;c}F^{-1A}_{\ .b} + \dot{F}^a_{.A}F^{-1A}_{\ .b;c} \\
&= F^{-1A}_{\ .b}\left[\frac{\mathrm{d}}{\mathrm{d}t}(F^a_{.A;c}) + F^a_{.A;d}L^d_{.c}\right] + \dot{F}^a_{.A}F^{-1A}_{\ .b;c} \\
&= F^{-1A}_{\ .b}F^{-1B}_{\ .c}\left[\frac{\mathrm{d}}{\mathrm{d}t}(F^a_{.A;d})F^d_{.B} + F^a_{.A;d}\dot{F}^d_{.B}\right] + \dot{F}^a_{.A}F^{-1A}_{\ .b;c} \\
&= F^{-1A}_{\ .b}F^{-1B}_{\ .c}\frac{\mathrm{d}}{\mathrm{d}t}\left(F^a_{.A;d}F^d_{.B}\right) + \dot{F}^a_{.A}F^{-1A}_{\ .b;c} \\
&= F^{-1A}_{\ .c}F^{-1B}_{\ .b}\frac{\mathrm{d}}{\mathrm{d}t}\left(F^a_{.A;d}F^d_{.B}\right) + \dot{F}^a_{.A}F^{-1A}_{\ .c;b} \\
&= v^a_{;cb}.
\end{aligned} \tag{3.185}$$

3.3.3 *Lie derivatives*

Consider a differentiable function $f(x,t)$, which may be a scalar, vector, or tensor field of higher order. In coordinates, the Lie derivative of f taken with respect to the spatial velocity field $\boldsymbol{v}(x,t)$ is

$$\begin{aligned}\mathcal{L}_v f^{a\ldots h}_{i\ldots r}(x,t) &= \left.\frac{\partial f^{a\ldots h}_{i\ldots r}}{\partial t}\right|_x + v^s f^{a\ldots h}_{i\ldots r\,;\,s} \\ &\quad - v^a_{;\,s} f^{s\cdots h}_{i\ldots r} - \cdots - v^h_{;\,s} f^{a\ldots s}_{i\ldots r} \\ &\quad + v^t_{;\,i} f^{a\ldots h}_{t\ldots r} + \cdots + v^t_{;\,r} f^{a\ldots h}_{i\ldots t} \\ &= \left.\frac{\partial f^{a\ldots h}_{i\ldots r}}{\partial t}\right|_x + v^s \partial_s f^{a\ldots h}_{i\ldots r} \\ &\quad - f^{s\ldots h}_{i\ldots r}\partial_s v^a - \cdots - f^{a\ldots s}_{i\ldots r}\partial_s v^h \\ &\quad + f^{a\ldots h}_{t\ldots r}\partial_i v^t + \cdots + f^{a\ldots h}_{i\ldots t}\partial_r v^t. \end{aligned} \tag{3.186}$$

Covariant derivatives and partial derivatives are interchangeable in (3.186) as a result of the symmetry of the coefficients of the Levi-Civita connection in configuration B. When f is a scalar field, the Lie derivative and material time derivative are equivalent:

$$\mathcal{L}_v f(x,t) = \left.\frac{\partial f}{\partial t}\right|_x + v^s f_{;s} = \left.\frac{\partial f}{\partial t}\right|_x + v^s \partial_s f = \dot{f}. \tag{3.187}$$

One notable example is the Lie derivative of the spatial metric tensor $g_{ab}(x)$:

$$\begin{aligned}(\mathcal{L}_v g)_{ab} &= \left.\frac{\partial g_{ab}}{\partial t}\right|_x + v^c g_{ab;c} + v^c_{;a} g_{cb} + v^c_{;b} g_{ac} \\ &= 2v^c_{;(b} g_{a)c} \\ &= 2L^c_{\cdot(b} g_{a)c} \\ &= L_{ab} + L_{ba} \\ &= 2L_{(ab)} \\ &= 2D_{ab}. \end{aligned} \tag{3.188}$$

3.3.4 *Reynolds transport theorem*

Consider a differentiable function $f(x,t)$, which may be a scalar, vector, or tensor field of higher order. Recalling from (3.173) that material time derivatives of referential and spatial volume elements are

$$\begin{aligned}&\frac{\mathrm{d}}{\mathrm{d}t}[\mathrm{d}V(X)] = 0, \\ &\frac{\mathrm{d}}{\mathrm{d}t}[\mathrm{d}v(x)] = \frac{\mathrm{d}}{\mathrm{d}t}[J(X,t)\mathrm{d}V(X)] = \dot{J}\mathrm{d}V = v^a_{;\,a}\mathrm{d}v, \end{aligned} \tag{3.189}$$

using definition (3.151) of the material time derivative, and applying the divergence theorem of Gauss, Reynolds transport theorem can be derived:

$$\begin{aligned}\frac{\mathrm{d}}{\mathrm{d}t}\int_v f\mathrm{d}v &= \int_v \dot{f}\mathrm{d}v + \int_v f\frac{\mathrm{d}}{\mathrm{d}t}(\mathrm{d}v) \\ &= \int_v \frac{\partial f}{\partial t}\Big|_x \mathrm{d}v + \int_v f_{;a}v^a\mathrm{d}v + \int_v f v^a_{;a}\mathrm{d}v \\ &= \int_v \frac{\partial f}{\partial t}\Big|_x \mathrm{d}v + \int_v (fv^a)_{;a}\mathrm{d}v \\ &= \frac{\partial}{\partial t}\int_v f\,\mathrm{d}v + \oint_s f v^a n_a\,\mathrm{d}s, \end{aligned} \tag{3.190}$$

where v is a simply connected spatial volume enclosed by surface s with unit outward normal $n_a(x)$. The rightmost term in the final equality of (3.190) represents the rate of change of f in volume v associated with the flux of f carried through the boundary s of v.

3.4 Coordinate Systems

Several common coordinate systems are described: Cartesian, cylindrical, and spherical systems. Metric tensors and connection coefficients are listed. Formulae in physical components for the gradient, Laplacian, divergence, and curl are provided. Physical components of spatial velocity gradient and acceleration are given. Representations of the deformation gradient are derived. Convected coordinates are discussed.

3.4.1 *Cartesian coordinates*

Spatial coordinates are considered first; analogous formulae apply for reference coordinates. In three spatial dimensions, Cartesian spatial coordinates are

$$(x^1, x^2, x^3) \to (x, y, z). \tag{3.191}$$

The squared length of a differential line element $\mathrm{d}\boldsymbol{x}$ is

$$|\mathrm{d}\boldsymbol{x}|^2 = \mathrm{d}\boldsymbol{x}\cdot\mathrm{d}\boldsymbol{x} = (\mathrm{d}x)^2 + (\mathrm{d}y)^2 + (\mathrm{d}z)^2. \tag{3.192}$$

The metric tensor and its inverse are

$$g_{ab} = \delta_{ab}, \qquad g^{ab} = \delta^{ab}. \tag{3.193}$$

Determinants of the metric tensor and its inverse are unity:

$$g = \det(g_{ab}) = 1, \qquad g^{-1} = \det(g^{ab}) = 1. \tag{3.194}$$

A differential volume element is

$$dv = \sqrt{g}\, dx^1 dx^2 dx^3 = dx\ dy\ dz. \tag{3.195}$$

Connection coefficients from (2.150) vanish identically since $\partial_c \delta_{ab} = 0$:

$$\overset{g}{\Gamma}{}^{\,\cdot\cdot a}_{bc} = 0 \qquad \forall a, b, c. \tag{3.196}$$

Thus, covariant and partial derivatives are equivalent:

$$\overset{g}{\nabla}_a(\cdot) = (\cdot);_a = \partial_a(\cdot). \tag{3.197}$$

Let $\boldsymbol{x}$ denote the position vector in Euclidean space:

$$\boldsymbol{x}(x, y, z) = x\boldsymbol{g}_x + y\boldsymbol{g}_y + z\boldsymbol{g}_z. \tag{3.198}$$

Natural basis vectors are

$$\boldsymbol{g}_a = \partial_a \boldsymbol{x}; \qquad \boldsymbol{g}_x = \partial_x \boldsymbol{x}, \quad \boldsymbol{g}_y = \partial_y \boldsymbol{x}, \quad \boldsymbol{g}_z = \partial_z \boldsymbol{x}. \tag{3.199}$$

Notice that basis vectors $\boldsymbol{g}_a$ are already dimensionless, and are equal to their representations $\boldsymbol{e}_a$ in physical components:

$$\boldsymbol{e}_x = \frac{\boldsymbol{g}_x}{\sqrt{g_{xx}}} = \boldsymbol{g}_x, \qquad \boldsymbol{e}_y = \frac{\boldsymbol{g}_y}{\sqrt{g_{yy}}} = \boldsymbol{g}_y, \qquad \boldsymbol{e}_z = \frac{\boldsymbol{g}_z}{\sqrt{g_{zz}}} = \boldsymbol{g}_z. \tag{3.200}$$

Therefore, the position vector from a fixed origin is

$$\boldsymbol{x}(x, y, z) = x\boldsymbol{e}_x + y\boldsymbol{e}_y + z\boldsymbol{e}_z. \tag{3.201}$$

Note also that since the physical basis is orthonormal,

$$\boldsymbol{e}^x = \boldsymbol{e}_x, \qquad \boldsymbol{e}^y = \boldsymbol{e}_y, \qquad \boldsymbol{e}^z = \boldsymbol{e}_z. \tag{3.202}$$

In physical components, the gradient of a scalar field $f(\boldsymbol{x})$ is

$$\overset{g}{\nabla} f = f_{;a}\ \boldsymbol{g}^a = \partial_a f\ \boldsymbol{e}_a = \partial_x f\ \boldsymbol{e}_x + \partial_y f\ \boldsymbol{e}_y + \partial_z f\ \boldsymbol{e}_z. \tag{3.203}$$

The Laplacian of f is

$$\overset{g}{\nabla}{}^2 f = \partial_a \partial_a f = \partial_x \partial_x f + \partial_y \partial_y f + \partial_z \partial_z f. \tag{3.204}$$

The divergence of a vector field $\boldsymbol{v}(\boldsymbol{x})$ is

$$\langle \overset{g}{\nabla}, \boldsymbol{v} \rangle = v^a_{;a} = \partial_a v^a = \partial_x v_x + \partial_y v_y + \partial_z v_z. \tag{3.205}$$

The curl of field $\boldsymbol{v}(\boldsymbol{x})$ is

$$\begin{aligned} \overset{g}{\nabla} \times \boldsymbol{v} &= \epsilon^{abc} v_{c;b} \boldsymbol{g}_a \\ &= e^{abc} \partial_b v_c \boldsymbol{e}_a \\ &= (\partial_y v_z - \partial_z v_y)\boldsymbol{e}_x + (\partial_z v_x - \partial_x v_z)\boldsymbol{e}_y + (\partial_x v_y - \partial_y v_x)\boldsymbol{e}_z. \end{aligned} \tag{3.206}$$

The divergence of a second-order tensor field $\boldsymbol{A}(\boldsymbol{x})$ is, in Cartesian coordinates,

$$\begin{aligned}\overset{g}{\nabla}\cdot\boldsymbol{A} &= \partial_a A_{ab}\boldsymbol{e}_b \\ &= (\partial_x A_{xx}+\partial_y A_{yx}+\partial_z A_{zx})\boldsymbol{e}_x \\ &\quad+(\partial_x A_{xy}+\partial_y A_{yy}+\partial_z A_{zy})\boldsymbol{e}_y \\ &\quad+(\partial_x A_{xz}+\partial_y A_{yz}+\partial_z A_{zz})\boldsymbol{e}_z. \end{aligned} \tag{3.207}$$

Now letting $\boldsymbol{v}(\boldsymbol{x},t)$ denote the spatial velocity field of (3.148), the velocity gradient of (3.162) in physical components is

$$[L_{ab}] = [\partial_b v_a] = \begin{bmatrix} \partial_x v_x & \partial_y v_x & \partial_z v_x \\ \partial_x v_y & \partial_y v_y & \partial_z v_y \\ \partial_x v_z & \partial_y v_z & \partial_z v_z \end{bmatrix}, \tag{3.208}$$

where the velocity vector in physical components is

$$\boldsymbol{v} = v_x\boldsymbol{e}_x + v_y\boldsymbol{e}_y + v_z\boldsymbol{e}_z = \dot{x}\ \boldsymbol{e}_x + \dot{y}\ \boldsymbol{e}_y + \dot{z}\ \boldsymbol{e}_z. \tag{3.209}$$

Finally, the spatial acceleration $\boldsymbol{a}(\boldsymbol{x},t)$ of (3.154) is

$$\begin{aligned}\boldsymbol{a} &= a_x\boldsymbol{e}_x + a_y\boldsymbol{e}_y + a_z\boldsymbol{e}_z \\ &= \dot{v}_x\boldsymbol{e}_x + \dot{v}_y\boldsymbol{e}_y + \dot{v}_z\boldsymbol{e}_z \\ &= \left(\frac{\partial v_x}{\partial t} + v_x\partial_x v_x + v_y\partial_y v_x + v_z\partial_z v_x\right)\boldsymbol{e}_x \\ &\quad+\left(\frac{\partial v_y}{\partial t} + v_x\partial_x v_y + v_y\partial_y v_y + v_z\partial_z v_y\right)\boldsymbol{e}_y \\ &\quad+\left(\frac{\partial v_z}{\partial t} + v_x\partial_x v_z + v_y\partial_y v_z + v_z\partial_z v_z\right)\boldsymbol{e}_z. \end{aligned} \tag{3.210}$$

Cartesian reference coordinates are

$$(X^1, X^2, X^3) \to (X, Y, Z). \tag{3.211}$$

The squared length of a differential line element $\mathrm{d}\boldsymbol{X}$ is

$$|\mathrm{d}\boldsymbol{X}|^2 = \mathrm{d}\boldsymbol{X}\cdot\mathrm{d}\boldsymbol{X} = (\mathrm{d}X)^2 + (\mathrm{d}Y)^2 + (\mathrm{d}Z)^2. \tag{3.212}$$

The metric tensor and its inverse are

$$G_{AB} = \delta_{AB}, \qquad G^{AB} = \delta^{AB}. \tag{3.213}$$

Determinants of the metric tensor and its inverse are unity:

$$G = \det(G_{ab}) = 1, \qquad G^{-1} = \det(G^{AB}) = 1. \tag{3.214}$$

A differential volume element is

$$\mathrm{d}V = \sqrt{G}\,\mathrm{d}X^1\mathrm{d}X^2\mathrm{d}X^3 = \mathrm{d}X\,\mathrm{d}Y\,\mathrm{d}Z. \tag{3.215}$$

Connection coefficients from (2.122) vanish identically since $\partial_C\delta_{AB} = 0$:

$$\overset{G}{\Gamma}{}_{BC}^{\;\cdot\cdot A} = 0 \qquad \forall A, B, C. \tag{3.216}$$

Thus, covariant and partial derivatives are equivalent:

$$\overset{G}{\nabla}_A(\cdot) = (\cdot)_{;A} = \partial_A(\cdot). \tag{3.217}$$

Let $\boldsymbol{X}$ denote the position vector in Euclidean space:

$$\boldsymbol{X}(X, Y, Z) = X\boldsymbol{G}_X + Y\boldsymbol{G}_Y + Z\boldsymbol{G}_Z. \tag{3.218}$$

Natural basis vectors are

$$\boldsymbol{G}_A = \partial_A\boldsymbol{X}; \qquad \boldsymbol{G}_X = \partial_X\boldsymbol{X}, \quad \boldsymbol{G}_Y = \partial_Y\boldsymbol{X}, \quad \boldsymbol{G}_Z = \partial_Z\boldsymbol{X}. \tag{3.219}$$

Basis vectors $\boldsymbol{G}_A$ are already dimensionless and are equal to their representations $\boldsymbol{E}_A$ in physical components:

$$\begin{aligned}\boldsymbol{E}_X &= \frac{\boldsymbol{G}_X}{\sqrt{G_{XX}}} = \boldsymbol{G}_X,\\ \boldsymbol{E}_Y &= \frac{\boldsymbol{G}_Y}{\sqrt{G_{YY}}} = \boldsymbol{G}_Y,\\ \boldsymbol{E}_Z &= \frac{\boldsymbol{G}_Z}{\sqrt{G_{ZZ}}} = \boldsymbol{G}_Z.\end{aligned} \tag{3.220}$$

Therefore, the position vector from a fixed origin is

$$\boldsymbol{X}(X, Y, Z) = X\boldsymbol{E}_X + Y\boldsymbol{E}_Y + Z\boldsymbol{E}_Z. \tag{3.221}$$

Since the physical basis is orthonormal,

$$\boldsymbol{E}^X = \boldsymbol{E}_X, \qquad \boldsymbol{E}^Y = \boldsymbol{E}_Y, \qquad \boldsymbol{E}^Z = \boldsymbol{E}_Z. \tag{3.222}$$

In Cartesian coordinates, deformation $\boldsymbol{x} = \boldsymbol{x}(\boldsymbol{X})$ is of the form

$$x = x(X, Y, Z), \qquad y = y(X, Y, Z), \qquad z = z(X, Y, Z). \tag{3.223}$$

Here coincident Cartesian frames are used such that $\boldsymbol{E}_A = \delta^a_A\boldsymbol{e}_a$:

$$\boldsymbol{E}_X = \boldsymbol{e}_x, \qquad \boldsymbol{E}_Y = \boldsymbol{e}_y, \qquad \boldsymbol{E}_Z = \boldsymbol{e}_z. \tag{3.224}$$

Mixed variant components of the shifter tensor between reference and spatial systems are then

$$g^a_A(\boldsymbol{x}, \boldsymbol{X}) = \langle \boldsymbol{g}^a, \boldsymbol{G}_A\rangle = \delta^a_A, \qquad g^A_a(\boldsymbol{x}, \boldsymbol{X}) = \langle \boldsymbol{g}_a, \boldsymbol{G}^A\rangle = \delta^A_a. \tag{3.225}$$

Referred to natural bases $\{\boldsymbol{g}_a, \boldsymbol{G}^A\}$, the deformation gradient is

$$\begin{aligned}\boldsymbol{F} &= \partial_A \boldsymbol{x} \otimes \boldsymbol{G}^A \\ &= \partial_A x^a \boldsymbol{g}_a \otimes \boldsymbol{G}^A \\ &= F^a_{.A} \boldsymbol{g}_a \otimes \boldsymbol{G}^A \\ &= \partial_X x\, \boldsymbol{g}_x \otimes \boldsymbol{G}^X + \partial_Y x\, \boldsymbol{g}_x \otimes \boldsymbol{G}^Y + \partial_Z x\, \boldsymbol{g}_x \otimes \boldsymbol{G}^Z \\ &\quad + \partial_X y\, \boldsymbol{g}_y \otimes \boldsymbol{G}^X + \partial_Y y\, \boldsymbol{g}_y \otimes \boldsymbol{G}^Y + \partial_Z y\, \boldsymbol{g}_y \otimes \boldsymbol{G}^Z \\ &\quad + \partial_X z\, \boldsymbol{g}_z \otimes \boldsymbol{G}^X + \partial_Y z\, \boldsymbol{g}_z \otimes \boldsymbol{G}^Y + \partial_Z z\, \boldsymbol{g}_z \otimes \boldsymbol{G}^Z. \end{aligned} \tag{3.226}$$

Components of $\boldsymbol{F}$ are the same in physical coordinates:

$$\begin{aligned}\boldsymbol{F} &= \partial_X x\, \boldsymbol{E}_X \otimes \boldsymbol{E}_X + \partial_Y x\, \boldsymbol{E}_X \otimes \boldsymbol{E}_Y + \partial_Z x\, \boldsymbol{E}_X \otimes \boldsymbol{E}_Z \\ &\quad + \partial_X y\, \boldsymbol{E}_Y \otimes \boldsymbol{E}_X + \partial_Y y\, \boldsymbol{E}_Y \otimes \boldsymbol{E}_Y + \partial_Z y\, \boldsymbol{E}_Y \otimes \boldsymbol{E}_Z \\ &\quad + \partial_X z\, \boldsymbol{E}_Z \otimes \boldsymbol{E}_X + \partial_Y z\, \boldsymbol{E}_Z \otimes \boldsymbol{E}_Y + \partial_Z z\, \boldsymbol{E}_Z \otimes \boldsymbol{E}_Z. \end{aligned} \tag{3.227}$$

Referential and deformed volume elements are related through the Jacobian determinant:

$$J = \mathrm{d}v/\mathrm{d}V = \det[F^a_{.A}]\sqrt{g/G} = \det[\partial_A x^a]\det[g^A_a] = \det[\partial_A x^a]. \tag{3.228}$$

Deformation tensor $\boldsymbol{C}$ of (3.86) is

$$\begin{aligned}\boldsymbol{C} &= C_{AB} \boldsymbol{E}_A \otimes \boldsymbol{E}_B \\ &= F^a_{.A} \delta_{ab} F^b_{.B} \boldsymbol{E}_A \otimes \boldsymbol{E}_B \\ &= [(\partial_X x)^2 + (\partial_X y)^2 + (\partial_X z)^2] \boldsymbol{E}_X \otimes \boldsymbol{E}_X \\ &\quad + [(\partial_X x)(\partial_Y x) + (\partial_X y)(\partial_Y y) + (\partial_X z)(\partial_Y z)] \boldsymbol{E}_X \otimes \boldsymbol{E}_Y \\ &\quad + [(\partial_X x)(\partial_Z x) + (\partial_X y)(\partial_Z y) + (\partial_X z)(\partial_Z z)] \boldsymbol{E}_X \otimes \boldsymbol{E}_Z \\ &\quad + [(\partial_X x)(\partial_Y x) + (\partial_X y)(\partial_Y y) + (\partial_X z)(\partial_Y z)] \boldsymbol{E}_Y \otimes \boldsymbol{E}_X \\ &\quad + [(\partial_Y x)^2 + (\partial_Y y)^2 + (\partial_Y z)^2] \boldsymbol{E}_Y \otimes \boldsymbol{E}_Y \\ &\quad + [(\partial_Y x)(\partial_Z x) + (\partial_Y y)(\partial_Z y) + (\partial_Y z)(\partial_Z z)] \boldsymbol{E}_Y \otimes \boldsymbol{E}_Z \\ &\quad + [(\partial_X x)(\partial_Z x) + (\partial_X y)(\partial_Z y) + (\partial_X z)(\partial_Z z)] \boldsymbol{E}_Z \otimes \boldsymbol{E}_X \\ &\quad + [(\partial_Y x)(\partial_Z x) + (\partial_Y y)(\partial_Z y) + (\partial_Y z)(\partial_Z z)] \boldsymbol{E}_Z \otimes \boldsymbol{E}_Y \\ &\quad + [(\partial_Z x)^2 + (\partial_Z y)^2 + (\partial_Z z)^2] \boldsymbol{E}_Z \otimes \boldsymbol{E}_Z. \end{aligned} \tag{3.229}$$

In Cartesian coordinates, inverse deformation $\boldsymbol{X} = \boldsymbol{X}(\boldsymbol{x})$ is of the form

$$X = X(x, y, z), \qquad Y = Y(x, y, z), \qquad Z = Z(x, y, z). \tag{3.230}$$

Referred to natural bases $\{\boldsymbol{G}_A, \boldsymbol{g}^a\}$, the inverse deformation gradient is

$$\begin{aligned}
\boldsymbol{F}^{-1} &= \partial_a \boldsymbol{X} \otimes \boldsymbol{g}^a \\
&= \partial_a X^A \boldsymbol{G}_A \otimes \boldsymbol{g}^a \\
&= F^{-1A}_{\ \ \bullet a} \boldsymbol{G}_A \otimes \boldsymbol{g}^a \\
&= \partial_x X \ \boldsymbol{G}_X \otimes \boldsymbol{g}^x + \partial_y X \ \boldsymbol{G}_X \otimes \boldsymbol{g}^y + \partial_z X \ \boldsymbol{G}_X \otimes \boldsymbol{g}^z \\
&\quad + \partial_x Y \ \boldsymbol{G}_Y \otimes \boldsymbol{g}^x + \partial_y Y \ \boldsymbol{G}_Y \otimes \boldsymbol{g}^y + \partial_z Y \ \boldsymbol{G}_Y \otimes \boldsymbol{g}^z \\
&\quad + \partial_x Z \ \boldsymbol{G}_Z \otimes \boldsymbol{g}^x + \partial_y Z \ \boldsymbol{G}_Z \otimes \boldsymbol{g}^y + \partial_z Z \ \boldsymbol{G}_Z \otimes \boldsymbol{g}^z. \qquad (3.231)
\end{aligned}$$

Components of $\boldsymbol{F}^{-1}$ are the same in physical coordinates:

$$\begin{aligned}
\boldsymbol{F}^{-1} &= \partial_x X \ \boldsymbol{e}_x \otimes \boldsymbol{e}_x + \partial_y X \ \boldsymbol{e}_x \otimes \boldsymbol{e}_y + \partial_z X \ \boldsymbol{e}_x \otimes \boldsymbol{e}_z \\
&\quad + \partial_x Y \ \boldsymbol{e}_y \otimes \boldsymbol{e}_x + \partial_y Y \ \boldsymbol{e}_y \otimes \boldsymbol{e}_y + \partial_z Y \ \boldsymbol{e}_y \otimes \boldsymbol{e}_z \\
&\quad + \partial_x Z \ \boldsymbol{e}_z \otimes \boldsymbol{e}_x + \partial_y Z \ \boldsymbol{e}_z \otimes \boldsymbol{e}_y + \partial_z Z \ \boldsymbol{e}_z \otimes \boldsymbol{e}_z. \qquad (3.232)
\end{aligned}$$

The inverse of the Jacobian determinant $J^{-1} = 1/J$ is

$$\begin{aligned}
J^{-1} &= \mathrm{d}V/\mathrm{d}v \\
&= \det[F^{-1A}_{\ \ \bullet a}]\sqrt{G/g} \\
&= \det[\partial_a X^A] \det[g^a_A] \\
&= \det[\partial_a X^A]. \qquad (3.233)
\end{aligned}$$

A spatial displacement vector field in Cartesian coordinates is

$$\begin{aligned}
\boldsymbol{u} &= \boldsymbol{x} - \boldsymbol{X} \\
&= x\boldsymbol{g}_x + y\boldsymbol{g}_y + z\boldsymbol{g}_z - X\boldsymbol{G}_X - Y\boldsymbol{G}_Y - Z\boldsymbol{G}_Z \\
&= (x - X)\boldsymbol{e}_x + (y - Y)\boldsymbol{e}_y + (z - Z)\boldsymbol{e}_z \\
&= u_x\boldsymbol{e}_x + u_y\boldsymbol{e}_y + u_z\boldsymbol{e}_z. \qquad (3.234)
\end{aligned}$$

3.4.2 *Cylindrical coordinates*

Spatial coordinates are considered first; analogous formulae for reference coordinates follow later. In three spatial dimensions, cylindrical spatial coordinates are

$$(x^1, x^2, x^3) \to (r, \theta, z); \qquad r \geq 0, \qquad \theta \in (-\pi, \pi]. \qquad (3.235)$$

The squared length of a differential line element $\mathrm{d}\boldsymbol{x}$ is

$$|\mathrm{d}\boldsymbol{x}|^2 = \mathrm{d}\boldsymbol{x} \cdot \mathrm{d}\boldsymbol{x} = (\mathrm{d}r)^2 + (r\,\mathrm{d}\theta)^2 + (\mathrm{d}z)^2. \qquad (3.236)$$

The metric tensor and its inverse are, in matrix form,

$$[g_{ab}] = \begin{bmatrix} g_{rr} & g_{r\theta} & g_{rz} \\ g_{\theta r} & g_{\theta\theta} & g_{\theta z} \\ g_{zr} & g_{z\theta} & g_{zz} \end{bmatrix} = \begin{bmatrix} 1 & 0 & 0 \\ 0 & r^2 & 0 \\ 0 & 0 & 1 \end{bmatrix},$$

$$[g^{ab}] = \begin{bmatrix} g^{rr} & g^{r\theta} & g^{rz} \\ g^{\theta r} & g^{\theta\theta} & g^{\theta z} \\ g^{zr} & g^{z\theta} & g^{zz} \end{bmatrix} = \begin{bmatrix} 1 & 0 & 0 \\ 0 & 1/r^2 & 0 \\ 0 & 0 & 1 \end{bmatrix}. \qquad (3.237)$$

Determinants of the metric tensor and its inverse are

$$g = \det(g_{ab}) = r^2, \qquad g^{-1} = \det(g^{ab}) = 1/r^2. \qquad (3.238)$$

A differential volume element is

$$\mathrm{d}v = \sqrt{g}\,\mathrm{d}x^1\mathrm{d}x^2\mathrm{d}x^3 = r\,\mathrm{d}r\,\mathrm{d}\theta\,\mathrm{d}z. \qquad (3.239)$$

Connection coefficients from (2.150) are

$$\overset{g}{\Gamma}{}^{..\theta}_{r\theta} = \overset{g}{\Gamma}{}^{..\theta}_{\theta r} = 1/r, \qquad \overset{g}{\Gamma}{}^{..r}_{\theta\theta} = -r, \qquad \overset{g}{\Gamma}{}^{..a}_{bc} = 0 \text{ otherwise}. \qquad (3.240)$$

Let $\boldsymbol{x}$ denote the position vector in Euclidean space. Natural basis vectors are

$$\boldsymbol{g}_a = \partial_a \boldsymbol{x}; \qquad \boldsymbol{g}_r(\theta) = \partial_r \boldsymbol{x}, \quad \boldsymbol{g}_\theta(r,\theta) = \partial_\theta \boldsymbol{x}, \quad \boldsymbol{g}_z = \partial_z \boldsymbol{x}. \qquad (3.241)$$

The position vector from a fixed origin is

$$\boldsymbol{x}(r,\theta,z) = r\boldsymbol{g}_r(\theta) + z\boldsymbol{g}_z, \qquad (3.242)$$

where $\boldsymbol{g}_\theta = r\partial_\theta \boldsymbol{g}_r$. Basis vectors $\boldsymbol{g}_a$ are not all dimensionless. Their dimensionless, unit length representations $\boldsymbol{e}_a$ in physical components are

$$\boldsymbol{e}_r = \frac{\boldsymbol{g}_r}{\sqrt{g_{rr}}} = \boldsymbol{g}_r, \qquad \boldsymbol{e}_\theta = \frac{\boldsymbol{g}_\theta}{\sqrt{g_{\theta\theta}}} = \frac{\boldsymbol{g}_\theta}{r}, \qquad \boldsymbol{e}_z = \frac{\boldsymbol{g}_z}{\sqrt{g_{zz}}} = \boldsymbol{g}_z. \qquad (3.243)$$

Since the physical basis is orthonormal,

$$\boldsymbol{e}^r = \boldsymbol{e}_r, \qquad \boldsymbol{e}^\theta = \boldsymbol{e}_\theta, \qquad \boldsymbol{e}^z = \boldsymbol{e}_z. \qquad (3.244)$$

The position vector from a fixed origin is then

$$\boldsymbol{x}(r,\theta,z) = r\boldsymbol{e}_r + z\boldsymbol{e}_z, \qquad (3.245)$$

noting that $\boldsymbol{e}_r = \boldsymbol{e}_r(\theta)$ and $\boldsymbol{e}_\theta = \partial_\theta \boldsymbol{e}_r$. Contravariant basis vectors are

$$\boldsymbol{g}^a = g^{ab}\boldsymbol{g}_b; \qquad \boldsymbol{g}^r(\theta) = \boldsymbol{g}_r, \quad \boldsymbol{g}^\theta(r,\theta) = (1/r^2)\boldsymbol{g}_\theta, \quad \boldsymbol{g}^z = \boldsymbol{g}_z. \qquad (3.246)$$

Cylindrical coordinates are related to Cartesian coordinates as

$$x = r\cos\theta, \qquad y = r\sin\theta; \qquad \boldsymbol{x} = x\boldsymbol{e}_x + y\boldsymbol{e}_y + z\boldsymbol{e}_z; \qquad (3.247)$$

$$\boldsymbol{e}_r = \cos\theta \boldsymbol{e}_x + \sin\theta \boldsymbol{e}_y, \quad \boldsymbol{e}_\theta = -\sin\theta \boldsymbol{e}_x + \cos\theta \boldsymbol{e}_y, \quad \boldsymbol{e}_z = \boldsymbol{e}_z. \tag{3.248}$$

In physical components, writing all indices in the subscript position, the gradient of a scalar field $f(\boldsymbol{x})$ is

$$\overset{g}{\nabla} f = \partial_r f \ \boldsymbol{e}_r + \frac{1}{r}\partial_\theta f \ \boldsymbol{e}_\theta + \partial_z f \ \boldsymbol{e}_z. \tag{3.249}$$

The Laplacian of f is

$$\overset{g}{\nabla}{}^2 f = \partial_r \partial_r f + \frac{1}{r}\partial_r f + \frac{1}{r^2}\partial_\theta \partial_\theta f + \partial_z \partial_z f. \tag{3.250}$$

The divergence of a vector field $\boldsymbol{v}(\boldsymbol{x})$ is

$$\langle \overset{g}{\nabla}, \boldsymbol{v} \rangle = \frac{1}{r}\partial_r (r v_r) + \frac{1}{r}\partial_\theta v_\theta + \partial_z v_z. \tag{3.251}$$

The curl of field $\boldsymbol{v}(\boldsymbol{x})$ is

$$\begin{aligned} \overset{g}{\nabla} \times \boldsymbol{v} = & \left[\frac{1}{r}\partial_\theta v_z - \partial_z v_\theta \right] \boldsymbol{e}_r \\ & + [\partial_z v_r - \partial_r v_z] \boldsymbol{e}_\theta + \frac{1}{r}[\partial_r (r v_\theta) - \partial_\theta v_r] \boldsymbol{e}_z. \end{aligned} \tag{3.252}$$

The divergence of a second-order tensor field $\boldsymbol{A}(\boldsymbol{x}) = A_{ab} \boldsymbol{e}_a \otimes \boldsymbol{e}_b$ is, in physical cylindrical coordinates [26],

$$\begin{aligned} \overset{g}{\nabla} \cdot \boldsymbol{A} &= \overset{g}{\nabla}_b A_{ba} \boldsymbol{e}_a \\ &= [(1/r)\partial_r (r A_{rr}) + (1/r)\partial_\theta A_{\theta r} + \partial_z A_{zr} - (1/r) A_{\theta\theta}] \boldsymbol{e}_r \\ &\quad + [(1/r)\partial_r (r A_{r\theta}) + (1/r)\partial_\theta A_{\theta\theta} + \partial_z A_{z\theta} + (1/r) A_{\theta r}] \boldsymbol{e}_\theta \\ &\quad + [(1/r)\partial_r (r A_{rz}) + (1/r)\partial_\theta A_{\theta z} + \partial_z A_{zz}] \boldsymbol{e}_z. \end{aligned} \tag{3.253}$$

Now letting $\boldsymbol{v}(\boldsymbol{x}, t)$ denote the spatial velocity field, the velocity gradient of (3.162) in physical components is

$$[L_{ab}] = [v_{a;b}] = \begin{bmatrix} \partial_r v_r & (1/r)(\partial_\theta v_r - v_\theta) & \partial_z v_r \\ \partial_r v_\theta & (1/r)(\partial_\theta v_\theta + v_r) & \partial_z v_\theta \\ \partial_r v_z & (1/r)\partial_\theta v_z & \partial_z v_z \end{bmatrix}, \tag{3.254}$$

where the velocity vector in physical components is

$$\begin{aligned} \boldsymbol{v} &= v_r \boldsymbol{e}_r + v_\theta \boldsymbol{e}_\theta + v_z \boldsymbol{e}_z \\ &= \dot{r} \ \boldsymbol{e}_r + r\dot{\theta} \ \boldsymbol{e}_\theta + \dot{z} \ \boldsymbol{e}_z \\ &= \frac{\mathrm{d}r}{\mathrm{d}t} \ \boldsymbol{e}_r + r\frac{\mathrm{d}\theta}{\mathrm{d}t} \ \boldsymbol{e}_\theta + \frac{\mathrm{d}z}{\mathrm{d}t} \ \boldsymbol{e}_z. \end{aligned} \tag{3.255}$$

Finally, the spatial acceleration $\boldsymbol{a}(\boldsymbol{x}, t)$ of (3.154) is

$$\begin{aligned}\boldsymbol{a} &= a_r \boldsymbol{e}_r + a_\theta \boldsymbol{e}_\theta + a_z \boldsymbol{e}_z \\ &= \dot{v}_r \boldsymbol{e}_r + \dot{v}_\theta \boldsymbol{e}_\theta + \dot{v}_z \boldsymbol{e}_z \\ &= \left(\frac{\partial v_r}{\partial t} + v_r \partial_r v_r + \frac{1}{r} v_\theta \partial_\theta v_r + v_z \partial_z v_r - \frac{1}{r} v_\theta v_\theta \right) \boldsymbol{e}_r \\ &+ \left(\frac{\partial v_\theta}{\partial t} + v_r \partial_r v_\theta + \frac{1}{r} v_\theta \partial_\theta v_\theta + v_z \partial_z v_\theta + \frac{1}{r} v_r v_\theta \right) \boldsymbol{e}_\theta \\ &+ \left(\frac{\partial v_z}{\partial t} + v_r \partial_r v_z + \frac{1}{r} v_\theta \partial_\theta v_z + v_z \partial_z v_z \right) \boldsymbol{e}_z. \end{aligned} \tag{3.256}$$

Reference cylindrical coordinates $\{X^A\}$ are

$$(X^1, X^2, X^3) \to (R, \Theta, Z); \qquad R \geq 0, \quad \Theta \in (-\pi, \pi]. \tag{3.257}$$

Let $\boldsymbol{X}$ denote the position vector measured from a fixed origin:

$$\boldsymbol{X}(R, \Theta, Z) = R\boldsymbol{G}_R(\Theta) + Z\boldsymbol{G}_Z. \tag{3.258}$$

Natural basis vectors are

$$\begin{aligned} &\boldsymbol{G}_A = \partial_A \boldsymbol{X}; \\ &\boldsymbol{G}_R(\Theta) = \partial_R \boldsymbol{X}, \quad \boldsymbol{G}_\Theta(R, \Theta) = \partial_\Theta \boldsymbol{X}, \quad \boldsymbol{G}_Z = \partial_Z \boldsymbol{X}. \end{aligned} \tag{3.259}$$

The squared length of a differential line element $\mathrm{d}\boldsymbol{X}$ is

$$|\mathrm{d}\boldsymbol{X}|^2 = \mathrm{d}\boldsymbol{X} \cdot \mathrm{d}\boldsymbol{X} = (\mathrm{d}R)^2 + (R\,\mathrm{d}\Theta)^2 + (\mathrm{d}Z)^2. \tag{3.260}$$

The metric tensor with components $G_{AB} = \boldsymbol{G}_A \cdot \boldsymbol{G}_B$ and its inverse are

$$[G_{AB}] = \begin{bmatrix} 1 & 0 & 0 \\ 0 & R^2 & 0 \\ 0 & 0 & 1 \end{bmatrix}, \quad [G^{AB}] = \begin{bmatrix} 1 & 0 & 0 \\ 0 & 1/R^2 & 0 \\ 0 & 0 & 1 \end{bmatrix}. \tag{3.261}$$

Contravariant basis vectors are

$$\begin{aligned} &\boldsymbol{G}^A = G^{AB} \boldsymbol{G}_B; \\ &\boldsymbol{G}^R(\Theta) = \boldsymbol{G}_R, \quad \boldsymbol{G}^\Theta(R, \Theta) = (1/R^2)\boldsymbol{G}_\Theta, \quad \boldsymbol{G}^Z = \boldsymbol{G}_Z. \end{aligned} \tag{3.262}$$

Physical (dimensionless unit) basis vectors are

$$\begin{aligned} &\boldsymbol{E}_A = \boldsymbol{G}_A / \sqrt{G_{AA}}; \\ &\boldsymbol{E}_R(\Theta) = \boldsymbol{G}_R, \quad \boldsymbol{E}_\Theta(\Theta) = (1/R)\boldsymbol{G}_\Theta, \quad \boldsymbol{E}_Z = \boldsymbol{G}_Z. \end{aligned} \tag{3.263}$$

Since cylindrical coordinates are orthogonal ($\boldsymbol{E}_A = \boldsymbol{E}^A$), there is no need to distinguish between contravariant and covariant physical components. Cylindrical coordinates are related to Cartesian coordinates as follows:

$$X = R\cos\Theta, \quad Y = R\sin\Theta; \qquad \boldsymbol{X} = X\boldsymbol{E}_X + Y\boldsymbol{E}_Y + Z\boldsymbol{E}_Z; \tag{3.264}$$

$$\begin{aligned}\boldsymbol{E}_R &= \cos\Theta\boldsymbol{E}_X + \sin\Theta\boldsymbol{E}_Y,\\ \boldsymbol{E}_\Theta &= -\sin\Theta\boldsymbol{E}_X + \cos\Theta\boldsymbol{E}_Y,\\ \boldsymbol{E}_Z &= \boldsymbol{E}_Z.\end{aligned} \tag{3.265}$$

Referential Christoffel symbols of the second kind are completely analogous to (3.240).

Here coincident Cartesian frames are used such that $\boldsymbol{E}_A = \delta^a_A \boldsymbol{e}_a$:

$$\boldsymbol{E}_X = \boldsymbol{e}_x, \qquad \boldsymbol{E}_Y = \boldsymbol{e}_y, \qquad \boldsymbol{E}_Z = \boldsymbol{e}_z. \tag{3.266}$$

Mixed variant components of the shifter tensor between reference and spatial systems are

$$g^a_A(\boldsymbol{x}, \boldsymbol{X}) = \langle \boldsymbol{g}^a, \boldsymbol{G}_A \rangle; \qquad \boldsymbol{g}^a(\boldsymbol{x}) = g^a_A \boldsymbol{G}^A(\boldsymbol{X}); \tag{3.267}$$

where $\langle \cdot, \cdot \rangle$ denotes the scalar product. Using (3.266), it follows that

$$\begin{aligned}g^r_R(\theta, \Theta) &= \langle \boldsymbol{g}^r(\theta), \boldsymbol{G}_R(\Theta) \rangle\\ &= \boldsymbol{e}_r(\theta) \cdot \boldsymbol{E}_R(\Theta)\\ &= (\cos\theta\boldsymbol{e}_x + \sin\theta\boldsymbol{e}_y) \cdot (\cos\Theta\boldsymbol{E}_X + \sin\Theta\boldsymbol{E}_Y)\\ &= \cos(\theta - \Theta).\end{aligned} \tag{3.268}$$

Applying similar calculations, the matrix of components of $g^a_A(r, \theta; R, \Theta)$ is

$$\begin{aligned}[g^a_A] &= \begin{bmatrix} g^r_R & g^r_\Theta & g^r_Z \\ g^\theta_R & g^\theta_\Theta & g^\theta_Z \\ g^z_R & g^z_\Theta & g^z_Z \end{bmatrix}\\ &= \begin{bmatrix} \cos(\theta - \Theta) & R\sin(\theta - \Theta) & 0 \\ -(1/r)\sin(\theta - \Theta) & (R/r)\cos(\theta - \Theta) & 0 \\ 0 & 0 & 1 \end{bmatrix}.\end{aligned} \tag{3.269}$$

The inverse of $[g^a_A]$ is $[g^A_a]$, such that $g^a_A g^A_b = \delta^a_b$:

$$\begin{aligned}[g^A_a] &= \begin{bmatrix} g^R_r & g^R_\theta & g^R_z \\ g^\Theta_r & g^\Theta_\theta & g^\Theta_z \\ g^Z_r & g^Z_\theta & g^Z_z \end{bmatrix}\\ &= \begin{bmatrix} \cos(\theta - \Theta) & -r\sin(\theta - \Theta) & 0 \\ (1/R)\sin(\theta - \Theta) & (r/R)\cos(\theta - \Theta) & 0 \\ 0 & 0 & 1 \end{bmatrix}.\end{aligned} \tag{3.270}$$

Noting $G = \det(G_{AB}) = R^2$, determinants obey

$$\det[g^a_A] = \sqrt{G/g} = R/r, \qquad \det[g^A_a] = \sqrt{g/G} = r/R. \tag{3.271}$$

In cylindrical coordinates, deformation $\boldsymbol{x} = \boldsymbol{x}(\boldsymbol{X})$ is of the form

$$r = r(R, \Theta, Z), \qquad \theta = \theta(R, \Theta, Z), \qquad z = z(R, \Theta, Z). \tag{3.272}$$

Referred to natural bases $\{\boldsymbol{g}_a, \boldsymbol{G}^A\}$, the deformation gradient (two-point tensor) is

$$\begin{aligned}
\boldsymbol{F} &= \partial_A \boldsymbol{x} \otimes \boldsymbol{G}^A \\
&= \partial_A x^a \boldsymbol{g}_a \otimes \boldsymbol{G}^A \\
&= F^a_{.A} \boldsymbol{g}_a \otimes \boldsymbol{G}^A \\
&= \partial_R r \ \boldsymbol{g}_r \otimes \boldsymbol{G}^R + \partial_\Theta r \ \boldsymbol{g}_r \otimes \boldsymbol{G}^\Theta + \partial_Z r \ \boldsymbol{g}_r \otimes \boldsymbol{G}^Z \\
&\quad + \partial_R \theta \ \boldsymbol{g}_\theta \otimes \boldsymbol{G}^R + \partial_\Theta \theta \ \boldsymbol{g}_\theta \otimes \boldsymbol{G}^\Theta + \partial_Z \theta \ \boldsymbol{g}_\theta \otimes \boldsymbol{G}^Z \\
&\quad + \partial_R z \ \boldsymbol{g}_z \otimes \boldsymbol{G}^R + \partial_\Theta z \ \boldsymbol{g}_z \otimes \boldsymbol{G}^\Theta + \partial_Z z \ \boldsymbol{g}_z \otimes \boldsymbol{G}^Z.
\end{aligned} \tag{3.273}$$

Referential and deformed volume elements

$$\begin{aligned}
\mathrm{d}V &= \sqrt{G}\mathrm{d}X^1\mathrm{d}X^2\mathrm{d}X^3 = R\mathrm{d}R\mathrm{d}\Theta\mathrm{d}Z, \\
\mathrm{d}v &= \sqrt{g}\mathrm{d}x^1\mathrm{d}x^2\mathrm{d}x^3 = r\mathrm{d}r\mathrm{d}\theta\mathrm{d}z
\end{aligned} \tag{3.274}$$

are related through the Jacobian determinant:

$$\begin{aligned}
J &= \mathrm{d}v/\mathrm{d}V \\
&= \det[F^a_{.A}]\sqrt{g/G} \\
&= \det[\partial_A x^a] \det[g^A_a] \\
&= \det[\partial_A x^a](r/R).
\end{aligned} \tag{3.275}$$

Using (3.243), (3.262), and (3.263) to convert natural basis vectors to cylindrical orthonormal bases $\{\boldsymbol{e}_a, \boldsymbol{E}_A\}$, the two-point deformation gradient in physical components is

$$\begin{aligned}
\boldsymbol{F} &= F_{\langle aA \rangle} \boldsymbol{e}_a \otimes \boldsymbol{E}_A \\
&= \partial_R r \ \boldsymbol{e}_r \otimes \boldsymbol{E}_R + (1/R)\partial_\Theta r \ \boldsymbol{e}_r \otimes \boldsymbol{E}_\Theta + \partial_Z r \ \boldsymbol{e}_r \otimes \boldsymbol{E}_Z \\
&\quad + r\partial_R \theta \ \boldsymbol{e}_\theta \otimes \boldsymbol{E}_R + (r/R)\partial_\Theta \theta \ \boldsymbol{e}_\theta \otimes \boldsymbol{E}_\Theta + r\partial_Z \theta \ \boldsymbol{e}_\theta \otimes \boldsymbol{E}_Z \\
&\quad + \partial_R z \ \boldsymbol{e}_z \otimes \boldsymbol{E}_R + (1/R)\partial_\Theta z \ \boldsymbol{e}_z \otimes \boldsymbol{E}_\Theta + \partial_Z z \ \boldsymbol{e}_z \otimes \boldsymbol{E}_Z,
\end{aligned} \tag{3.276}$$

where physical scalar components are written in angled brackets. The Jacobian determinant is simply

$$J = \det[F_{\langle aA \rangle}]. \tag{3.277}$$

Deformation tensor $\boldsymbol{C}$ of (3.86) is, in curvilinear coordinates,

$$\begin{aligned}
\boldsymbol{C} &= C_{AB}\,\boldsymbol{G}^A \otimes \boldsymbol{G}^B \\
&= F^a_{\cdot A} g_{ab} F^b_{\cdot B}\,\boldsymbol{G}^A \otimes \boldsymbol{G}^B \\
&= [(\partial_R r)^2 + r^2(\partial_R \theta)^2 + (\partial_R z)^2]\,\boldsymbol{G}^R \otimes \boldsymbol{G}^R \\
&\quad + [(\partial_R r)(\partial_\Theta r) + r^2(\partial_R \theta)(\partial_\Theta \theta) + (\partial_R z)(\partial_\Theta z)]\,\boldsymbol{G}^R \otimes \boldsymbol{G}^\Theta \\
&\quad + [(\partial_R r)(\partial_Z r) + r^2(\partial_R \theta)(\partial_Z \theta) + (\partial_R z)(\partial_Z z)]\,\boldsymbol{G}^R \otimes \boldsymbol{G}^Z \\
&\quad + [(\partial_R r)(\partial_\Theta r) + r^2(\partial_R \theta)(\partial_\Theta \theta) + (\partial_R z)(\partial_\Theta z)]\,\boldsymbol{G}^\Theta \otimes \boldsymbol{G}^R \\
&\quad + [(\partial_\Theta r)^2 + r^2(\partial_\Theta \theta)^2 + (\partial_\Theta z)^2]\,\boldsymbol{G}^\Theta \otimes \boldsymbol{G}^\Theta \\
&\quad + [(\partial_\Theta r)(\partial_Z r) + r^2(\partial_\Theta \theta)(\partial_Z \theta) + (\partial_\Theta z)(\partial_Z z)]\,\boldsymbol{G}^\Theta \otimes \boldsymbol{G}^Z \\
&\quad + [(\partial_R r)(\partial_Z r) + r^2(\partial_R \theta)(\partial_Z \theta) + (\partial_R z)(\partial_Z z)]\,\boldsymbol{G}^Z \otimes \boldsymbol{G}^R \\
&\quad + [(\partial_\Theta r)(\partial_Z r) + r^2(\partial_\Theta \theta)(\partial_Z \theta) + (\partial_\Theta z)(\partial_Z z)]\,\boldsymbol{G}^Z \otimes \boldsymbol{G}^\Theta \\
&\quad + [(\partial_Z r)^2 + r^2(\partial_Z \theta)^2 + (\partial_Z z)^2]\,\boldsymbol{G}^Z \otimes \boldsymbol{G}^Z. \qquad (3.278)
\end{aligned}$$

In physical components,

$$\begin{aligned}
\boldsymbol{C} &= C_{\langle AB\rangle}\,\boldsymbol{E}_A \otimes \boldsymbol{E}_B \\
&= F_{\langle aA\rangle} F_{\langle aB\rangle}\,\boldsymbol{E}_A \otimes \boldsymbol{E}_B \\
&= \quad [(\partial_R r)^2 + r^2(\partial_R \theta)^2 + (\partial_R z)^2]\,\boldsymbol{E}_R \otimes \boldsymbol{E}_R \\
&\quad + (1/R)[(\partial_R r)(\partial_\Theta r) + r^2(\partial_R \theta)(\partial_\Theta \theta) \\
&\qquad\qquad + (\partial_R z)(\partial_\Theta z)]\,\boldsymbol{E}_R \otimes \boldsymbol{E}_\Theta \\
&\quad + [(\partial_R r)(\partial_Z r) + r^2(\partial_R \theta)(\partial_Z \theta) \\
&\qquad\qquad + (\partial_R z)(\partial_Z z)]\,\boldsymbol{E}_R \otimes \boldsymbol{E}_Z \\
&\quad + (1/R)[(\partial_R r)(\partial_\Theta r) + r^2(\partial_R \theta)(\partial_\Theta \theta) \\
&\qquad\qquad + (\partial_R z)(\partial_\Theta z)]\,\boldsymbol{E}_\Theta \otimes \boldsymbol{E}_R \\
&\quad + (1/R^2)[(\partial_\Theta r)^2 + r^2(\partial_\Theta \theta)^2 \\
&\qquad\qquad + (\partial_\Theta z)^2]\,\boldsymbol{E}_\Theta \otimes \boldsymbol{E}_\Theta \\
&\quad + (1/R)[(\partial_\Theta r)(\partial_Z r) + r^2(\partial_\Theta \theta)(\partial_Z \theta) \\
&\qquad\qquad + (\partial_\Theta z)(\partial_Z z)]\,\boldsymbol{E}_\Theta \otimes \boldsymbol{E}_Z \\
&\quad + [(\partial_R r)(\partial_Z r) + r^2(\partial_R \theta)(\partial_Z \theta) \\
&\qquad\qquad + (\partial_R z)(\partial_Z z)]\,\boldsymbol{E}_Z \otimes \boldsymbol{E}_R \\
&\quad + (1/R)[(\partial_\Theta r)(\partial_Z r) + r^2(\partial_\Theta \theta)(\partial_Z \theta) \\
&\qquad\qquad + (\partial_\Theta z)(\partial_Z z)]\,\boldsymbol{E}_Z \otimes \boldsymbol{E}_\Theta \\
&\quad + [(\partial_Z r)^2 + r^2(\partial_Z \theta)^2 + (\partial_Z z)^2]\,\boldsymbol{E}_Z \otimes \boldsymbol{E}_Z. \qquad (3.279)
\end{aligned}$$

Using the shifter, the deformation gradient can be expressed completely with respect to Lagrangian bases:

$$\begin{aligned}\boldsymbol{F} &= \partial_A x^a (g_a^B \boldsymbol{G}_B) \otimes \boldsymbol{G}^A \\ &= [\partial_A x^a g_a^B]\, \boldsymbol{G}_B \otimes \boldsymbol{G}^A \\ &= [\partial_R r \cos(\theta - \Theta) - r\partial_R \theta \sin(\theta - \Theta)]\, \boldsymbol{G}_R \otimes \boldsymbol{G}^R \\ &\quad + [\partial_\Theta r \cos(\theta - \Theta) - r\partial_\Theta \theta \sin(\theta - \Theta)]\, \boldsymbol{G}_R \otimes \boldsymbol{G}^\Theta \\ &\quad + [\partial_Z r \cos(\theta - \Theta) - r\partial_Z \theta \sin(\theta - \Theta)]\, \boldsymbol{G}_R \otimes \boldsymbol{G}^Z \\ &\quad + [(1/R)\partial_R r \sin(\theta - \Theta) + (r/R)\partial_R \theta \cos(\theta - \Theta)]\, \boldsymbol{G}_\Theta \otimes \boldsymbol{G}^R \\ &\quad + [(1/R)\partial_\Theta r \sin(\theta - \Theta) + (r/R)\partial_\Theta \theta \cos(\theta - \Theta)]\, \boldsymbol{G}_\Theta \otimes \boldsymbol{G}^\Theta \\ &\quad + [(1/R)\partial_Z r \sin(\theta - \Theta) + (r/R)\partial_Z \theta \cos(\theta - \Theta)]\, \boldsymbol{G}_\Theta \otimes \boldsymbol{G}^Z \\ &\quad + [\partial_R z]\, \boldsymbol{G}_Z \otimes \boldsymbol{G}^R + [\partial_\Theta z]\, \boldsymbol{G}_Z \otimes \boldsymbol{G}^\Theta + [\partial_Z z]\, \boldsymbol{G}_Z \otimes \boldsymbol{G}^Z. \end{aligned} \tag{3.280}$$

In physical components this becomes

$$\begin{aligned}\boldsymbol{F} &= F_{\langle AB \rangle} \boldsymbol{E}_A \otimes \boldsymbol{E}_B \\ &= [\partial_R r \cos(\theta - \Theta) - r\partial_R \theta \sin(\theta - \Theta)] \boldsymbol{E}_R \otimes \boldsymbol{E}_R \\ &\quad + [(1/R)\partial_\Theta r \cos(\theta - \Theta) - (r/R)\partial_\Theta \theta \sin(\theta - \Theta)] \boldsymbol{E}_R \otimes \boldsymbol{E}_\Theta \\ &\quad + [\partial_Z r \cos(\theta - \Theta) - r\partial_Z \theta \sin(\theta - \Theta)] \boldsymbol{E}_R \otimes \boldsymbol{E}_Z \\ &\quad + [\partial_R r \sin(\theta - \Theta) + r\partial_R \theta \cos(\theta - \Theta)] \boldsymbol{E}_\Theta \otimes \boldsymbol{E}_R \\ &\quad + [(1/R)\partial_\Theta r \sin(\theta - \Theta) + (r/R)\partial_\Theta \theta \cos(\theta - \Theta)] \boldsymbol{E}_\Theta \otimes \boldsymbol{E}_\Theta \\ &\quad + [\partial_Z r \sin(\theta - \Theta) + r\partial_Z \theta \cos(\theta - \Theta)] \boldsymbol{E}_\Theta \otimes \boldsymbol{E}_Z \\ &\quad + [\partial_R z] \boldsymbol{E}_Z \otimes \boldsymbol{E}_R + [(1/R)\partial_\Theta z] \boldsymbol{E}_Z \otimes \boldsymbol{E}_\Theta + [\partial_Z z] \boldsymbol{E}_Z \otimes \boldsymbol{E}_Z. \end{aligned} \tag{3.281}$$

Similarly, expressing $\boldsymbol{F}$ with respect to natural spatial bases,

$$\begin{aligned}\boldsymbol{F} &= \partial_A x^a \boldsymbol{g}_a \otimes (g_b^A \boldsymbol{g}^b) \\ &= [\partial_A x^a g_b^A] \boldsymbol{g}_a \otimes \boldsymbol{g}^b \\ &= [\partial_R r \cos(\theta - \Theta) + (1/R)\partial_\Theta r \sin(\theta - \Theta)] \boldsymbol{g}_r \otimes \boldsymbol{g}^r \\ &\quad + [\partial_R \theta \cos(\theta - \Theta) + (1/R)\partial_\Theta \theta \sin(\theta - \Theta)] \boldsymbol{g}_\theta \otimes \boldsymbol{g}^r \\ &\quad + [\partial_R z \cos(\theta - \Theta) + (1/R)\partial_\Theta z \sin(\theta - \Theta)] \boldsymbol{g}_z \otimes \boldsymbol{g}^r \\ &\quad + [-r\partial_R r \sin(\theta - \Theta) + (r/R)\partial_\Theta r \cos(\theta - \Theta)] \boldsymbol{g}_r \otimes \boldsymbol{g}^\theta \\ &\quad + [-r\partial_R \theta \sin(\theta - \Theta) + (r/R)\partial_\Theta \theta \cos(\theta - \Theta)] \boldsymbol{g}_\theta \otimes \boldsymbol{g}^\theta \\ &\quad + [-r\partial_R z \sin(\theta - \Theta) + (r/R)\partial_\Theta z \cos(\theta - \Theta)] \boldsymbol{g}_z \otimes \boldsymbol{g}^\theta \\ &\quad + [\partial_Z r] \boldsymbol{g}_r \otimes \boldsymbol{g}^z + [\partial_Z \theta] \boldsymbol{g}_\theta \otimes \boldsymbol{g}^z + [\partial_Z z] \boldsymbol{g}_z \otimes \boldsymbol{g}^z. \end{aligned} \tag{3.282}$$

In spatial physical components,

$$\begin{aligned}
\boldsymbol{F} &= F_{\langle ab\rangle}\boldsymbol{e}_a \otimes \boldsymbol{e}_b \\
&= [\partial_R r\cos(\theta-\Theta) + (1/R)\partial_\Theta r\sin(\theta-\Theta)]\boldsymbol{e}_r\otimes\boldsymbol{e}_r \\
&\quad + [r\partial_R\theta\cos(\theta-\Theta) + (r/R)\partial_\Theta\theta\sin(\theta-\Theta)]\boldsymbol{e}_\theta\otimes\boldsymbol{e}_r \\
&\quad + [\partial_R z\cos(\theta-\Theta) + (1/R)\partial_\Theta z\sin(\theta-\Theta)]\boldsymbol{e}_z\otimes\boldsymbol{e}_r \\
&\quad + [-\partial_R r\sin(\theta-\Theta) + (1/R)\partial_\Theta r\cos(\theta-\Theta)]\boldsymbol{e}_r\otimes\boldsymbol{e}_\theta \\
&\quad + [-r\partial_R\theta\sin(\theta-\Theta) + (r/R)\partial_\Theta\theta\cos(\theta-\Theta)]\boldsymbol{e}_\theta\otimes\boldsymbol{e}_\theta \\
&\quad + [-\partial_R z\sin(\theta-\Theta) + (1/R)\partial_\Theta z\cos(\theta-\Theta)]\boldsymbol{e}_z\otimes\boldsymbol{e}_\theta \\
&\quad + [\partial_Z r]\boldsymbol{e}_r\otimes\boldsymbol{e}_z + [r\partial_Z\theta]\boldsymbol{e}_\theta\otimes\boldsymbol{e}_z + [\partial_Z z]\boldsymbol{e}_z\otimes\boldsymbol{e}_z. \qquad (3.283)
\end{aligned}$$

Letting $\boldsymbol{F}\to\tilde{\boldsymbol{F}}$, (3.280) becomes an example of possibly anholonomic deformation referred to coincident curvilinear coordinate frames in reference and intermediate configurations, as proposed in [8] and introduced later in (4.166) of this book. With $\boldsymbol{F}\to\bar{\boldsymbol{F}}$, (3.282) becomes an example of coincident curvilinear systems in current and intermediate configurations, as in [8] and (4.180) later in this text. In particular, coincident spatial and intermediate cylindrical coordinate systems are examined more in Section 4.2.5 of the present book.

In cylindrical coordinates, inverse deformation $\boldsymbol{X}=\boldsymbol{X}(\boldsymbol{x})$ is of the form

$$R = R(r,\theta,z), \qquad \Theta=\Theta(r,\theta,z), \qquad Z = Z(r,\theta,z). \qquad (3.284)$$

Referred to natural bases $\{\boldsymbol{G}_A, \boldsymbol{g}^a\}$, the inverse deformation gradient is

$$\begin{aligned}
\boldsymbol{F}^{-1} &= \partial_a\boldsymbol{X}\otimes\boldsymbol{g}^a \\
&= \partial_a X^A\boldsymbol{G}_A\otimes\boldsymbol{g}^a \\
&= F^{-1A}_{\ \ \bullet a}\boldsymbol{G}_A\otimes\boldsymbol{g}^a \\
&= \partial_r R\ \boldsymbol{G}_R\otimes\boldsymbol{g}^r + \partial_\theta R\ \boldsymbol{G}_R\otimes\boldsymbol{g}^\theta + \partial_z R\ \boldsymbol{G}_R\otimes\boldsymbol{g}^z \\
&\quad + \partial_r\Theta\ \boldsymbol{G}_\Theta\otimes\boldsymbol{g}^r + \partial_\theta\Theta\ \boldsymbol{G}_\Theta\otimes\boldsymbol{g}^\theta + \partial_z\Theta\ \boldsymbol{G}_\Theta\otimes\boldsymbol{g}^z \\
&\quad + \partial_r Z\ \boldsymbol{G}_Z\otimes\boldsymbol{g}^r + \partial_\theta Z\ \boldsymbol{G}_Z\otimes\boldsymbol{g}^\theta + \partial_z Z\ \boldsymbol{G}_Z\otimes\boldsymbol{g}^z. \qquad (3.285)
\end{aligned}$$

The inverse of the Jacobian determinant $J^{-1}=1/J$ is

$$\begin{aligned}
J^{-1} &= \mathrm{d}V/\mathrm{d}v \\
&= \det[F^{-1A}_{\ \ \bullet a}]\sqrt{G/g} \\
&= \det[\partial_a X^A]\det[g^a_A] \\
&= \det[\partial_a X^A](R/r). \qquad (3.286)
\end{aligned}$$

With respect to orthonormal bases $\{\boldsymbol{E}_A, \boldsymbol{e}_a\}$, the two-point inverse deformation gradient is

$$\begin{aligned}
\boldsymbol{F}^{-1} &= F^{-1}_{\langle Aa\rangle}\boldsymbol{E}_A \otimes \boldsymbol{e}_a \\
&= \partial_r R\, \boldsymbol{E}_R \otimes \boldsymbol{e}_r + (1/r)\partial_\theta R\, \boldsymbol{E}_R \otimes \boldsymbol{e}_\theta + \partial_z R\, \boldsymbol{E}_R \otimes \boldsymbol{e}_z \\
&\quad + R\partial_r \Theta\, \boldsymbol{E}_\Theta \otimes \boldsymbol{e}_r + (R/r)\partial_\theta \Theta\, \boldsymbol{E}_\Theta \otimes \boldsymbol{e}_\theta + R\partial_z \Theta\, \boldsymbol{E}_\Theta \otimes \boldsymbol{e}_z \\
&\quad + \partial_r Z\, \boldsymbol{E}_Z \otimes \boldsymbol{e}_r + (1/r)\partial_\theta Z\, \boldsymbol{E}_Z \otimes \boldsymbol{e}_\theta + \partial_z Z\, \boldsymbol{E}_Z \otimes \boldsymbol{e}_z, \qquad (3.287)
\end{aligned}$$

with inverse Jacobian determinant

$$J^{-1} = \det[F^{-1}_{\langle Aa\rangle}]. \qquad (3.288)$$

A spatial displacement vector field in physical cylindrical coordinates can be constructed using the shifter:

$$\begin{aligned}
\boldsymbol{u} &= \boldsymbol{x} - \boldsymbol{X} \\
&= r\boldsymbol{g}_r + z\boldsymbol{g}_z - R\boldsymbol{G}_R - Z\boldsymbol{G}_Z \\
&= r\boldsymbol{g}_r - Rg^a_R\boldsymbol{g}_a + (z - Z)\boldsymbol{G}_Z \\
&= [r - R\cos(\theta - \Theta)]\boldsymbol{e}_r + [R\sin(\theta - \Theta)]\boldsymbol{e}_\theta + [z - Z]\boldsymbol{e}_z \\
&= u_r\boldsymbol{e}_r + u_\theta\boldsymbol{e}_\theta + u_z\boldsymbol{e}_z. \qquad (3.289)
\end{aligned}$$

3.4.3 *Spherical coordinates*

Spatial coordinates are considered first here; analogous formulae apply for reference coordinates as will be shown later. In three spatial dimensions, spherical spatial coordinates are

$$(x^1, x^2, x^3) \to (r, \theta, \phi); \qquad r \geq 0, \quad \theta \in [0, \pi], \quad \phi \in (-\pi, \pi]. \qquad (3.290)$$

The squared length of a differential line element $\mathrm{d}\boldsymbol{x}$ is

$$|\mathrm{d}\boldsymbol{x}|^2 = \mathrm{d}\boldsymbol{x} \cdot \mathrm{d}\boldsymbol{x} = (\mathrm{d}r)^2 + (r\,\mathrm{d}\theta)^2 + (r\,\sin\theta\,\mathrm{d}\phi)^2. \qquad (3.291)$$

In matrix form, the metric tensor and its inverse are, respectively,

$$[g_{ab}] = \begin{bmatrix} g_{rr} & g_{r\theta} & g_{r\phi} \\ g_{\theta r} & g_{\theta\theta} & g_{\theta\phi} \\ g_{\phi r} & g_{\phi\theta} & g_{\phi\phi} \end{bmatrix} = \begin{bmatrix} 1 & 0 & 0 \\ 0 & r^2 & 0 \\ 0 & 0 & r^2\sin^2\theta \end{bmatrix},$$

$$[g^{ab}] = \begin{bmatrix} g^{rr} & g^{r\theta} & g^{r\phi} \\ g^{\theta r} & g^{\theta\theta} & g^{\theta\phi} \\ g^{\phi r} & g^{\phi\theta} & g^{\phi\phi} \end{bmatrix} = \begin{bmatrix} 1 & 0 & 0 \\ 0 & 1/r^2 & 0 \\ 0 & 0 & 1/(r^2\sin^2\theta) \end{bmatrix}. \qquad (3.292)$$

Determinants of the metric tensor and its inverse are

$$g = \det(g_{ab}) = r^4\sin^2\theta, \qquad g^{-1} = \det(g^{ab}) = 1/(r^4\sin^2\theta). \qquad (3.293)$$

A differential volume element is

$$\mathrm{d}v = \sqrt{g}\,\mathrm{d}x^1\mathrm{d}x^2\mathrm{d}x^3 = r^2 \sin\theta\,\mathrm{d}r\,\mathrm{d}\theta\,\mathrm{d}\phi. \tag{3.294}$$

Connection coefficients from (2.150) are

$$\begin{aligned}
&\overset{g}{\Gamma}{}_{r\theta}^{\cdot\cdot\theta} = \overset{g}{\Gamma}{}_{\theta r}^{\cdot\cdot\theta} = \overset{g}{\Gamma}{}_{r\phi}^{\cdot\cdot\phi} = \overset{g}{\Gamma}{}_{\phi r}^{\cdot\cdot\phi} = 1/r, &&\overset{g}{\Gamma}{}_{\theta\theta}^{\cdot\cdot r} = -r,\\
&\overset{g}{\Gamma}{}_{\phi\phi}^{\cdot\cdot r} = -r\sin^2\theta, &&\overset{g}{\Gamma}{}_{\phi\phi}^{\cdot\cdot\theta} = -\sin\theta\,\cos\theta,\\
&\overset{g}{\Gamma}{}_{\theta\phi}^{\cdot\cdot\phi} = \overset{g}{\Gamma}{}_{\phi\theta}^{\cdot\cdot\phi} = \cot\theta, &&\overset{g}{\Gamma}{}_{bc}^{\cdot\cdot a} = 0 \text{ otherwise.}
\end{aligned} \tag{3.295}$$

Let $\boldsymbol{x}$ denote the position vector in Euclidean space. Natural basis vectors $\boldsymbol{g}_a = \partial_a \boldsymbol{x}$ are

$$\boldsymbol{g}_r(\theta,\phi) = \partial_r \boldsymbol{x}, \quad \boldsymbol{g}_\theta(r,\theta,\phi) = \partial_\theta \boldsymbol{x}, \quad \boldsymbol{g}_\phi(r,\theta,\phi) = \partial_\phi \boldsymbol{x}. \tag{3.296}$$

The position vector from a fixed origin is

$$\boldsymbol{x}(r,\theta,\phi) = r\boldsymbol{g}_r(\theta,\phi). \tag{3.297}$$

Contravariant basis vectors are

$$\begin{aligned}
&\boldsymbol{g}^a = g^{ab}\boldsymbol{g}_b;\\
&\boldsymbol{g}^r = \boldsymbol{g}_r, \qquad \boldsymbol{g}^\theta = (1/r^2)\boldsymbol{g}_\theta, \qquad \boldsymbol{g}^\phi = [1/(r^2\sin^2\theta)]\boldsymbol{g}_\phi.
\end{aligned} \tag{3.298}$$

Basis vectors $\boldsymbol{g}_a$ are not all dimensionless. Their dimensionless, unit length representations $\boldsymbol{e}_a$ in physical components are

$$\begin{aligned}
\boldsymbol{e}_r &= \frac{\boldsymbol{g}_r}{\sqrt{g_{rr}}} = \boldsymbol{g}_r,\\
\boldsymbol{e}_\theta &= \frac{\boldsymbol{g}_\theta}{\sqrt{g_{\theta\theta}}} = \frac{\boldsymbol{g}_\theta}{r},\\
\boldsymbol{e}_\phi &= \frac{\boldsymbol{g}_\phi}{\sqrt{g_{\phi\phi}}} = \frac{\boldsymbol{g}_\phi}{r\sin\theta}.
\end{aligned} \tag{3.299}$$

Since the physical basis is orthonormal,

$$\boldsymbol{e}^r = \boldsymbol{e}_r, \qquad \boldsymbol{e}^\theta = \boldsymbol{e}_\theta, \qquad \boldsymbol{e}^\phi = \boldsymbol{e}_\phi. \tag{3.300}$$

The position vector from a fixed origin is then

$$\boldsymbol{x}(r,\theta,\phi) = r\boldsymbol{e}_r(\theta,\phi). \tag{3.301}$$

Spherical coordinates are related to Cartesian coordinates as

$$\begin{aligned}
&x = r\sin\theta\cos\phi, \quad y = r\sin\theta\sin\phi, \quad z = r\cos\theta;\\
&\boldsymbol{x} = x\boldsymbol{e}_x + y\boldsymbol{e}_y + z\boldsymbol{e}_z;
\end{aligned} \tag{3.302}$$

$$\begin{aligned} \boldsymbol{e}_r &= \sin\theta\cos\phi\boldsymbol{e}_x + \sin\theta\sin\phi\boldsymbol{e}_y + \cos\theta\boldsymbol{e}_z, \\ \boldsymbol{e}_\theta &= \cos\theta\cos\phi\boldsymbol{e}_x + \cos\theta\sin\phi\boldsymbol{e}_y - \sin\theta\boldsymbol{e}_z, \\ \boldsymbol{e}_\phi &= -\sin\phi\boldsymbol{e}_x + \cos\phi\boldsymbol{e}_y. \end{aligned} \tag{3.303}$$

In physical components, writing all indices in the subscript position, the gradient of a scalar field $f(\boldsymbol{x})$ is

$$\overset{g}{\nabla} f = \partial_r f \; \boldsymbol{e}_r + \frac{1}{r}\partial_\theta f \; \boldsymbol{e}_\theta + \frac{1}{r\,\sin\theta}\partial_\phi f \; \boldsymbol{e}_\phi. \tag{3.304}$$

The Laplacian of f is

$$\overset{g}{\nabla}{}^2 f = \frac{1}{r^2}\partial_r(r^2\partial_r f) + \frac{1}{r^2\,\sin\theta}\partial_\theta(\sin\theta\partial_\theta f) + \frac{1}{r^2\sin^2\theta}\partial_\phi\partial_\phi f. \tag{3.305}$$

The divergence of a vector field $\boldsymbol{v}(\boldsymbol{x})$ is

$$\langle \overset{g}{\nabla}, \boldsymbol{v}\rangle = \frac{1}{r^2}\partial_r(r^2 v_r) + \frac{1}{r\,\sin\theta}[\partial_\theta(v_\theta\,\sin\theta) + \partial_\phi v_\phi]. \tag{3.306}$$

The curl of field $\boldsymbol{v}(\boldsymbol{x})$ is

$$\begin{aligned} \overset{g}{\nabla} \times \boldsymbol{v} = &\frac{1}{r\,\sin\theta}[\partial_\theta(v_\phi\sin\theta) - \partial_\phi v_\theta]\boldsymbol{e}_r \\ &+ \frac{1}{r}\left[\frac{1}{\sin\theta}\partial_\phi v_r - \partial_r(r v_\phi)\right]\boldsymbol{e}_\theta \\ &+ \frac{1}{r}[\partial_r(r v_\theta) - \partial_\theta v_r]\boldsymbol{e}_\phi. \end{aligned} \tag{3.307}$$

The divergence of a second-order tensor field $\boldsymbol{A}(\boldsymbol{x}) = A_{ab}\boldsymbol{e}_a \otimes \boldsymbol{e}_b$ in physical spherical coordinates is [26]

$$\begin{aligned} \overset{g}{\nabla} \cdot \boldsymbol{A} &= \overset{g}{\nabla}_b A_{ba}\boldsymbol{e}_a \\ &= \{(1/r^2)\partial_r(r^2 A_{rr}) + [1/(r\sin\theta)]\partial_\theta(\sin\theta A_{\theta r}) \\ &\quad + [1/(r\sin\theta)]\partial_\phi A_{\phi r} - (1/r)(A_{\theta\theta} + A_{\phi\phi})\}\boldsymbol{e}_r \\ &\quad + \{(1/r^2)\partial_r(r^2 A_{r\theta}) + [1/(r\sin\theta)]\partial_\theta(\sin\theta A_{\theta\theta}) \\ &\quad + [1/(r\sin\theta)]\partial_\phi A_{\phi\theta} + (1/r)(A_{\theta r} - \cot\theta A_{\phi\phi})\}\boldsymbol{e}_\theta \\ &\quad + \{(1/r^2)\partial_r(r^2 A_{r\phi}) + (1/r)\partial_\theta(\sin\theta A_{\theta\phi}) \\ &\quad + [1/(r\sin\theta)]\partial_\phi A_{\phi\phi} + (1/r)(A_{\phi r} + A_{\phi\theta})\}\boldsymbol{e}_\phi. \end{aligned} \tag{3.308}$$

Now letting $\boldsymbol{v}(\boldsymbol{x}, t)$ denote the spatial velocity field, the velocity gradient of (3.162) in physical components is

$$\begin{aligned} [L_{ab}] &= [v_{a;b}] \\ &= \begin{bmatrix} \partial_r v_r & (1/r)(\partial_\theta v_r - v_\theta) & (1/r\,\sin\theta)\partial_\phi v_r - (1/r)v_\phi \\ \partial_r v_\theta & (1/r)(\partial_\theta v_\theta + v_r) & (1/r\,\sin\theta)\partial_\phi v_\theta - (1/r)v_\phi\cot\phi \\ \partial_r v_\phi & (1/r)\partial_\theta v_\phi & (1/r\,\sin\theta)\partial_\phi v_\phi + (1/r)(v_\theta\cot\theta + v_r) \end{bmatrix}, \end{aligned} \tag{3.309}$$

where the velocity vector in physical components is

$$\begin{aligned} \boldsymbol{v} &= v_r \boldsymbol{e}_r + v_\theta \boldsymbol{e}_\theta + v_\phi \boldsymbol{e}_\phi \\ &= \dot{r}\ \boldsymbol{e}_r + r\dot{\theta}\ \boldsymbol{e}_\theta + r\dot{\phi}\sin\theta\ \boldsymbol{e}_\phi \\ &= \frac{\mathrm{d}r}{\mathrm{d}t}\ \boldsymbol{e}_r + r\frac{\mathrm{d}\theta}{\mathrm{d}t}\ \boldsymbol{e}_\theta + r\frac{\mathrm{d}\phi}{\mathrm{d}t}\sin\theta\ \boldsymbol{e}_\phi. \end{aligned} \tag{3.310}$$

Finally, the spatial acceleration $\boldsymbol{a}(\boldsymbol{x}, t)$ of (3.154) is

$$\begin{aligned} \boldsymbol{a} &= a_r \boldsymbol{e}_r + a_\theta \boldsymbol{e}_\theta + a_\phi \boldsymbol{e}_\phi \\ &= \dot{v}_r \boldsymbol{e}_r + \dot{v}_\theta \boldsymbol{e}_\theta + \dot{v}_\phi \boldsymbol{e}_\phi \\ &= \Bigg[\frac{\partial v_r}{\partial t} + v_r \partial_r v_r + \frac{1}{r} v_\theta \partial_\theta v_r \\ &\qquad + \frac{1}{r\sin\theta} v_\phi \partial_\phi v_r - \frac{1}{r}(v_\theta v_\theta + v_\phi v_\phi)\Bigg] \boldsymbol{e}_r \\ &\quad + \Bigg[\frac{\partial v_\theta}{\partial t} + v_r \partial_r v_\theta + \frac{1}{r} v_\theta \partial_\theta v_\theta \\ &\qquad + \frac{1}{r\sin\theta} v_\phi \partial_\phi v_\theta + \frac{1}{r}(v_r v_\theta - v_\phi v_\phi \cot\theta)\Bigg] \boldsymbol{e}_\theta \\ &\quad + \Bigg[\frac{\partial v_\phi}{\partial t} + v_r \partial_r v_\phi + \frac{1}{r} v_\theta \partial_\theta v_\phi \\ &\qquad + \frac{1}{r\sin\theta} v_\phi \partial_\phi v_\phi + \frac{1}{r}(v_r v_\phi + v_\theta v_\phi \cot\theta)\Bigg] \boldsymbol{e}_\phi. \end{aligned} \tag{3.311}$$

Reference spherical coordinates $\{X^A\}$ are

$$(X^1, X^2, X^3) \to (R, \Theta, \Phi); \qquad R \geq 0, \quad \Theta \in [0, \pi] \quad \Phi \in (-\pi, \pi]. \tag{3.312}$$

Let $\boldsymbol{X}$ denote the position vector measured from a fixed origin:

$$\boldsymbol{X}(R, \Theta, \Phi) = R\boldsymbol{G}_R(\Theta, \Phi). \tag{3.313}$$

Natural basis vectors $\boldsymbol{G}_A - \partial_A \boldsymbol{X}$ are

$$\begin{aligned} \boldsymbol{G}_R(\Theta, \Phi) &= \partial_R \boldsymbol{X}, \\ \boldsymbol{G}_\Theta(R, \Theta, \Phi) &= \partial_\Theta \boldsymbol{X}, \\ \boldsymbol{G}_\Phi(R, \Theta, \Phi) &= \partial_\Phi \boldsymbol{X}. \end{aligned} \tag{3.314}$$

The squared length of a differential line element $\mathrm{d}\boldsymbol{X}$ is

$$|\mathrm{d}\boldsymbol{X}|^2 = \mathrm{d}\boldsymbol{X} \cdot \mathrm{d}\boldsymbol{X} = (\mathrm{d}R)^2 + (R\,\mathrm{d}\Theta)^2 + (R\,\sin\Theta\,\mathrm{d}\Phi)^2. \tag{3.315}$$

The metric tensor with components $G_{AB} = \boldsymbol{G}_A \cdot \boldsymbol{G}_B$ and its inverse are respectively

$$[G_{AB}] = \begin{bmatrix} 1 & 0 & 0 \\ 0 & R^2 & 0 \\ 0 & 0 & R^2 \sin^2 \Theta \end{bmatrix},$$

$$[G^{AB}] = \begin{bmatrix} 1 & 0 & 0 \\ 0 & 1/R^2 & 0 \\ 0 & 0 & 1/(R^2 \sin^2 \Theta) \end{bmatrix}. \quad (3.316)$$

Contravariant basis vectors are

$$\boldsymbol{G}^A = G^{AB} \boldsymbol{G}_B;$$

$$\boldsymbol{G}^R = \boldsymbol{G}_R, \qquad \boldsymbol{G}^\Theta = \frac{\boldsymbol{G}_\Theta}{R^2}, \qquad \boldsymbol{G}^\Phi = \frac{\boldsymbol{G}_\Phi}{R^2 \sin^2 \Theta}. \quad (3.317)$$

Physical (dimensionless unit) basis vectors are

$$\boldsymbol{E}_A = \boldsymbol{G}_A / \sqrt{G_{AA}};$$

$$\boldsymbol{E}_R = \boldsymbol{G}_R, \qquad \boldsymbol{E}_\Theta = \frac{\boldsymbol{G}_\Theta}{R}, \qquad \boldsymbol{E}_\Phi = \frac{\boldsymbol{G}_\Phi}{R \sin \Theta}. \quad (3.318)$$

Since spherical coordinates are orthogonal ($\boldsymbol{E}_A = \boldsymbol{E}^A$), there is no need to distinguish between contravariant and covariant physical components. Spherical coordinates are related to Cartesian coordinates as follows:

$$X = R \sin \Theta \cos \Phi, \quad Y = R \sin \Theta \sin \Phi, \quad Z = R \cos \Theta;$$

$$\boldsymbol{X} = X \boldsymbol{E}_X + Y \boldsymbol{E}_Y + Z \boldsymbol{E}_Z; \quad (3.319)$$

$$\boldsymbol{E}_R = \sin \Theta \cos \Phi \boldsymbol{E}_X + \sin \Theta \sin \Phi \boldsymbol{E}_Y + \cos \Theta \boldsymbol{E}_Z,$$

$$\boldsymbol{E}_\Theta = \cos \Theta \cos \Phi \boldsymbol{E}_X + \cos \Theta \sin \Phi \boldsymbol{E}_Y - \sin \Theta \boldsymbol{E}_Z,$$

$$\boldsymbol{E}_\Phi = - \sin \Phi \boldsymbol{E}_X + \cos \Phi \boldsymbol{E}_Y. \quad (3.320)$$

Referential Christoffel symbols of the second kind are completely analogous to (3.295).

Using (3.266), the matrix of shifter components $g^a_A(r, \theta, \phi; R, \Theta, \Phi)$ is

$$[g^a_A] = \begin{bmatrix} g^r_R & g^r_\Theta & g^r_\Phi \\ g^\theta_R & g^\theta_\Theta & g^\theta_\Phi \\ g^\phi_R & g^\phi_\Theta & g^\phi_\Phi \end{bmatrix}; \quad (3.321)$$

$$g^r_R = \sin \theta \sin \Theta \cos(\phi - \Phi) + \cos \theta \cos \Theta, \quad (3.322)$$

$$g^r_\Theta = R[\sin \theta \cos \Theta \cos(\phi - \Phi) - \cos \theta \sin \Theta], \quad (3.323)$$

$$g^r_\Phi = R\sin\theta\sin\Theta\sin(\phi - \Phi), \tag{3.324}$$

$$g^\theta_R = (1/r)[\cos\theta\sin\Theta\cos(\phi - \Phi) - \sin\theta\cos\Theta], \tag{3.325}$$

$$g^\theta_\Theta = (R/r)[\cos\theta\cos\Theta\cos(\phi - \Phi) + \sin\theta\sin\Theta], \tag{3.326}$$

$$g^\theta_\Phi = (R/r)\cos\theta\sin\Theta\sin(\phi - \Phi), \tag{3.327}$$

$$g^\phi_R = -[\sin\Theta/(r\sin\theta)]\sin(\phi - \Phi), \tag{3.328}$$

$$g^\phi_\Theta = -[(R\cos\Theta)/(r\sin\theta)]\sin(\phi - \Phi), \tag{3.329}$$

$$g^\phi_\Phi = [(R\sin\Theta)/(r\sin\theta)]\cos(\phi - \Phi). \tag{3.330}$$

Letting $G = \det(G_{AB}) = R^4\sin^2\Theta$, determinants obey

$$\det[g^a_A] = \sqrt{G/g} = \frac{R^2\sin\Theta}{r^2\sin\theta}, \quad \det[g^A_a] = \sqrt{g/G} = \frac{r^2\sin\theta}{R^2\sin\Theta}. \tag{3.331}$$

In spherical coordinates, deformation $\boldsymbol{x} = \boldsymbol{x}(\boldsymbol{X})$ is of the form

$$r = r(R, \Theta, \Phi), \qquad \theta = \theta(R, \Theta, \Phi), \qquad \phi = \phi(R, \Theta, \Phi). \tag{3.332}$$

Referred to natural bases $\{\boldsymbol{g}_a, \boldsymbol{G}^A\}$, the deformation gradient is

$$\begin{aligned}
\boldsymbol{F} &= \partial_A\boldsymbol{x} \otimes \boldsymbol{G}^A \\
&= \partial_A x^a \boldsymbol{g}_a \otimes \boldsymbol{G}^A \\
&= F^a_{\cdot A}\boldsymbol{g}_a \otimes \boldsymbol{G}^A \\
&= \partial_R r\ \boldsymbol{g}_r \otimes \boldsymbol{G}^R + \partial_\Theta r\ \boldsymbol{g}_r \otimes \boldsymbol{G}^\Theta + \partial_\Phi r\ \boldsymbol{g}_r \otimes \boldsymbol{G}^\Phi \\
&\quad + \partial_R\theta\ \boldsymbol{g}_\theta \otimes \boldsymbol{G}^R + \partial_\Theta\theta\ \boldsymbol{g}_\theta \otimes \boldsymbol{G}^\Theta + \partial_\Phi\theta\ \boldsymbol{g}_\theta \otimes \boldsymbol{G}^\Phi \\
&\quad + \partial_R\phi\ \boldsymbol{g}_\phi \otimes \boldsymbol{G}^R + \partial_\Theta\phi\ \boldsymbol{g}_\phi \otimes \boldsymbol{G}^\Theta + \partial_\Phi\phi\ \boldsymbol{g}_\phi \otimes \boldsymbol{G}^\Phi.
\end{aligned} \tag{3.333}$$

Referential and deformed volume elements

$$\begin{aligned}
\mathrm{d}V &= \sqrt{G}\mathrm{d}X^1\mathrm{d}X^2\mathrm{d}X^3 = R^2\sin\Theta\mathrm{d}R\mathrm{d}\Theta\mathrm{d}\Phi, \\
\mathrm{d}v &= \sqrt{g}\mathrm{d}x^1\mathrm{d}x^2\mathrm{d}x^3 = r^2\sin\theta\mathrm{d}r\mathrm{d}\theta\mathrm{d}\phi
\end{aligned} \tag{3.334}$$

are related through the Jacobian determinant:

$$\begin{aligned}
J &= \frac{\mathrm{d}v}{\mathrm{d}V} \\
&= \det[F^a_{\cdot A}]\sqrt{g/G} \\
&= \det[\partial_A x^a]\det[g^A_a] \\
&= \det[\partial_A x^a]\frac{r^2\sin\theta}{R^2\sin\Theta}.
\end{aligned} \tag{3.335}$$

Using (3.299), (3.317), and (3.318) to convert natural basis vectors to spherical orthonormal bases $\{\boldsymbol{e}_a, \boldsymbol{E}_A\}$, the two-point deformation gradient in physical components is

$$\begin{aligned}
\boldsymbol{F} &= F_{\langle aA\rangle}\boldsymbol{e}_a \otimes \boldsymbol{E}_A \\
&= \partial_R r\ \boldsymbol{e}_r \otimes \boldsymbol{E}_R + (1/R)\partial_\Theta r\ \boldsymbol{e}_r \otimes \boldsymbol{E}_\Theta \\
&\quad + [1/(R\sin\Theta)]\partial_\Phi r\ \boldsymbol{e}_r \otimes \boldsymbol{E}_\Phi + r\partial_R\theta\ \boldsymbol{e}_\theta \otimes \boldsymbol{E}_R \\
&\quad + (r/R)\partial_\Theta\theta\ \boldsymbol{e}_\theta \otimes \boldsymbol{E}_\Theta + [r/(R\sin\Theta)]\partial_\Phi\theta\ \boldsymbol{e}_\theta \otimes \boldsymbol{E}_\Phi \\
&\quad + r\sin\theta\partial_R\phi\ \boldsymbol{e}_\phi \otimes \boldsymbol{E}_R + (r/R)\sin\theta\partial_\Theta\phi\ \boldsymbol{e}_\phi \otimes \boldsymbol{E}_\Theta \\
&\quad + [r\sin\theta/(R\sin\Theta)]\partial_\Phi\phi\ \boldsymbol{e}_\phi \otimes \boldsymbol{E}_\Phi,
\end{aligned} \tag{3.336}$$

where physical scalar components are written in angled brackets. The Jacobian determinant is simply

$$J = \det[F_{\langle aA\rangle}]. \tag{3.337}$$

Deformation tensor $\boldsymbol{C}$ of (3.86) is, in spherical curvilinear coordinates,

$$\begin{aligned}
\boldsymbol{C} &= C_{AB}\boldsymbol{G}^A \otimes \boldsymbol{G}^B \\
&= F^a_{.A} g_{ab} F^b_{.B}\boldsymbol{G}^A \otimes \boldsymbol{G}^B \\
&= [(\partial_R r)^2 + r^2(\partial_R\theta)^2 + r^2\sin^2\theta(\partial_R\phi)^2]\boldsymbol{G}^R \otimes \boldsymbol{G}^R \\
&\quad + [(\partial_R r)(\partial_\Theta r) + r^2(\partial_R\theta)(\partial_\Theta\theta) \\
&\qquad + r^2\sin^2\theta(\partial_R\phi)(\partial_\Theta\phi)]\boldsymbol{G}^R \otimes \boldsymbol{G}^\Theta \\
&\quad + [(\partial_R r)(\partial_\Phi r) + r^2(\partial_R\theta)(\partial_\Phi\theta) \\
&\qquad + r^2\sin^2\theta(\partial_R\phi)(\partial_\Phi\phi)]\boldsymbol{G}^R \otimes \boldsymbol{G}^\Phi \\
&\quad + [(\partial_R r)(\partial_\Theta r) + r^2(\partial_R\theta)(\partial_\Theta\theta) \\
&\qquad + r^2\sin^2\theta(\partial_R\phi)(\partial_\Theta\phi)]\boldsymbol{G}^\Theta \otimes \boldsymbol{G}^R \\
&\quad + [(\partial_\Theta r)^2 + r^2(\partial_\Theta\theta)^2 \\
&\qquad + r^2\sin^2\theta(\partial_\Theta\phi)^2]\boldsymbol{G}^\Theta \otimes \boldsymbol{G}^\Theta \\
&\quad + [(\partial_\Theta r)(\partial_\Phi r) + r^2(\partial_\Theta\theta)(\partial_\Phi\theta) \\
&\qquad + r^2\sin^2\theta(\partial_\Theta\phi)(\partial_\Phi\phi)]\boldsymbol{G}^\Theta \otimes \boldsymbol{G}^\Phi \\
&\quad + [(\partial_R r)(\partial_\Phi r) + r^2(\partial_R\theta)(\partial_\Phi\theta) \\
&\qquad + r^2\sin^2\theta(\partial_R\phi)(\partial_\Phi\phi)]\boldsymbol{G}^\Phi \otimes \boldsymbol{G}^R \\
&\quad + [(\partial_\Theta r)(\partial_\Phi r) + r^2(\partial_\Theta\theta)(\partial_\Phi\theta) \\
&\qquad + r^2\sin^2\theta(\partial_\Theta\phi)(\partial_\Phi\phi)]\boldsymbol{G}^\Phi \otimes \boldsymbol{G}^\Theta \\
&\quad + [(\partial_\Phi r)^2 + r^2(\partial_\Phi\theta)^2 + r^2\sin^2\theta(\partial_\Phi\phi)^2]\boldsymbol{G}^\Phi \otimes \boldsymbol{G}^\Phi.
\end{aligned} \tag{3.338}$$

In physical components, this tensor becomes

$$\begin{aligned}
\boldsymbol{C} &= C_{\langle AB\rangle}\boldsymbol{E}_A\otimes\boldsymbol{E}_B\\
&= F_{\langle aA\rangle}F_{\langle aB\rangle}\boldsymbol{E}_A\otimes\boldsymbol{E}_B\\
&= [(\partial_R r)^2 + r^2(\partial_R\theta)^2 + r^2\sin^2\theta(\partial_R\phi)^2]\,\boldsymbol{G}^R\otimes\boldsymbol{G}^R\\
&\quad + (1/R)[(\partial_R r)(\partial_\Theta r) + r^2(\partial_R\theta)(\partial_\Theta\theta)\\
&\qquad + r^2\sin^2\theta(\partial_R\phi)(\partial_\Theta\phi)]\,\boldsymbol{G}^R\otimes\boldsymbol{G}^\Theta\\
&\quad + [1/(R\sin\Theta)][(\partial_R r)(\partial_\Phi r) + r^2(\partial_R\theta)(\partial_\Phi\theta)\\
&\qquad + r^2\sin^2\theta(\partial_R\phi)(\partial_\Phi\phi)]\,\boldsymbol{G}^R\otimes\boldsymbol{G}^\Phi\\
&\quad + (1/R)[(\partial_R r)(\partial_\Theta r) + r^2(\partial_R\theta)(\partial_\Theta\theta)\\
&\qquad + r^2\sin^2\theta(\partial_R\phi)(\partial_\Theta\phi)]\,\boldsymbol{G}^\Theta\otimes\boldsymbol{G}^R\\
&\quad + (1/R^2)[(\partial_\Theta r)^2 + r^2(\partial_\Theta\theta)^2\\
&\qquad + r^2\sin^2\theta(\partial_\Theta\phi)^2]\,\boldsymbol{G}^\Theta\otimes\boldsymbol{G}^\Theta\\
&\quad + [1/(R^2\sin\Theta)][(\partial_\Theta r)(\partial_\Phi r) + r^2(\partial_\Theta\theta)(\partial_\Phi\theta)\\
&\qquad + r^2\sin^2\theta(\partial_\Theta\phi)(\partial_\Phi\phi)]\,\boldsymbol{G}^\Theta\otimes\boldsymbol{G}^\Phi\\
&\quad + [1/(R\sin\Theta)][(\partial_R r)(\partial_\Phi r) + r^2(\partial_R\theta)(\partial_\Phi\theta)\\
&\qquad + r^2\sin^2\theta(\partial_R\phi)(\partial_\Phi\phi)]\,\boldsymbol{G}^\Phi\otimes\boldsymbol{G}^R\\
&\quad + [1/(R^2\sin\Theta)][(\partial_\Theta r)(\partial_\Phi r) + r^2(\partial_\Theta\theta)(\partial_\Phi\theta)\\
&\qquad + r^2\sin^2\theta(\partial_\Theta\phi)(\partial_\Phi\phi)]\,\boldsymbol{G}^\Phi\otimes\boldsymbol{G}^\Theta\\
&\quad + [1/(R^2\sin^2\Theta)][(\partial_\Phi r)^2 + r^2(\partial_\Phi\theta)^2\\
&\qquad + r^2\sin^2\theta(\partial_\Phi\phi)^2]\,\boldsymbol{G}^\Phi\otimes\boldsymbol{G}^\Phi.
\end{aligned}\tag{3.339}$$

Inverse deformation $\boldsymbol{X} = \boldsymbol{X}(\boldsymbol{x})$ is of the form

$$R = R(r,\theta,\phi),\qquad \Theta = \Theta(r,\theta,\phi),\qquad \Phi = \Phi(r,\theta,\phi).\tag{3.340}$$

Referred to natural bases $\{\boldsymbol{G}_A, \boldsymbol{g}^a\}$, the inverse deformation gradient is

$$\begin{aligned}
\boldsymbol{F}^{-1} &= \partial_a\boldsymbol{X}\otimes\boldsymbol{g}^a\\
&= \partial_a X^A\boldsymbol{G}_A\otimes\boldsymbol{g}^a\\
&= F^{-1A}_{\ \bullet a}\boldsymbol{G}_A\otimes\boldsymbol{g}^a\\
&= \partial_r R\ \boldsymbol{G}_R\otimes\boldsymbol{g}^r + \partial_\theta R\ \boldsymbol{G}_R\otimes\boldsymbol{g}^\theta + \partial_\phi R\ \boldsymbol{G}_R\otimes\boldsymbol{g}^\phi\\
&\quad + \partial_r\Theta\ \boldsymbol{G}_\Theta\otimes\boldsymbol{g}^r + \partial_\theta\Theta\ \boldsymbol{G}_\Theta\otimes\boldsymbol{g}^\theta + \partial_\phi\Theta\ \boldsymbol{G}_\Theta\otimes\boldsymbol{g}^\phi\\
&\quad + \partial_r\Phi\ \boldsymbol{G}_\Phi\otimes\boldsymbol{g}^r + \partial_\theta\Phi\ \boldsymbol{G}_\Phi\otimes\boldsymbol{g}^\theta + \partial_\phi\Phi\ \boldsymbol{G}_\Phi\otimes\boldsymbol{g}^\phi.
\end{aligned}\tag{3.341}$$

The inverse of the Jacobian determinant $J^{-1} = 1/J$ is

$$\begin{aligned} J^{-1} &= \mathrm{d}V/\mathrm{d}v \\ &= \det[F^{-1}{}^{A}_{.a}]\sqrt{G/g} \\ &= \det[\partial_a X^A]\det[g^a_A] \\ &= \det[\partial_a X^A][R^2 \sin\Theta/(r^2 \sin\theta)]. \end{aligned} \tag{3.342}$$

With respect to orthonormal bases $\{\boldsymbol{e}_a, \boldsymbol{E}_A\}$, the inverse deformation gradient is

$$\begin{aligned} \boldsymbol{F}^{-1} &= F^{-1}_{\langle Aa\rangle}\boldsymbol{E}_A \otimes \boldsymbol{e}_a \\ &= \partial_r R\, \boldsymbol{E}_R \otimes \boldsymbol{e}_r + (1/r)\partial_\theta R\, \boldsymbol{E}_R \otimes \boldsymbol{e}_\theta \\ &\quad + [1/(r\sin\theta)]\partial_\phi R\, \boldsymbol{E}_R \otimes \boldsymbol{e}_\phi + R\partial_r\Theta\, \boldsymbol{E}_\Theta \otimes \boldsymbol{e}_r \\ &\quad + (R/r)\partial_\theta\Theta\, \boldsymbol{E}_\Theta \otimes \boldsymbol{e}_\theta + [R/(r\sin\theta)]\partial_\phi\Theta\, \boldsymbol{E}_\Theta \otimes \boldsymbol{e}_\phi \\ &\quad + R\sin\Theta\partial_r\Phi\, \boldsymbol{E}_\Phi \otimes \boldsymbol{e}_r + (R/r)\sin\Theta\partial_\theta\Phi\, \boldsymbol{E}_\Phi \otimes \boldsymbol{e}_\theta \\ &\quad + [R\sin\Theta/(r\sin\theta)]\partial_\phi\Phi\, \boldsymbol{E}_\Phi \otimes \boldsymbol{e}_\phi \end{aligned} \tag{3.343}$$

with inverse Jacobian determinant

$$J^{-1} = \det[F^{-1}_{\langle Aa\rangle}]. \tag{3.344}$$

A spatial displacement vector field in physical spherical coordinates can be constructed using the shifter:

$$\begin{aligned} \boldsymbol{u} &= \boldsymbol{x} - \boldsymbol{X} \\ &= r\boldsymbol{g}_r - R\boldsymbol{G}_R \\ &= r\boldsymbol{g}_r - Rg^a_R\boldsymbol{g}_a \\ &= [r - Rg^r_R]\boldsymbol{e}_r + [-Rrg^\theta_R]\boldsymbol{e}_\theta + [-Rr\sin\theta g^\phi_R]\boldsymbol{e}_\phi \\ &= u_r\boldsymbol{e}_r + u_\theta\boldsymbol{e}_\theta + u_\phi\boldsymbol{e}_\phi, \end{aligned} \tag{3.345}$$

where g^r_R, g^θ_R, and g^ϕ_R are given by (3.322), (3.325), and (3.328).

3.4.4 *Convected coordinates*

Let $\boldsymbol{x} \in B$ and $\boldsymbol{X} \in B_0$ denote position vectors in Euclidean space. First, let convected coordinates ξ^A be attached to material particles of the body such that

$$\boldsymbol{X} = \boldsymbol{X}(\xi^A), \qquad \xi^A = \delta^A_B X^B, \qquad \partial(\cdot)/\partial\xi^A = \partial(\cdot)/\partial X^A. \tag{3.346}$$

Reference basis vectors $\boldsymbol{G}_A(X)$ are equivalent to those in (2.4):

$$\boldsymbol{G}_A = \frac{\partial \boldsymbol{X}}{\partial X^A} = \frac{\partial \boldsymbol{X}}{\partial \xi^A} = G_{AB}\boldsymbol{G}^B, \qquad G_{AB} = \boldsymbol{G}_A \cdot \boldsymbol{G}_B. \tag{3.347}$$

Spatial position can be written

$$\boldsymbol{x} = \boldsymbol{x}(X^A, t) = \boldsymbol{x}(\xi^A, t). \quad (3.348)$$

Convected spatial basis vectors $\boldsymbol{g}'_A(X, t)$ are time dependent:

$$\boldsymbol{g}'_A = \frac{\partial \boldsymbol{x}}{\partial X^A} = \frac{\partial \boldsymbol{x}}{\partial \xi^A} = g'_{AB}\boldsymbol{g}'^B, \qquad g'_{AB} = \boldsymbol{g}'_A \cdot \boldsymbol{g}'_B. \quad (3.349)$$

Deformation gradient $\boldsymbol{F}(X, t)$ is

$$\begin{aligned}\boldsymbol{F} &= F^a_{.A}\boldsymbol{g}_a \otimes \boldsymbol{G}^A \\ &= \frac{\partial x^a}{\partial X^A}\boldsymbol{g}_a \otimes \boldsymbol{G}^A \\ &= \frac{\partial x^a}{\partial \xi^A}\frac{\partial \boldsymbol{x}}{\partial x^a} \otimes \boldsymbol{G}^A \\ &= \frac{\partial \boldsymbol{x}}{\partial \xi^A} \otimes \boldsymbol{G}^A \\ &= \boldsymbol{g}'_A \otimes \boldsymbol{G}^A \\ &= \delta^A_B \boldsymbol{g}'_A \otimes \boldsymbol{G}^B, \end{aligned} \quad (3.350)$$

implying that convected basis vectors and their associated metric obey

$$\boldsymbol{g}'_A = F^a_{.A}\boldsymbol{g}_a, \qquad g'_{AB} = F^a_{.A}\boldsymbol{g}_a \cdot F^b_{.B}\boldsymbol{g}_b = F^a_{.A} g_{ab} F^b_{.B} = C_{AB}, \quad (3.351)$$

where referential deformation tensor C_{AB} is defined in (3.86). Reciprocal convected basis vectors satisfy the relationships

$$\boldsymbol{g}'^A = g'^{AB}\boldsymbol{g}'_B = C^{-1\,AB} F^a_{.B}\boldsymbol{g}_a = F^{-1A}_{\quad .a}\boldsymbol{g}^a, \quad (3.352)$$

such that

$$\langle \boldsymbol{g}'^A, \boldsymbol{g}'_B \rangle = F^{-1A}_{\quad .a} F^b_{.B} \langle \boldsymbol{g}^a, \boldsymbol{g}_b \rangle = F^{-1A}_{\quad .a} F^a_{.B} = \delta^A_B. \quad (3.353)$$

To second order in $\mathrm{d}\boldsymbol{X}$,

$$|\mathrm{d}\boldsymbol{x}|^2 = \mathrm{d}\boldsymbol{x} \cdot \mathrm{d}\boldsymbol{x} = g'_{AB}\mathrm{d}X^A \mathrm{d}X^B = \langle \mathrm{d}\boldsymbol{X}, \boldsymbol{g}'\mathrm{d}\boldsymbol{X} \rangle, \quad (3.354)$$

where $\boldsymbol{g}' = g'_{AD}\boldsymbol{G}^A \otimes \boldsymbol{G}^B$.

Now consider an alternative representation where convected coordinates ζ^a are attached to spatial points of the body:

$$\boldsymbol{x} = \boldsymbol{x}(\zeta^a), \qquad \zeta^a = \delta^a_b x^b, \qquad \partial(\cdot)/\partial \zeta^a = \partial(\cdot)/\partial x^a. \quad (3.355)$$

Spatial basis vectors $\boldsymbol{g}_a(x)$ are equivalent to those in (2.4):

$$\boldsymbol{g}_a = \frac{\partial \boldsymbol{x}}{\partial x^a} = \frac{\partial \boldsymbol{x}}{\partial \zeta^a} = g_{ab}\boldsymbol{g}^b, \qquad g_{ab} = \boldsymbol{g}_a \cdot \boldsymbol{g}_b. \quad (3.356)$$

Reference position can be written

$$\boldsymbol{X} = \boldsymbol{X}(x^a, t) = \boldsymbol{X}(\zeta^a, t). \tag{3.357}$$

Convected reference basis vectors $\boldsymbol{G}'_a(x,t)$ are time dependent:

$$\boldsymbol{G}'_a = \frac{\partial \boldsymbol{X}}{\partial x^a} = \frac{\partial \boldsymbol{X}}{\partial \zeta^a} = G'_{ab}\boldsymbol{G}'^b, \qquad G'_{ab} = \boldsymbol{G}'_a \cdot \boldsymbol{G}'_b. \tag{3.358}$$

Inverse deformation gradient $\boldsymbol{F}^{-1}(X,t)$ is

$$\begin{aligned}\boldsymbol{F}^{-1} &= F^{-1A}_{\ \ \ \boldsymbol{.}a}\boldsymbol{G}_A \otimes \boldsymbol{g}^a \\ &= \frac{\partial X^A}{\partial x^a}\boldsymbol{G}_A \otimes \boldsymbol{g}^a \\ &= \frac{\partial X^A}{\partial \zeta^a}\frac{\partial \boldsymbol{X}}{\partial X^A} \otimes \boldsymbol{g}^a \\ &= \frac{\partial \boldsymbol{X}}{\partial \zeta^a} \otimes \boldsymbol{g}^a \\ &= \boldsymbol{G}'_a \otimes \boldsymbol{g}^a \\ &= \delta^a_b \boldsymbol{G}'_a \otimes \boldsymbol{g}^b,\end{aligned} \tag{3.359}$$

implying that convected basis vectors and their associated metric obey

$$\boldsymbol{G}'_a = F^{-1A}_{\ \ \ \boldsymbol{.}a}\boldsymbol{G}_A, \tag{3.360}$$

$$G'_{ab} = F^{-1A}_{\ \ \ \boldsymbol{.}a}\boldsymbol{G}_A \cdot F^{-1B}_{\ \ \ \boldsymbol{.}b}\boldsymbol{G}_B = F^{-1A}_{\ \ \ \boldsymbol{.}a}G_{AB}F^{-1B}_{\ \ \ \boldsymbol{.}b} = B^{-1}_{ab}, \tag{3.361}$$

where spatial deformation tensor B^{ab} is defined in (3.87). Reciprocal convected basis vectors satisfy the relationships

$$\boldsymbol{G}'^a = G'^{ab}\boldsymbol{G}'_b = B^{ab}F^{-1A}_{\ \ \ \boldsymbol{.}b}\boldsymbol{G}_A = F^a_{\boldsymbol{.}A}\boldsymbol{G}^A, \tag{3.362}$$

such that

$$\langle \boldsymbol{G}'^a, \boldsymbol{G}'_b \rangle = F^a_{\boldsymbol{.}A}F^{-1B}_{\ \ \ \boldsymbol{.}b}\langle \boldsymbol{G}^A, \boldsymbol{G}_B \rangle = F^a_{\boldsymbol{.}A}F^{-1A}_{\ \ \ \boldsymbol{.}b} = \delta^a_b. \tag{3.363}$$

To second order in $\mathrm{d}\boldsymbol{x}$,

$$|\mathrm{d}\boldsymbol{X}|^2 = \mathrm{d}\boldsymbol{X} \cdot \mathrm{d}\boldsymbol{X} = G'_{ab}\mathrm{d}x^a\mathrm{d}x^b = \langle \mathrm{d}\boldsymbol{x}, \boldsymbol{G}'\mathrm{d}\boldsymbol{x} \rangle, \tag{3.364}$$

where the convected metric tensor is $\boldsymbol{G}' = G'_{ab}\boldsymbol{g}^a \otimes \boldsymbol{g}^b$.

Chapter 4

Geometry of Anholonomic Deformation

In this chapter, a mathematical framework describing geometry associated with non-integrable deformation fields or deformation mappings is developed. In other words, the description is directed towards anholonomic differential geometry.

4.1 Anholonomic Spaces and Geometric Interpretation

The deformation gradient is split multiplicatively into two terms, neither of which is necessarily integrable. Geometric consequences of such a construction are considered, including rules for differentiation, coordinate systems, metric tensors, and connection coefficients convected from reference or current configurations to anholonomic space.

4.1.1 *Two-term decomposition of deformation gradient*

Consider a multiplicative split of the (total) deformation gradient $\boldsymbol{F}$ into two terms:

$$\boldsymbol{F} = \bar{\boldsymbol{F}}\tilde{\boldsymbol{F}}, \tag{4.1}$$

or in indicial notation, letting Greek index $\alpha = 1, \ldots n$, where n is the dimension of Euclidean space,

$$\frac{\partial \varphi^a}{\partial X^A} = \partial_A x^a = F^a_{.A} = \bar{F}^a_{.\alpha}\tilde{F}^\alpha_{.A}. \tag{4.2}$$

In coordinates, terms on the right of (4.1) can be written

$$\bar{\boldsymbol{F}}(X,t) = \bar{F}^a_{.\alpha}\boldsymbol{g}_a \otimes \tilde{\boldsymbol{g}}^\alpha, \qquad \tilde{\boldsymbol{F}}(X,t) = \tilde{F}^\alpha_{.A}\tilde{\boldsymbol{g}}_\alpha \otimes \boldsymbol{G}^A. \tag{4.3}$$

Basis vectors $\tilde{\boldsymbol{g}}_\alpha$ and their reciprocals $\tilde{\boldsymbol{g}}^\alpha$ will be described in detail later. Both $\bar{\boldsymbol{F}}$ and $\tilde{\boldsymbol{F}}$ are second-order, two-point tensor fields with positive determinants (assuming here that $n = 3$):

$$\det \bar{\boldsymbol{F}} = \frac{1}{6} e_{abc} e^{\alpha\beta\chi} \bar{F}^a_{\cdot\alpha} \bar{F}^b_{\cdot\beta} \bar{F}^c_{\cdot\chi} > 0,$$

$$\det \tilde{\boldsymbol{F}} = \frac{1}{6} e_{\alpha\beta\chi} e^{ABC} \tilde{F}^\alpha_{\cdot A} \tilde{F}^\beta_{\cdot B} \tilde{F}^\chi_{\cdot C} > 0. \tag{4.4}$$

Third-order permutation symbols obey, analogously to (2.8) and (2.9),

$$\begin{aligned} e^{\alpha\beta\chi} &= e_{\alpha\beta\chi} = 0 \quad \text{when any two indices are equal,} \\ e^{\alpha\beta\chi} &= e_{\alpha\beta\chi} = +1 \quad \text{for } \alpha\beta\chi = 123, 231, 312, \\ e^{\alpha\beta\chi} &= e_{\alpha\beta\chi} = -1 \quad \text{for } \alpha\beta\chi = 132, 213, 321. \end{aligned} \tag{4.5}$$

While it would be possible to prescribe, as an alternative to (4.4), that both $\det \bar{\boldsymbol{F}} < 0$ and $\det \tilde{\boldsymbol{F}} < 0$, while still maintaining condition (3.6) requiring $\det \boldsymbol{F} > 0$, such a possibility is not considered in the present work.

Noting that $(\boldsymbol{AB})^{-1} = \boldsymbol{B}^{-1}\boldsymbol{A}^{-1}$ for two generic non-singular matrices $\boldsymbol{A}$ and $\boldsymbol{B}$, inversion of (4.1)-(4.3) leads to

$$\boldsymbol{F}^{-1} = \tilde{\boldsymbol{F}}^{-1} \bar{\boldsymbol{F}}^{-1}; \tag{4.6}$$

$$\frac{\partial \Phi^A}{\partial x^a} = \partial_a X^A = F^{-1A}_{\cdot a} = \tilde{F}^{-1A}_{\cdot\alpha} \bar{F}^{-1\alpha}_{\cdot a}; \tag{4.7}$$

$$\tilde{\boldsymbol{F}}^{-1}(x,t) = \tilde{F}^{-1A}_{\cdot\alpha} \boldsymbol{G}_A \otimes \tilde{\boldsymbol{g}}^\alpha, \qquad \bar{\boldsymbol{F}}^{-1}(x,t) = \bar{F}^{-1\alpha}_{\cdot a} \tilde{\boldsymbol{g}}_\alpha \otimes \boldsymbol{g}^a. \tag{4.8}$$

Furthermore, from the definition of the inverse,

$$\tilde{F}^\alpha_{\cdot B} \tilde{F}^{-1A}_{\cdot\alpha} = \delta^A_B, \qquad \bar{F}^a_{\cdot\alpha} \bar{F}^{-1\alpha}_{\cdot b} = \delta^a_b; \tag{4.9}$$

$$\tilde{F}^\alpha_{\cdot A} \tilde{F}^{-1A}_{\cdot\beta} = \bar{F}^a_{\cdot\beta} \bar{F}^{-1\alpha}_{\cdot a} = \delta^\alpha_\beta; \tag{4.10}$$

with Kronecker delta symbols, similarly to (2.6),

$$\delta^\alpha_\beta = 1 \quad \forall \alpha = \beta, \qquad \delta^\alpha_\beta = 0 \quad \forall \alpha \neq \beta. \tag{4.11}$$

In geometrically nonlinear crystal mechanics [7], $\bar{\boldsymbol{F}}$ is often associated with lattice (*i.e.*, thermoelastic) deformation, and $\tilde{\boldsymbol{F}}$ is often associated with plastic (*i.e.*, dislocation slip-enabled) deformation, but the physical meaning of each of these tensors is unimportant in the present context. The target space of $\tilde{\boldsymbol{F}}$ and $\bar{\boldsymbol{F}}^{-1}$ is intermediate configuration $\tilde{B}$. Figure 4.1 illustrates deformation mappings among reference, intermediate, and current configurations entering (4.1).

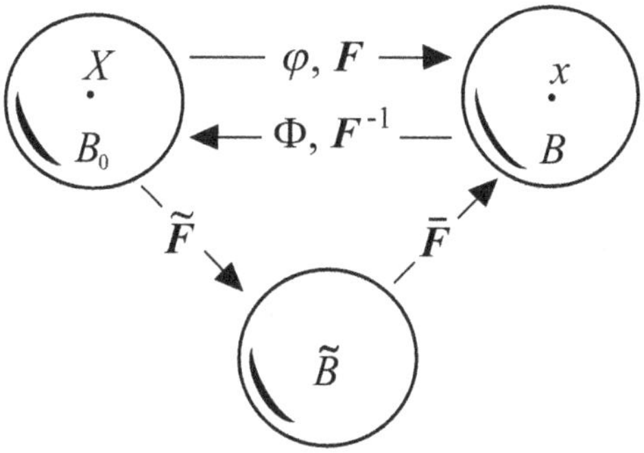

Fig. 4.1 Mappings among reference, intermediate, and current configurations of a deformable body.

4.1.2 *Anholonomicity conditions and partial differentiation*

It is assumed that $\tilde{\boldsymbol{F}}(X,t)$ is at least twice differentiable with respect to referential coordinates X^A. Single-valued coordinates $\tilde{x}^\alpha(X,t)$ referred to intermediate configuration $\tilde{B}$ that are continuous and at least three times differentiable with respect to X^A exist if and only if the following integrability conditions apply [28]:

$$\partial_A \tilde{F}^{\alpha}_{.B} = \partial_B \tilde{F}^{\alpha}_{.A} \Leftrightarrow \frac{\partial^2 \tilde{x}^\alpha}{\partial X^A \partial X^B} = \frac{\partial^2 \tilde{x}^\alpha}{\partial X^B \partial X^A}. \tag{4.12}$$

When (4.12) applies for all $X \in B_0$ throughout a simply connected body, intermediate configuration $\tilde{B}$ is said to be holonomic, and then like reference configuration B_0, can be treated as a Euclidean n-dimensional space:

$$\partial_{[A} \tilde{F}^{\alpha}_{.B]} = 0 \Leftrightarrow \tilde{F}^{\alpha}_{.A} = \partial_A \tilde{x}^\alpha. \tag{4.13}$$

Otherwise, single-valued coordinates $\tilde{x}^\alpha(X,t)$ continuously differentiable with respect to X^A do not exist in intermediate configuration $\tilde{B}$, which is then termed an anholonomic space:

$$\partial_{[A} \tilde{F}^{\alpha}_{.B]} \neq 0 \leftrightarrow \tilde{x}^\alpha(X,t) \text{ anholonomic}. \tag{4.14}$$

The physical domain of $\tilde{\boldsymbol{F}}$ must be simply connected to ensure that (4.12) are sufficient conditions for the existence of a uniform covering $\tilde{x}^\alpha$ of that domain, in which case both $\tilde{x}^\alpha$ and X^A are single-valued for all $X \in B_0$. That conditions (4.13) are necessary for the existence of $\tilde{x}^\alpha(X,t)$ follows directly from (2.17). Sufficiency of the left equality in (4.13) for the existence of $\tilde{x}^\alpha(X,t)$ is discussed further in the context of Stokes's theorem in Section 5.2.3. When (4.13) [(4.14)] applies only over local simply connected regions of B_0, deformation map $\tilde{\boldsymbol{F}}$ can be designated as holonomic [anholonomic] only over those regions.

Regardless of which of (4.13) or (4.14) applies, partial differentiation with respect to intermediate coordinates is defined as follows:

$$\partial_\alpha(\cdot) = \partial_A(\cdot)\tilde{F}^{-1A}_{\cdot\alpha}. \tag{4.15}$$

For many arguments including (4.13), once differentiability of mapping $\tilde{\boldsymbol{F}}$ is sufficient; however, subsequent identities involving curvature, for example, require that $\tilde{\boldsymbol{F}}$ be at least twice differentiable with respect to X^A.

Arguments in (4.12)-(4.15) can be repeated for $\bar{\boldsymbol{F}}^{-1}(x,t)$, which is assumed at least twice differentiable with respect to spatial coordinates x^a. Single-valued coordinates $\tilde{x}^\alpha(x,t)$ referred to intermediate configuration $\tilde{B}$ that are continuous and at least three times differentiable with respect to x^a exist if and only if

$$\partial_a \bar{F}^{-1\alpha}_{\cdot b} = \partial_b \bar{F}^{-1\alpha}_{\cdot a} \Leftrightarrow \frac{\partial^2 \tilde{x}^\alpha}{\partial x^a \partial x^b} = \frac{\partial^2 \tilde{x}^\alpha}{\partial x^b \partial x^a}. \tag{4.16}$$

When (4.16) applies globally over B, intermediate configuration $\tilde{B}$ is holonomic:

$$\partial_{[a}\bar{F}^{-1\alpha}_{\cdot b]} = 0 \Leftrightarrow \bar{F}^{-1\alpha}_{\cdot a} = \partial_a \tilde{x}^\alpha. \tag{4.17}$$

Otherwise,

$$\partial_{[a}\bar{F}^{-1\alpha}_{\cdot b]} \neq 0 \Leftrightarrow \tilde{x}^\alpha(x,t) \text{ anholonomic}. \tag{4.18}$$

Partial differentiation with respect to intermediate coordinates obeys

$$\partial_\alpha(\cdot) = \partial_a(\cdot)\bar{F}^a_{\cdot\alpha}. \tag{4.19}$$

Verification that (4.15) and (4.19) are equivalent is straightforward using (3.8) or (3.9) and (4.2):

$$\begin{aligned}\partial_\alpha(\cdot) &= \partial_A(\cdot)\tilde{F}^{-1A}_{\cdot\alpha}\\ &= \partial_a(\cdot)\,\partial_A x^a\,\tilde{F}^{-1A}_{\cdot\alpha}\\ &= \partial_a(\cdot)F^a_{\cdot A}\tilde{F}^{-1A}_{\cdot\alpha}\\ &= \partial_a(\cdot)\bar{F}^a_{\cdot\alpha}.\end{aligned} \tag{4.20}$$

Second partial anholonomic derivatives obey the relations

$$\begin{aligned}\partial_\alpha[\partial_\beta(\cdot)] &= \partial_A[\partial_B(\cdot)\tilde{F}^{-1B}_{\cdot\beta}]\tilde{F}^{-1A}_{\cdot\alpha}\\ &= \partial_A[\partial_B(\cdot)]\tilde{F}^{-1B}_{\cdot\beta}\tilde{F}^{-1A}_{\cdot\alpha} + \partial_B(\cdot)\partial_A(\tilde{F}^{-1B}_{\cdot\beta})\tilde{F}^{-1A}_{\cdot\alpha}\\ &= \partial_B[\partial_A(\cdot)]\tilde{F}^{-1B}_{\cdot\beta}\tilde{F}^{-1A}_{\cdot\alpha} + \partial_A(\cdot)\partial_B(\tilde{F}^{-1A}_{\cdot\beta})\tilde{F}^{-1B}_{\cdot\alpha}\end{aligned}$$

$$
\begin{aligned}
&= \partial_B[\partial_A(\cdot)\tilde{F}^{-1A}_{\ \ \ \cdot\alpha}]\tilde{F}^{-1B}_{\ \ \ \cdot\beta} \\
&\quad - \partial_A(\cdot)\partial_B(\tilde{F}^{-1A}_{\ \ \ \cdot\alpha})\tilde{F}^{-1B}_{\ \ \ \cdot\beta} + \partial_A(\cdot)\partial_B(\tilde{F}^{-1A}_{\ \ \ \cdot\beta})\tilde{F}^{-1B}_{\ \ \ \cdot\alpha} \\
&= \partial_\beta[\partial_\alpha(\cdot)] - \partial_A(\cdot)\partial_\beta\tilde{F}^{-1A}_{\ \ \ \cdot\alpha} + \partial_A(\cdot)\partial_\alpha\tilde{F}^{-1A}_{\ \ \ \cdot\beta} \\
&= \partial_\beta[\partial_\alpha(\cdot)] + 2\partial_A(\cdot)\partial_{[\alpha}\tilde{F}^{-1A}_{\ \ \ \cdot\beta]}.
\end{aligned} \tag{4.21}
$$

Following similar arguments,

$$
\partial_\alpha[\partial_\beta(\cdot)] = \partial_\beta[\partial_\alpha(\cdot)] + 2\partial_a(\cdot)\partial_{[\alpha}\bar{F}^a_{\cdot\beta]}. \tag{4.22}
$$

In general, second partial anholonomic differentiation is not symmetric. Explicitly,

$$
\begin{aligned}
\partial_{[\alpha}[\partial_{\beta]}(\cdot)] &= \frac{1}{2}\{\partial_\alpha[\partial_\beta(\cdot)] - \partial_\beta[\partial_\alpha(\cdot)]\} \\
&= \partial_A(\cdot)\partial_{[\alpha}\tilde{F}^{-1A}_{\ \ \ \cdot\beta]} \\
&= \partial_a(\cdot)\partial_{[\alpha}\bar{F}^a_{\cdot\beta]}.
\end{aligned} \tag{4.23}
$$

However, when (4.13) and (4.17) apply such that $\tilde{B}$ is holonomic, second partial anholonomic differentiation is symmetric:

$$
\begin{aligned}
\tilde{F}^{-1A}_{\ \ \ \cdot\beta} &= \partial_\beta X^A \\
\Rightarrow \partial_{[\alpha}\tilde{F}^{-1A}_{\ \ \ \cdot\beta]} &= \partial_{[\alpha}\partial_{\beta]}X^A = 0 \\
\Rightarrow \partial_\alpha[\partial_\beta(\cdot)] &= \partial_\beta[\partial_\alpha(\cdot)],
\end{aligned} \tag{4.24}
$$

$$
\begin{aligned}
\bar{F}^a_{\cdot\beta} &= \partial_\beta x^a \\
\Rightarrow \partial_{[\alpha}\bar{F}^a_{\cdot\beta]} &= \partial_{[\alpha}\partial_{\beta]}x^a = 0 \\
\Rightarrow \partial_\alpha[\partial_\beta(\cdot)] &= \partial_\beta[\partial_\alpha(\cdot)].
\end{aligned} \tag{4.25}
$$

Even when (4.13) and (4.17) apply globally on B_0 and B respectively, it follows from the inverse function theorem that coordinate functions $X^A(\tilde{x}, t)$ and $x^a(\tilde{x}, t)$ may only be available locally [25].

4.1.3 *Anholonomic basis vectors and metric tensors*

By definition, basis vectors and their reciprocals in intermediate configuration $\tilde{B}$ obey

$$
\langle \tilde{\boldsymbol{g}}^\alpha, \tilde{\boldsymbol{g}}_\beta \rangle = \delta^\alpha_\beta. \tag{4.26}
$$

A symmetric metric tensor $\tilde{\boldsymbol{g}}$ on $\tilde{B}$ is defined in components in terms of a scalar or dot product as

$$
\tilde{g}_{\alpha\beta} = \tilde{\boldsymbol{g}}_\alpha \cdot \tilde{\boldsymbol{g}}_\beta = \tilde{\boldsymbol{g}}_\beta \cdot \tilde{\boldsymbol{g}}_\alpha = \tilde{g}_{\beta\alpha} = \tilde{g}_{(\alpha\beta)}. \tag{4.27}
$$

The dot product of two generic contravariant vectors $\boldsymbol{V} = V^\alpha \tilde{\boldsymbol{g}}_\alpha$ and $\boldsymbol{W} = W^\alpha \tilde{\boldsymbol{g}}_\alpha$ is computed as

$$\begin{aligned}\boldsymbol{V} \cdot \boldsymbol{W} &= V^\alpha \tilde{\boldsymbol{g}}_\alpha \cdot W^\beta \tilde{\boldsymbol{g}}_\beta \\ &= V^\alpha W^\beta (\tilde{\boldsymbol{g}}_\alpha \cdot \tilde{\boldsymbol{g}}_\beta) \\ &= V^\alpha \tilde{g}_{\alpha\beta} W^\beta \\ &= V^\alpha W_\alpha \\ &= V_\alpha W^\alpha. \end{aligned} \tag{4.28}$$

As indicated, the metric tensor can be used to lower contravariant indices in the usual manner:

$$V_\alpha = V^\beta \tilde{g}_{\alpha\beta}, \qquad \tilde{\boldsymbol{g}}_\alpha = \tilde{g}_{\alpha\beta} \tilde{\boldsymbol{g}}^\beta. \tag{4.29}$$

Metric $\tilde{\boldsymbol{g}}$ is assumed to be positive definite, with positive determinant $\tilde{g}$ over any volume (*i.e.*, excluding possible points, lines, or surfaces where $\tilde{g}$ may be zero or undefined):

$$\tilde{g} = \det \tilde{\boldsymbol{g}} = \det(\tilde{g}_{\alpha\beta}) = \frac{1}{6} e^{\alpha\beta\chi} e^{\delta\epsilon\phi} \tilde{g}_{\alpha\delta}\, \tilde{g}_{\beta\epsilon}\, \tilde{g}_{\chi\phi} > 0. \tag{4.30}$$

The inverse $\tilde{\boldsymbol{g}}^{-1}$ with components $\tilde{g}^{\alpha\beta}$ on $\tilde{B}$ obeys, by definition,

$$\tilde{g}^{\alpha\beta} = \tilde{\boldsymbol{g}}^\alpha \cdot \tilde{\boldsymbol{g}}^\beta = \tilde{\boldsymbol{g}}^\beta \cdot \tilde{\boldsymbol{g}}^\alpha = \tilde{g}^{\beta\alpha} = \tilde{g}^{(\alpha\beta)}, \tag{4.31}$$

$$\tilde{g}^{\alpha\chi} \tilde{g}_{\chi\beta} = \delta^\alpha_\beta. \tag{4.32}$$

Inverse metric (4.31) enables the dot product of generic covariant vectors $\boldsymbol{\alpha} = \alpha_\alpha \tilde{\boldsymbol{g}}^\alpha$ and $\boldsymbol{\beta} = \beta_\alpha \tilde{\boldsymbol{g}}^\alpha$ on $\tilde{B}$:

$$\boldsymbol{\alpha} \cdot \boldsymbol{\beta} = \alpha_\alpha \tilde{\boldsymbol{g}}^\alpha \cdot \beta_\beta \tilde{\boldsymbol{g}}^\beta = \alpha_\alpha \beta_\beta (\tilde{\boldsymbol{g}}^\alpha \cdot \tilde{\boldsymbol{g}}^\beta) = \alpha_\alpha \tilde{g}^{\alpha\beta} \beta_\beta = \alpha^\alpha \beta_\alpha = \alpha_\alpha \beta^\alpha. \tag{4.33}$$

Components of $\tilde{\boldsymbol{g}}^{-1}$ can be used to raise indices as indicated above:

$$\alpha^\alpha = \alpha_\beta \tilde{g}^{\alpha\beta}, \qquad \tilde{\boldsymbol{g}}^\alpha = \tilde{g}^{\alpha\beta} \tilde{\boldsymbol{g}}_\beta. \tag{4.34}$$

Analogously to (4.30),

$$\tilde{g}^{-1} = 1/\tilde{g} = \det \tilde{\boldsymbol{g}}^{-1} = \det(\tilde{g}^{\alpha\beta}) = \frac{1}{6} e_{\alpha\beta\chi} e_{\delta\epsilon\phi} \tilde{g}^{\alpha\delta}\, \tilde{g}^{\beta\epsilon}\, \tilde{g}^{\chi\phi} > 0. \tag{4.35}$$

Permutation tensors in configuration $\tilde{B}$ are defined according to

$$\epsilon^{\alpha\beta\chi} = \frac{1}{\sqrt{\tilde{g}}}\, e^{\alpha\beta\chi}, \quad \epsilon_{\alpha\beta\chi} = \sqrt{\tilde{g}}\, e_{\alpha\beta\chi}, \tag{4.36}$$

where permutation symbols are defined in (4.5). Shifter tensors can also be introduced among basis vectors in intermediate and reference or spatial configurations:

$$\begin{aligned} g^\alpha_A &= \langle \tilde{\boldsymbol{g}}^\alpha, \boldsymbol{G}_A \rangle, \qquad & g^A_\alpha &= \langle \tilde{\boldsymbol{g}}_\alpha, \boldsymbol{G}^A \rangle; \\ g^\alpha_a &= \langle \tilde{\boldsymbol{g}}^\alpha, \boldsymbol{g}_a \rangle, \qquad & g^a_\alpha &= \langle \tilde{\boldsymbol{g}}_\alpha, \boldsymbol{g}^a \rangle; \end{aligned} \tag{4.37}$$

$$g^{\alpha A} = \tilde{\boldsymbol{g}}^{\alpha} \cdot \boldsymbol{G}^{A} = g^{A\alpha}, \quad g_{\alpha A} = \tilde{\boldsymbol{g}}_{\alpha} \cdot \boldsymbol{G}_{A} = g_{A\alpha}; \tag{4.38}$$

$$g^{\alpha a} = \tilde{\boldsymbol{g}}^{\alpha} \cdot \boldsymbol{g}^{a} = g^{a\alpha}, \quad g_{\alpha a} = \tilde{\boldsymbol{g}}_{\alpha} \cdot \boldsymbol{g}_{a} = g_{a\alpha}; \tag{4.39}$$

$$g^{A}_{\alpha} g^{\alpha}_{B} = \delta^{A}_{B}, \quad g^{a}_{\alpha} g^{\alpha}_{b} = \delta^{a}_{b}, \quad g^{\alpha}_{A} g^{A}_{\beta} = g^{\alpha}_{a} g^{a}_{\beta} = \delta^{\alpha}_{\beta}; \tag{4.40}$$

$$\det(g^{A}_{\alpha}) = 1/\det(g^{\alpha}_{A}) = \sqrt{\det(\tilde{g}_{\alpha\beta})/\det(G_{AB})} = \sqrt{\tilde{g}/G}; \tag{4.41}$$

$$\det(g^{\alpha}_{a}) = 1/\det(g^{a}_{\alpha}) = \sqrt{\det(g_{ab})/\det(\tilde{g}_{\alpha\beta})} = \sqrt{g/\tilde{g}}. \tag{4.42}$$

The following rules apply for shifting of basis vectors:

$$\tilde{\boldsymbol{g}}_{\alpha} = g^{A}_{\alpha} \boldsymbol{G}_{A} = g^{a}_{\alpha} \boldsymbol{g}_{a}, \quad \tilde{\boldsymbol{g}}^{\alpha} = g^{\alpha}_{A} \boldsymbol{G}^{A} = g^{\alpha}_{a} \boldsymbol{g}^{a}; \tag{4.43}$$

$$\boldsymbol{G}_{A} = g^{\alpha}_{A} \tilde{\boldsymbol{g}}_{\alpha}, \quad \boldsymbol{G}^{A} = g^{A}_{\alpha} \tilde{\boldsymbol{g}}^{\alpha}; \quad \boldsymbol{g}_{a} = g^{\alpha}_{a} \tilde{\boldsymbol{g}}_{\alpha}, \quad \boldsymbol{g}^{a} = g^{a}_{\alpha} \tilde{\boldsymbol{g}}^{\alpha}. \tag{4.44}$$

Furthermore, since

$$\boldsymbol{g}_{a} = g^{\alpha}_{a} \tilde{\boldsymbol{g}}_{\alpha} = g^{\alpha}_{a} g^{A}_{\alpha} \boldsymbol{G}_{A} = g^{A}_{a} \boldsymbol{G}_{A} \Leftrightarrow g^{A}_{a} = g^{\alpha}_{a} g^{A}_{\alpha}, \tag{4.45}$$

it follows from the product rule of determinants that

$$\det(g^{A}_{a}) = [\det(g^{\alpha}_{a})][\det(g^{A}_{\alpha})] = [\sqrt{g/\tilde{g}}][\sqrt{\tilde{g}/G}] = \sqrt{g/G}, \tag{4.46}$$

in agreement with (2.54).

First consider the case when (4.13) applies, such that $\tilde{B}$ can be regarded as a Euclidean space and a position vector $\tilde{\boldsymbol{x}}$ can be assigned to any point $\tilde{x}(X,t) \in \tilde{B}$. In that case, natural basis vectors can be defined in the usual manner, similarly to (2.4):

$$\tilde{\boldsymbol{g}}_{\alpha}(\tilde{x}) = \partial_{\alpha} \tilde{\boldsymbol{x}} = \frac{\partial \tilde{\boldsymbol{x}}}{\partial \tilde{x}^{\alpha}}. \tag{4.47}$$

Metric tensor components corresponding to (4.47) are

$$\tilde{g}_{\alpha\beta}(\tilde{x}) = \partial_{\alpha} \tilde{\boldsymbol{x}} \cdot \partial_{\beta} \tilde{\boldsymbol{x}}. \tag{4.48}$$

Any time-independent Euclidean coordinate system (*e.g.*, curvilinear or Cartesian coordinates) can be used for $\tilde{x}^{\alpha}$ in this case.

Next consider the case when (4.14) applies, such that single-valued coordinates $\tilde{x}^{\alpha}(X,t)$ continuously differentiable with respect to X^{A} do not exist in $\tilde{B}$. In that case, a one-to-one correspondence between a material particle X and an intermediate point $\tilde{x}$ is not applicable. Intermediate basis vectors associated with a given material particle should then be assigned to X rather than $\tilde{x}$, *i.e.*, $\tilde{\boldsymbol{g}}_{\alpha} = \tilde{\boldsymbol{g}}_{\alpha}(X) = \tilde{\boldsymbol{g}}_{\alpha}[X(x,t)]$. Any time-independent coordinate system can be used for $\tilde{\boldsymbol{g}}_{\alpha}(X)$ in this case. [A different notation

$\tilde{\boldsymbol{g}}'_\alpha(X,t)$ is introduced later for convected intermediate basis vectors that may vary explicitly with time t at fixed X.] However, it is often practical to select identical coordinate systems in configurations B_0 and $\tilde{B}$ [7]:

$$\tilde{\boldsymbol{g}}_\alpha(X) = \delta^A_\alpha \boldsymbol{G}_A = \delta^A_\alpha \partial_A \boldsymbol{X} = \delta^A_\alpha \frac{\partial \boldsymbol{X}}{\partial X^A} = \delta^A_\alpha \tilde{F}^\beta_{.A} \partial_\beta \boldsymbol{X}. \tag{4.49}$$

Metric tensor components corresponding to (4.49) are

$$\tilde{g}_{\alpha\beta}(X) = \tilde{\boldsymbol{g}}_\alpha \cdot \tilde{\boldsymbol{g}}_\beta = \delta^A_\alpha \boldsymbol{G}_A \cdot \delta^B_\beta \boldsymbol{G}_B = \delta^A_\alpha \delta^B_\beta G_{AB}, \tag{4.50}$$

where $G_{AB}(X)$ is the referential metric tensor of (2.22). Mixed Kronecker delta symbols are equivalent to shifter tensors for coincident coordinate systems and have the usual meaning, *i.e.*,

$$\delta^A_\alpha = 1 \quad \forall A = \alpha, \qquad \delta^A_\alpha = 0 \quad \forall A \neq \alpha. \tag{4.51}$$

The determinant of the metric is simply

$$\tilde{g}(X) = \det(\tilde{g}_{\alpha\beta}) = \det(\delta^A_\alpha)\det(\delta^B_\beta)\det(G_{AB}) = \det(G_{AB}) = G. \tag{4.52}$$

Explicitly, shifter tensor components are

$$g^\alpha_A = \langle \tilde{\boldsymbol{g}}^\alpha, \boldsymbol{G}_A \rangle = \langle \delta^\alpha_B \boldsymbol{G}^B, \boldsymbol{G}_A \rangle = \delta^\alpha_B \delta^B_A = \delta^\alpha_A, \tag{4.53}$$

with the determinant in (4.41) reducing to

$$\det(g^A_\alpha) = 1/\det(g^\alpha_A) = \sqrt{\tilde{g}/G} = 1. \tag{4.54}$$

Now consider the case when (4.18) applies, such that single-valued coordinates $\tilde{x}^\alpha(x,t)$ continuously differentiable with respect to x^a do not exist in $\tilde{B}$. In that case, a one-to-one correspondence between a spatial point x and an intermediate point $\tilde{x}$ is not applicable. Intermediate basis vectors associated with a given spatial point should then be assigned to x rather than $\tilde{x}$, *i.e.*, $\tilde{\boldsymbol{g}}_\alpha = \tilde{\boldsymbol{g}}_\alpha(x) = \tilde{\boldsymbol{g}}_\alpha[x(X,t)]$. Any time-independent coordinate system can be used for $\tilde{\boldsymbol{g}}_\alpha(x)$ in this case. [A different notation $\bar{\boldsymbol{g}}'_\alpha(x,t)$ is introduced later for convected intermediate basis vectors that may vary explicitly with time t at fixed x.] However, a pragmatic choice corresponds to identical coordinate systems in configurations B and $\tilde{B}$ [7]:

$$\tilde{\boldsymbol{g}}_\alpha(x) = \delta^a_\alpha \boldsymbol{g}_a = \delta^a_\alpha \partial_a \boldsymbol{x} = \delta^a_\alpha \frac{\partial \boldsymbol{x}}{\partial x^a} = \delta^a_\alpha \bar{F}^{-1\beta}_{.a} \partial_\beta \boldsymbol{x}. \tag{4.55}$$

When (4.18) applies, the choice of basis vectors in (4.49) is still possible; it will be shown later in (4.99) that condition (4.18) implies (4.14) and vice-versa. Choices (4.49) and (4.55) differ but are not contradictory. In particular, basis vectors in (4.49) and (4.55) are related by the shifter g^A_a of (2.47) since from (2.57),

$$\{\delta^a_\alpha \boldsymbol{g}_a[x(X,t)]\} = \delta^a_\alpha g^A_a(x,X)\boldsymbol{G}_A = \delta^a_\alpha g^A_a \delta^\beta_A \{\delta^B_\beta \boldsymbol{G}_B(X)\}. \tag{4.56}$$

Metric tensor components corresponding to (4.55) are

$$\tilde{g}_{\alpha\beta}(x) = \tilde{\boldsymbol{g}}_\alpha \cdot \tilde{\boldsymbol{g}}_\beta = \delta^a_\alpha \boldsymbol{g}_a \cdot \delta^b_\beta \boldsymbol{g}_b = \delta^a_\alpha \delta^b_\beta g_{ab}, \tag{4.57}$$

where $g_{ab}(x)$ is the spatial metric tensor of (2.23). Mixed Kronecker delta symbols satisfy

$$\delta^a_\alpha = 1 \quad \forall a = \alpha, \qquad \delta^a_\alpha = 0 \quad \forall a \neq \alpha. \tag{4.58}$$

The determinant of the metric is

$$\tilde{g}(x) = \det(\tilde{g}_{\alpha\beta}) = \det(\delta^a_\alpha)\det(\delta^b_\beta)\det(g_{ab}) = \det(g_{ab}) = g. \tag{4.59}$$

Explicitly, shifter tensor components are

$$g^a_\alpha = \langle \tilde{\boldsymbol{g}}_\alpha, \boldsymbol{g}^a \rangle = \langle \delta^b_\alpha \boldsymbol{g}_b, \boldsymbol{g}^a \rangle = \delta^b_\alpha \delta^a_b = \delta^a_\alpha, \tag{4.60}$$

with determinant of (4.42) reducing to

$$\det(g^\alpha_a) = 1/\det(g^a_\alpha) = \sqrt{g/\tilde{g}} = 1. \tag{4.61}$$

Finally, the simplest choice of coordinate system for configuration $\tilde{B}$ is a Cartesian system with constant basis vectors $\boldsymbol{e}_\alpha$:

$$\tilde{\boldsymbol{g}}_\alpha = \boldsymbol{e}_\alpha, \qquad \tilde{g}_{\alpha\beta} = \boldsymbol{e}_\alpha \cdot \boldsymbol{e}_\beta = \delta_{\alpha\beta}, \qquad \tilde{g} = \det(\delta_{\alpha\beta}) = 1. \tag{4.62}$$

When anholonomicity conditions (4.14) or (4.18) apply, such a Cartesian frame is assigned externally to configuration $\tilde{B}$, since in that case $\tilde{B}$ is not a Euclidean space.

The most general kind of anholonomic basis vectors may depend on both time and location, *i.e.*,

$$\tilde{\boldsymbol{g}}_\alpha = \tilde{\boldsymbol{g}}_\alpha(X, t) \quad \text{or} \quad \tilde{\boldsymbol{g}}_\alpha = \tilde{\boldsymbol{g}}_\alpha(x, t), \tag{4.63}$$

where neither (4.49) nor (4.55) need apply. Such general basis vectors need not be continuously differentiable with respect to X^A or x^a. This generic definition, used implicitly elsewhere [7], places no restrictions on the choice of intermediate basis vectors, besides the requirement that at each point (X or x) they exist and are linearly independent (*i.e.*, the definition of a basis). However, those defined explicitly in (4.49), (4.55), or (4.62) appear to be most practical for typical applications of the general theory [8].

Analogously to (2.20), the following perhaps more intuitive modern notation can be introduced for basis vectors and their reciprocals:

$$\tilde{\boldsymbol{g}}_\alpha = \frac{\boldsymbol{\partial}}{\boldsymbol{\partial}\tilde{x}^\alpha} = \boldsymbol{\partial}_\alpha, \qquad \tilde{\boldsymbol{g}}^\alpha = \boldsymbol{d}\tilde{x}^\alpha. \tag{4.64}$$

The above notation is used sparingly in the present work. Quantities in (4.64) are not gradient vectors or true differentials unless configuration $\tilde{B}$ is holonomic. With this notation (4.26) becomes

$$\langle \boldsymbol{\partial}_\beta, \boldsymbol{d}\tilde{x}^\alpha \rangle = \delta^\alpha_\beta. \tag{4.65}$$

4.1.4 Convected anholonomic connection coefficients

Arbitrary connection coefficients $\Gamma^{\cdot\cdot A}_{BC}$ in reference configuration B_0 can be mapped to coefficients $\hat{\Gamma}^{\cdot\cdot\alpha}_{\beta\chi}$ in configuration $\tilde{B}$ via [7,28]

$$\begin{aligned}\hat{\Gamma}^{\cdot\cdot\alpha}_{\beta\chi} &= \tilde{F}^{\alpha}_{\cdot A}\tilde{F}^{-1B}_{\cdot\beta}\tilde{F}^{-1C}_{\cdot\chi}\Gamma^{\cdot\cdot A}_{BC} + \tilde{F}^{\alpha}_{\cdot A}\partial_{\beta}\tilde{F}^{-1A}_{\cdot\chi} \\ &= \tilde{F}^{\alpha}_{\cdot A}\tilde{F}^{-1B}_{\cdot\beta}\tilde{F}^{-1C}_{\cdot\chi}\Gamma^{\cdot\cdot A}_{BC} - \tilde{F}^{-1B}_{\cdot\beta}\tilde{F}^{-1A}_{\cdot\chi}\partial_{B}\tilde{F}^{\alpha}_{\cdot A},\end{aligned} \tag{4.66}$$

where (4.15) applies, as does the following identity:

$$\begin{aligned}0 &= \partial_{\beta}\delta^{\alpha}_{\chi} \\ &= \partial_{\beta}(\tilde{F}^{\alpha}_{\cdot A}\tilde{F}^{-1A}_{\cdot\chi}) \\ &= \tilde{F}^{-1A}_{\cdot\chi}\partial_{\beta}\tilde{F}^{\alpha}_{\cdot A} + \tilde{F}^{\alpha}_{\cdot A}\partial_{\beta}\tilde{F}^{-1A}_{\cdot\chi} \\ &= \tilde{F}^{-1A}_{\cdot\chi}\tilde{F}^{-1B}_{\cdot\beta}\partial_{B}\tilde{F}^{\alpha}_{\cdot A} + \tilde{F}^{\alpha}_{\cdot A}\tilde{F}^{-1B}_{\cdot\beta}\partial_{B}\tilde{F}^{-1A}_{\cdot\chi}.\end{aligned} \tag{4.67}$$

Torsion tensor components $T^{\cdot\cdot A}_{BC}$ of (2.73) in configuration B_0 map correspondingly using (4.66) as

$$\begin{aligned}\hat{T}^{\cdot\cdot\alpha}_{\beta\chi} &= \tilde{F}^{\alpha}_{\cdot A}\tilde{F}^{-1B}_{\cdot\beta}\tilde{F}^{-1C}_{\cdot\chi}T^{\cdot\cdot A}_{BC} \\ &= \tilde{F}^{\alpha}_{\cdot A}\tilde{F}^{-1B}_{\cdot\beta}\tilde{F}^{-1C}_{\cdot\chi}\Gamma^{\cdot\cdot A}_{[BC]} \\ &= \hat{\Gamma}^{\cdot\cdot\alpha}_{[\beta\chi]} + \tilde{F}^{-1A}_{\cdot\beta}\tilde{F}^{-1B}_{\cdot\chi}\partial_{[A}\tilde{F}^{\alpha}_{\cdot B]} \\ &= \hat{\Gamma}^{\cdot\cdot\alpha}_{[\beta\chi]} + \hat{\kappa}^{\cdot\cdot\alpha}_{\beta\chi} \\ &= \hat{T}^{\cdot\cdot\alpha}_{[\beta\chi]},\end{aligned} \tag{4.68}$$

where components of the anholonomic object $\hat{\boldsymbol{\kappa}}$ are defined as [28]

$$\hat{\kappa}^{\cdot\cdot\alpha}_{\beta\chi} = \tilde{F}^{-1A}_{\cdot\beta}\tilde{F}^{-1B}_{\cdot\chi}\partial_{[A}\tilde{F}^{\alpha}_{\cdot B]} = \hat{\kappa}^{\cdot\cdot\alpha}_{[\beta\chi]}. \tag{4.69}$$

Note also the identities

$$\begin{aligned}\hat{\kappa}^{\cdot\cdot\chi}_{\alpha\beta} &= \tilde{F}^{-1A}_{\cdot\alpha}\tilde{F}^{-1B}_{\cdot\beta}\partial_{[A}\tilde{F}^{\chi}_{\cdot B]} \\ &= \tilde{F}^{-1A}_{\cdot[\alpha}\tilde{F}^{-1B}_{\cdot\beta]}\partial_{A}\tilde{F}^{\chi}_{\cdot B} \\ &= -\tilde{F}^{-1A}_{\cdot[\alpha}\tilde{F}^{\chi}_{\cdot|B|}\partial_{A}\tilde{F}^{-1B}_{\cdot\beta]} \\ &= -\tilde{F}^{\chi}_{\cdot B}\partial_{[\alpha}\tilde{F}^{-1B}_{\cdot\beta]}.\end{aligned} \tag{4.70}$$

When $\tilde{\boldsymbol{F}}(X,t)$ is an integrable function of X^A and hence (4.13) applies, then $\hat{\kappa}^{\cdot\cdot\alpha}_{\beta\chi} = 0$. Riemann-Christoffel curvature tensor components $R^{\cdot\cdot\cdot A}_{BCD}$ of (2.75) in configuration B_0 map as [7,28]

$$\begin{aligned}\hat{R}^{\cdot\cdot\cdot\alpha}_{\beta\chi\delta} &= \tilde{F}^{\alpha}_{\cdot A}\tilde{F}^{-1B}_{\cdot\beta}\tilde{F}^{-1C}_{\cdot\chi}\tilde{F}^{-1D}_{\cdot\delta}R^{\cdot\cdot\cdot A}_{BCD} \\ &= 2\partial_{[\beta}\hat{\Gamma}^{\cdot\cdot\alpha}_{\chi]\delta} + 2\hat{\Gamma}^{\cdot\cdot\alpha}_{[\beta|\epsilon|}\hat{\Gamma}^{\cdot\cdot\epsilon}_{\chi]\delta} + 2\hat{\kappa}^{\cdot\cdot\epsilon}_{\beta\chi}\hat{\Gamma}^{\cdot\cdot\alpha}_{\epsilon\delta} \\ &= \hat{R}^{\cdot\cdot\cdot\alpha}_{[\beta\chi]\delta}.\end{aligned} \tag{4.71}$$

Contraction over the final two indices of (4.71) leads to

$$\begin{aligned}\hat{R}_{\beta\chi\alpha}^{\cdot\cdot\cdot\alpha} &= 2\partial_{[\beta}\hat{\Gamma}_{\chi]\alpha}^{\cdot\cdot\alpha} + 2\hat{\Gamma}_{[\beta|\epsilon|}^{\cdot\cdot\alpha}\hat{\Gamma}_{\chi]\alpha}^{\cdot\cdot\epsilon} + 2\hat{\kappa}_{\beta\chi}^{\cdot\cdot\epsilon}\hat{\Gamma}_{\epsilon\alpha}^{\cdot\cdot\alpha} \\ &= 2\partial_{[\beta}\hat{\Gamma}_{\chi]\alpha}^{\cdot\cdot\alpha} + 2\hat{\kappa}_{\beta\chi}^{\cdot\cdot\epsilon}\hat{\Gamma}_{\epsilon\alpha}^{\cdot\cdot\alpha}\end{aligned} \tag{4.72}$$

instead of (2.83). Definitions (4.68) and (4.71) are consistent with the final equalities in (2.76) and (2.77) [28]:

$$\begin{aligned}\nabla_{[\beta}\nabla_{\chi]}V^{\alpha} &= \nabla_{[\beta}(\partial_{\chi]}V^{\alpha} + \hat{\Gamma}_{\chi]\delta}^{\cdot\cdot\alpha}V^{\delta}) \\ &= \partial_{[\beta}(\partial_{\chi]}V^{\alpha}) + \partial_{[\beta}\hat{\Gamma}_{\chi]\delta}^{\cdot\cdot\alpha}V^{\delta} \\ &\quad + \hat{\Gamma}_{[\beta|\delta|}^{\cdot\cdot\alpha}\partial_{\chi]}V^{\delta} + \hat{\Gamma}_{[\chi|\delta|}^{\cdot\cdot\alpha}\partial_{\beta]}V^{\delta} \\ &\quad + \hat{\Gamma}_{[\beta|\delta|}^{\cdot\cdot\alpha}\hat{\Gamma}_{\chi]\epsilon}^{\cdot\cdot\delta}V^{\epsilon} - \hat{\Gamma}_{[\beta\chi]}^{\cdot\cdot\delta}\partial_{\delta}V^{\alpha} \\ &\quad - \hat{\Gamma}_{[\beta\chi]}^{\cdot\cdot\epsilon}\hat{\Gamma}_{\epsilon\delta}^{\cdot\cdot\alpha}V^{\delta} \\ &= \partial_{[\beta}\hat{\Gamma}_{\chi]\delta}^{\cdot\cdot\alpha}V^{\delta} + \hat{\Gamma}_{[\beta|\delta|}^{\cdot\cdot\alpha}\hat{\Gamma}_{\chi]\epsilon}^{\cdot\cdot\delta}V^{\epsilon} - \hat{\Gamma}_{[\beta\chi]}^{\cdot\cdot\delta}\partial_{\delta}V^{\alpha} \\ &\quad - \hat{\Gamma}_{[\beta\chi]}^{\cdot\cdot\delta}\hat{\Gamma}_{\delta\epsilon}^{\cdot\cdot\alpha}V^{\epsilon} + \partial_{[\beta}(\partial_{\chi]}V^{\alpha}) \\ &= \partial_{[\beta}\hat{\Gamma}_{\chi]\delta}^{\cdot\cdot\alpha}V^{\delta} + \hat{\Gamma}_{[\beta|\delta|}^{\cdot\cdot\alpha}\hat{\Gamma}_{\chi]\epsilon}^{\cdot\cdot\delta}V^{\epsilon} - \hat{\Gamma}_{[\beta\chi]}^{\cdot\cdot\delta}\partial_{\delta}V^{\alpha} \\ &\quad - \hat{\Gamma}_{[\beta\chi]}^{\cdot\cdot\delta}\hat{\Gamma}_{\delta\epsilon}^{\cdot\cdot\alpha}V^{\epsilon} - \hat{\kappa}_{\beta\chi}^{\cdot\cdot\delta}\partial_{\delta}V^{\alpha} \\ &= \partial_{[\beta}\hat{\Gamma}_{\chi]\delta}^{\cdot\cdot\alpha}V^{\delta} + \hat{\Gamma}_{[\beta|\delta|}^{\cdot\cdot\alpha}\hat{\Gamma}_{\chi]\epsilon}^{\cdot\cdot\delta}V^{\epsilon} - \hat{\Gamma}_{[\beta\chi]}^{\cdot\cdot\delta}\partial_{\delta}V^{\alpha} \\ &\quad - \hat{\Gamma}_{[\beta\chi]}^{\cdot\cdot\delta}\hat{\Gamma}_{\delta\epsilon}^{\cdot\cdot\alpha}V^{\epsilon} + (\hat{\Gamma}_{[\beta\chi]}^{\cdot\cdot\delta} - \hat{T}_{\beta\chi}^{\cdot\cdot\delta})\partial_{\delta}V^{\alpha} \\ &= \frac{1}{2}\hat{R}_{\beta\chi\delta}^{\cdot\cdot\cdot\alpha}V^{\delta} - \hat{T}_{\beta\chi}^{\cdot\cdot\delta}\nabla_{\delta}V^{\alpha},\end{aligned} \tag{4.73}$$

$$\begin{aligned}\nabla_{[\beta}\nabla_{\chi]}\alpha_{\delta} &= \nabla_{[\beta}(\partial_{\chi]}\alpha_{\delta} - \hat{\Gamma}_{\chi]\delta}^{\cdot\cdot\alpha}\alpha_{\alpha}) \\ &= \partial_{[\beta}(\partial_{\chi]}\alpha_{\delta}) - \partial_{[\beta}\hat{\Gamma}_{\chi]\delta}^{\cdot\cdot\alpha}\alpha_{\alpha} - \hat{\Gamma}_{[\beta|\delta|}^{\cdot\cdot\alpha}\partial_{\chi]}\alpha_{\alpha} \\ &\quad - \hat{\Gamma}_{[\chi|\delta|}^{\cdot\cdot\alpha}\partial_{\beta]}\alpha_{\alpha} - \hat{\Gamma}_{[\beta\chi]}^{\cdot\cdot\alpha}\partial_{\alpha}\alpha_{\delta} \\ &\quad + \hat{\Gamma}_{[\beta\chi]}^{\cdot\cdot\alpha}\hat{\Gamma}_{\alpha\delta}^{\cdot\cdot\epsilon}\alpha_{\epsilon} + \hat{\Gamma}_{[\beta|\delta|}^{\cdot\cdot\alpha}\hat{\Gamma}_{\chi]\alpha}^{\cdot\cdot\epsilon}\alpha_{\epsilon} \\ &= -\frac{1}{2}\hat{R}_{\beta\chi\delta}^{\cdot\cdot\cdot\alpha}\alpha_{\alpha} - \hat{T}_{\beta\chi}^{\cdot\cdot\alpha}\nabla_{\alpha}\alpha_{\delta},\end{aligned} \tag{4.74}$$

where V^{α} and α_{δ} are respective anholonomic components of twice differentiable vector and covector fields, and covariant differentiation with respect to anholonomic coordinates is defined formally later in (4.138). For example, for a twice differentiable scalar field A, the second covariant derivative is

$$\nabla_{\alpha}(\nabla_{\beta}A) = \partial_{\alpha}(\partial_{\beta}A) - \hat{\Gamma}_{\alpha\beta}^{\cdot\cdot\chi}\partial_{\chi}A. \tag{4.75}$$

Skew components of second partial derivatives in (4.23) can be written as follows in terms of the anholonomic object:

$$\begin{aligned}
\partial_{[\alpha}[\partial_{\beta]}(\cdot)] &= \frac{1}{2}\{\partial_\alpha[\partial_\beta(\cdot)] - \partial_\beta[\partial_\alpha(\cdot)]\} \\
&= \partial_A(\cdot)\partial_{[\alpha}\tilde{F}^{-1A}_{\cdot\beta]} \\
&= \partial_\chi(\cdot)\tilde{F}^{-1B}_{\cdot[\alpha}\partial_{|B|}(\tilde{F}^{-1A}_{\cdot\beta]})\tilde{F}^\chi_{\cdot A} \\
&= -\partial_\chi(\cdot)\tilde{F}^{-1B}_{\cdot[\alpha}\tilde{F}^{-1A}_{\cdot\beta]}\partial_B\tilde{F}^\chi_{\cdot A} \\
&= \partial_\chi(\cdot)\tilde{F}^{-1B}_{\cdot[\beta}\tilde{F}^{-1A}_{\cdot\alpha]}\partial_B\tilde{F}^\chi_{\cdot A} \\
&= \hat{\kappa}^{\cdot\cdot\chi}_{\beta\alpha}\partial_\chi(\cdot) \\
&= -\hat{\kappa}^{\cdot\cdot\chi}_{\alpha\beta}\partial_\chi(\cdot).
\end{aligned} \tag{4.76}$$

Note that (4.76) has already been used in derivations (4.73) and (4.74). The skew second covariant derivative of scalar field A is—from (4.68) and (4.75)—related to the torsion as

$$\begin{aligned}
\nabla_{[\alpha}(\nabla_{\beta]}A) &= \partial_{[\alpha}(\partial_{\beta]}A) - \hat{\Gamma}^{\cdot\cdot\chi}_{[\alpha\beta]}\partial_\chi A \\
&= -(\hat{\kappa}^{\cdot\cdot\chi}_{\alpha\beta} + \hat{\Gamma}^{\cdot\cdot\chi}_{[\alpha\beta]})\partial_\chi A \\
&= -\hat{T}^{\cdot\cdot\chi}_{\alpha\beta}\partial_\chi A.
\end{aligned} \tag{4.77}$$

An identity holds for the skew derivative of the anholonomic object [28]:

$$\partial_{[\epsilon}\hat{\kappa}^{\cdot\cdot\chi}_{\beta\alpha]} = 2\hat{\kappa}^{\cdot\cdot\phi}_{[\epsilon\beta}\hat{\kappa}^{\cdot\cdot\chi}_{\alpha]\phi}. \tag{4.78}$$

Let $\tilde{g}_{\alpha\beta}$ and its inverse $\tilde{g}^{\alpha\beta}$ denote symmetric, three times differentiable, invertible, but otherwise arbitrary second-order tensors. Define the covariant derivative

$$\hat{M}^{\cdot\cdot\alpha}_{\beta\chi} = \tilde{g}^{\alpha\delta}\hat{M}_{\beta\chi\delta} = -\tilde{g}^{\alpha\delta}\nabla_\beta\tilde{g}_{\chi\delta} = \tilde{g}_{\chi\delta}\nabla_\beta\tilde{g}^{\alpha\delta}, \tag{4.79}$$

the partial derivative

$$\hat{\chi}_{\alpha\beta\chi} = \frac{1}{2}\partial_\alpha\tilde{g}_{\beta\chi} = \hat{\chi}_{\alpha\chi\beta}, \tag{4.80}$$

the fully covariant components of the anholonomic object

$$\hat{\kappa}_{\beta\chi\delta} = \hat{\kappa}^{\cdot\cdot\alpha}_{\beta\chi}\tilde{g}_{\alpha\delta} = -\hat{\kappa}_{\chi\beta\delta}, \tag{4.81}$$

and the torsion

$$\hat{T}_{\beta\chi\delta} = \hat{T}^{\cdot\cdot\alpha}_{\beta\chi}\tilde{g}_{\alpha\delta} = -\hat{T}_{\chi\beta\delta}. \tag{4.82}$$

Arbitrary anholonomic connection coefficients $\hat{\Gamma}^{\cdot\cdot\alpha}_{\beta\chi}$ can then be written as follows, analogously to (2.94) [28]:

$$\hat{\Gamma}^{\cdot\cdot\alpha}_{\beta\chi} = \frac{1}{2}\tilde{g}^{\alpha\delta}(2\hat{\chi}_{\{\beta\delta\chi\}} - 2\hat{T}_{\{\beta\delta\chi\}} + \hat{M}_{\{\beta\delta\chi\}} + 2\hat{\kappa}_{\{\beta\delta\chi\}}). \tag{4.83}$$

Relationships analogous to (4.66)–(4.71) can be used to map connection coefficients and torsion and curvature tensors referred to spatial coordinates to intermediate space $\tilde{B}$ by replacing referential indices with spatial indices (*e.g.*, $X^A \to x^a$) and replacing components $\tilde{F}^{\alpha}_{.A}$ with $\bar{F}^{-1\alpha}_{.a}$.

A convected coordinate representation of $\tilde{\boldsymbol{F}}(X,t)$ can be used to verify transformation (4.66) when $\Gamma^{..A}_{BC} = \overset{G}{\Gamma}{}^{..A}_{BC}$ corresponds to the Levi-Civita connection of (2.122) for Euclidean space on B_0. Convected anholonomic basis vectors and their reciprocals are defined as

$$\tilde{\boldsymbol{g}}'_{\alpha}(X,t) = \tilde{F}^{-1A}_{.\alpha}\boldsymbol{G}_A, \qquad \tilde{\boldsymbol{g}}'^{\alpha}(X,t) = \tilde{F}^{\alpha}_{.A}\boldsymbol{G}^A; \tag{4.84}$$

$$\langle \tilde{\boldsymbol{g}}'_{\alpha}, \tilde{\boldsymbol{g}}'^{\beta}\rangle = \tilde{F}^{-1A}_{.\alpha}\tilde{F}^{\beta}_{.B}\langle \boldsymbol{G}_A, \boldsymbol{G}^B\rangle = \tilde{F}^{-1A}_{.\alpha}\tilde{F}^{\beta}_{.B}\delta^B_A = \tilde{F}^{-1A}_{.\alpha}\tilde{F}^{\beta}_{.A} = \delta^{\beta}_{\alpha}. \tag{4.85}$$

The deformation map of the second of (4.3) can then be expressed as

$$\tilde{\boldsymbol{F}}(X,t) = \tilde{F}^{\alpha}_{.A}\tilde{\boldsymbol{g}}_{\alpha} \otimes \boldsymbol{G}^A = \tilde{\boldsymbol{g}}_{\alpha} \otimes \tilde{F}^{\alpha}_{.A}\boldsymbol{G}^A = \tilde{\boldsymbol{g}}_{\alpha} \otimes \tilde{\boldsymbol{g}}'^{\alpha}. \tag{4.86}$$

Components of the metric tensor corresponding to basis vectors in (4.84) are

$$\begin{aligned}
\tilde{\boldsymbol{g}}'_{\alpha} \cdot \tilde{\boldsymbol{g}}'_{\beta} &= \tilde{F}^{-1A}_{.\alpha}\boldsymbol{G}_A \cdot \tilde{F}^{-1B}_{.\beta}\boldsymbol{G}_B \\
&= \tilde{F}^{-1A}_{.\alpha}\tilde{F}^{-1B}_{.\beta}\boldsymbol{G}_A \cdot \boldsymbol{G}_B \\
&= \tilde{F}^{-1A}_{.\alpha}G_{AB}\tilde{F}^{-1B}_{.\beta}.
\end{aligned} \tag{4.87}$$

Taking the partial anholonomic derivative of $\tilde{\boldsymbol{g}}'_{\alpha}$ using (4.15) results in

$$\begin{aligned}
\partial_{\beta}\tilde{\boldsymbol{g}}'_{\alpha} &= \partial_{\beta}(\tilde{F}^{-1A}_{.\alpha}\boldsymbol{G}_A) \\
&= \tilde{F}^{-1A}_{.\alpha}\partial_{\beta}\boldsymbol{G}_A + \partial_{\beta}(\tilde{F}^{-1A}_{.\alpha})\boldsymbol{G}_A \\
&= \tilde{F}^{-1A}_{.\alpha}\tilde{F}^{-1B}_{.\beta}\partial_B\boldsymbol{G}_A + \partial_{\beta}(\tilde{F}^{-1A}_{.\alpha})\tilde{F}^{\chi}_{.A}\tilde{\boldsymbol{g}}'_{\chi} \\
&= \tilde{F}^{-1A}_{.\alpha}\tilde{F}^{-1B}_{.\beta}\overset{G}{\Gamma}{}^{..C}_{BA}\boldsymbol{G}_C - \partial_{\beta}(\tilde{F}^{\chi}_{.A})\tilde{F}^{-1A}_{.\alpha}\tilde{\boldsymbol{g}}'_{\chi} \\
&= \tilde{F}^{-1A}_{.\alpha}\tilde{F}^{-1B}_{.\beta}\tilde{F}^{\chi}_{.C}\overset{G}{\Gamma}{}^{..C}_{BA}\tilde{\boldsymbol{g}}'_{\chi} - \tilde{F}^{-1A}_{.\alpha}\tilde{F}^{-1B}_{.\beta}\partial_B\tilde{F}^{\chi}_{.A}\tilde{\boldsymbol{g}}'_{\chi} \\
&= \left(\tilde{F}^{\chi}_{.C}\tilde{F}^{-1B}_{.\beta}\tilde{F}^{-1A}_{.\alpha}\overset{G}{\Gamma}{}^{..C}_{BA} - \tilde{F}^{-1B}_{.\beta}\tilde{F}^{-1A}_{.\alpha}\partial_B\tilde{F}^{\chi}_{.A}\right)\tilde{\boldsymbol{g}}'_{\chi} \\
&= \tilde{\Gamma}^{..\chi}_{\beta\alpha}\tilde{\boldsymbol{g}}'_{\chi},
\end{aligned} \tag{4.88}$$

in agreement with (4.66). The torsion tensor of (4.68) vanishes identically in this case from (2.122):

$$\tilde{T}^{..\alpha}_{\beta\chi} = \tilde{F}^{\alpha}_{.A}\tilde{F}^{-1B}_{.\beta}\tilde{F}^{-1C}_{.\chi}\overset{G}{\Gamma}{}^{..A}_{[BC]} = \tilde{\Gamma}^{..\alpha}_{[\beta\chi]} + \tilde{\kappa}^{..\alpha}_{\beta\chi} = 0, \tag{4.89}$$

where the anholonomic object is the same as that of (4.69):

$$\tilde{\kappa}^{..\alpha}_{\beta\chi} = \tilde{F}^{-1A}_{.\beta}\tilde{F}^{-1B}_{.\chi}\partial_{[A}\tilde{F}^{\alpha}_{.B]}. \tag{4.90}$$

Following an analogous approach with $\bar{\boldsymbol{F}}^{-1}(x,t)$, convected anholonomic basis vectors and their reciprocals are defined as

$$\bar{\boldsymbol{g}}'_\alpha(x,t) = \bar{F}^a_{.\alpha}\boldsymbol{g}_a, \qquad \bar{\boldsymbol{g}}'^\alpha(x,t) = \bar{F}^{-1\alpha}_{.a}\boldsymbol{g}^a; \tag{4.91}$$

$$\begin{aligned}\langle \bar{\boldsymbol{g}}'_\alpha, \bar{\boldsymbol{g}}'^\beta\rangle &= \bar{F}^a_{.\alpha}\bar{F}^{-1\beta}_{.b}\langle \boldsymbol{g}_a, \boldsymbol{g}^b\rangle \\ &= \bar{F}^a_{.\alpha}\bar{F}^{-1\beta}_{.b}\delta^b_a \\ &= \bar{F}^a_{.\alpha}\bar{F}^{-1\beta}_{.a} \\ &= \delta^\beta_\alpha.\end{aligned} \tag{4.92}$$

The deformation map of the first of (4.3) can then be expressed using this convected basis as

$$\bar{\boldsymbol{F}}(X,t) = \bar{F}^a_{.\alpha}\boldsymbol{g}_a \otimes \tilde{\boldsymbol{g}}^\alpha = \bar{\boldsymbol{g}}'_\alpha \otimes \tilde{\boldsymbol{g}}^\alpha. \tag{4.93}$$

Components of the metric tensor corresponding to basis vectors in (4.91) are

$$\begin{aligned}\bar{\boldsymbol{g}}'_\alpha \cdot \bar{\boldsymbol{g}}'_\beta &= \bar{F}^a_{.\alpha}\boldsymbol{g}_a \cdot \bar{F}^b_{.\beta}\boldsymbol{g}_b \\ &= \bar{F}^a_{.\alpha}\bar{F}^b_{.\beta}\boldsymbol{g}_a \cdot \boldsymbol{g}_b \\ &= \bar{F}^a_{.\alpha}g_{ab}\bar{F}^b_{.\beta}.\end{aligned} \tag{4.94}$$

Taking the partial anholonomic derivative of $\bar{\boldsymbol{g}}'_\alpha$ using (4.19) results in

$$\begin{aligned}\partial_\beta\bar{\boldsymbol{g}}'_\alpha &= \partial_\beta(\bar{F}^a_{.\alpha}\boldsymbol{g}_a) \\ &= \bar{F}^a_{.\alpha}\boldsymbol{g}_a\partial_\beta\boldsymbol{g}_a + \partial_\beta(\bar{F}^a_{.\alpha})\boldsymbol{g}_a \\ &= \bar{F}^a_{.\alpha}\bar{F}^b_{.\beta}\partial_b\boldsymbol{g}_a + \partial_\beta(\bar{F}^a_{.\alpha})\bar{F}^{-1\chi}_{.a}\bar{\boldsymbol{g}}'_\chi \\ &= \bar{F}^a_{.\alpha}\bar{F}^b_{.\beta}\,{}^g\Gamma^{..c}_{ba}\boldsymbol{g}_c - \partial_\beta(\bar{F}^{-1\chi}_{.a})\bar{F}^a_{.\alpha}\bar{\boldsymbol{g}}'_\chi \\ &= \bar{F}^a_{.\alpha}\bar{F}^b_{.\beta}\bar{F}^{-1\chi}_{.c}\,{}^g\Gamma^{..c}_{ba}\bar{\boldsymbol{g}}'_\chi - \bar{F}^a_{.\alpha}\bar{F}^b_{.\beta}\partial_b\bar{F}^{-1\chi}_{.a}\bar{\boldsymbol{g}}'_\chi \\ &= \left(\bar{F}^{-1\chi}_{.c}\bar{F}^b_{.\beta}\bar{F}^a_{.\alpha}\,{}^g\Gamma^{..c}_{ba} - \bar{F}^b_{.\beta}\bar{F}^a_{.\alpha}\partial_b\bar{F}^{-1\chi}_{.a}\right)\bar{\boldsymbol{g}}'_\chi \\ &= \bar{\Gamma}^{..\chi}_{\beta\alpha}\bar{\boldsymbol{g}}'_\chi.\end{aligned} \tag{4.95}$$

The torsion tensor of (4.68) vanishes identically in this case from (2.150):

$$\bar{T}^{..\alpha}_{\beta\chi} = \bar{F}^{-1\alpha}_{.a}\bar{F}^b_{.\beta}\bar{F}^c_{.\chi}\,{}^g\Gamma^{..a}_{[bc]} = \bar{\Gamma}^{..\alpha}_{[\beta\chi]} + \bar{\kappa}^{..\alpha}_{\beta\chi} = 0, \tag{4.96}$$

where the corresponding anholonomic object is

$$\bar{\kappa}^{..\alpha}_{\beta\chi} = \bar{F}^a_{.\beta}\bar{F}^b_{.\chi}\partial_{[a}\bar{F}^{-1\alpha}_{.b]}. \tag{4.97}$$

Connection coefficients $\tilde{\Gamma}^{\cdot\cdot\chi}_{\beta\alpha}$ of (4.88) and $\bar{\Gamma}^{\cdot\cdot\chi}_{\beta\alpha}$ of (4.95) are generally unequal. However, following from decomposition (4.1) and (4.2), their skew covariant components are equal [7]:

$$
\begin{aligned}
\bar{\Gamma}^{\cdot\cdot\chi}_{[\beta\alpha]} &= \bar{F}^{b}_{\cdot[\beta}\bar{F}^{a}_{\cdot\alpha]}\bar{F}^{-1\chi}_{\cdot c}\overset{g}{\Gamma}{}^{\cdot\cdot c}_{ba} - \bar{F}^{b}_{\cdot[\beta}\bar{F}^{a}_{\cdot\alpha]}\partial_b\bar{F}^{-1\chi}_{\cdot a} \\
&= \bar{F}^{b}_{\cdot\beta}\bar{F}^{a}_{\cdot\alpha}\bar{F}^{-1\chi}_{\cdot c}\overset{g}{\Gamma}{}^{\cdot\cdot c}_{[ba]} - \bar{F}^{b}_{\cdot\beta}\bar{F}^{a}_{\cdot\alpha}\partial_{[b}\bar{F}^{-1\chi}_{\cdot a]} \\
&= -\bar{F}^{b}_{\cdot\beta}\bar{F}^{a}_{\cdot\alpha}\partial_{[b}\bar{F}^{-1\chi}_{\cdot a]} \\
&= -\bar{F}^{b}_{\cdot[\beta}\bar{F}^{a}_{\cdot\alpha]}\partial_b\bar{F}^{-1\chi}_{\cdot a} \\
&= -F^{b}_{\cdot B}F^{a}_{\cdot A}\tilde{F}^{-1B}_{\cdot[\beta}\tilde{F}^{-1A}_{\cdot\alpha]}\partial_b(\tilde{F}^{\chi}_{\cdot C}F^{-1C}_{\cdot a}) \\
&= -F^{b}_{\cdot B}F^{a}_{\cdot A}\tilde{F}^{-1B}_{\cdot[\beta}\tilde{F}^{-1A}_{\cdot\alpha]}(F^{-1C}_{\cdot a}\partial_b\tilde{F}^{\chi}_{\cdot C} + \tilde{F}^{\chi}_{\cdot C}\partial_b F^{-1C}_{\cdot a}) \\
&= -\tilde{F}^{-1B}_{\cdot[\beta}\tilde{F}^{-1A}_{\cdot\alpha]}(F^{b}_{\cdot B}\partial_b\tilde{F}^{\chi}_{\cdot A} + F^{b}_{\cdot B}F^{a}_{\cdot A}\tilde{F}^{\chi}_{\cdot C}\partial_b F^{-1C}_{\cdot a}) \\
&= -\tilde{F}^{-1B}_{\cdot[\beta}\tilde{F}^{-1A}_{\cdot\alpha]}(\partial_B\tilde{F}^{\chi}_{\cdot A} + F^{b}_{\cdot B}F^{a}_{\cdot A}\tilde{F}^{\chi}_{\cdot C}\partial_b F^{-1C}_{\cdot a}) \\
&= -\tilde{F}^{-1B}_{\cdot[\beta}\tilde{F}^{-1A}_{\cdot\alpha]}\partial_B\tilde{F}^{\chi}_{\cdot A} - \tilde{F}^{-1B}_{\cdot[\beta}\tilde{F}^{-1A}_{\cdot\alpha]}F^{b}_{\cdot B}F^{a}_{\cdot A}\tilde{F}^{\chi}_{\cdot C}\partial_b F^{-1C}_{\cdot a} \\
&= -\tilde{F}^{-1B}_{\cdot[\beta}\tilde{F}^{-1A}_{\cdot\alpha]}\partial_B\tilde{F}^{\chi}_{\cdot A} - \tilde{F}^{-1B}_{\cdot\beta}\tilde{F}^{-1A}_{\cdot\alpha}F^{b}_{\cdot[B}F^{a}_{\cdot A]}\tilde{F}^{\chi}_{\cdot C}\partial_b F^{-1C}_{\cdot a} \\
&= -\tilde{F}^{-1B}_{\cdot[\beta}\tilde{F}^{-1A}_{\cdot\alpha]}\partial_B\tilde{F}^{\chi}_{\cdot A} - \tilde{F}^{-1B}_{\cdot\beta}\tilde{F}^{-1A}_{\cdot\alpha}F^{b}_{\cdot B}F^{a}_{\cdot A}\tilde{F}^{\chi}_{\cdot C}\partial_{[b}F^{-1C}_{\cdot a]} \\
&= -\tilde{F}^{-1B}_{\cdot[\beta}\tilde{F}^{-1A}_{\cdot\alpha]}\partial_B\tilde{F}^{\chi}_{\cdot A} - \tilde{F}^{-1B}_{\cdot\beta}\tilde{F}^{-1A}_{\cdot\alpha}F^{b}_{\cdot B}F^{a}_{\cdot A}\tilde{F}^{\chi}_{\cdot C}\partial_{[b}\partial_{a]}X^{C} \\
&= -\tilde{F}^{-1B}_{\cdot[\beta}\tilde{F}^{-1A}_{\cdot\alpha]}\partial_B\tilde{F}^{\chi}_{\cdot A} \\
&= -\tilde{F}^{-1B}_{\cdot[\beta}\tilde{F}^{-1A}_{\cdot\alpha]}\partial_B\tilde{F}^{\chi}_{\cdot A} + \tilde{F}^{-1B}_{\cdot\beta}\tilde{F}^{-1A}_{\cdot\alpha}\tilde{F}^{\chi}_{\cdot C}\overset{G}{\Gamma}{}^{\cdot\cdot C}_{[BA]} \\
&= -\tilde{F}^{-1B}_{\cdot[\beta}\tilde{F}^{-1A}_{\cdot\alpha]}\partial_B\tilde{F}^{\chi}_{\cdot A} + \tilde{F}^{-1B}_{\cdot[\beta}\tilde{F}^{-1A}_{\cdot\alpha]}\tilde{F}^{\chi}_{\cdot C}\overset{G}{\Gamma}{}^{\cdot\cdot C}_{BA} \\
&= \tilde{\Gamma}^{\cdot\cdot\chi}_{[\beta\alpha]}.
\end{aligned}
\tag{4.98}
$$

Thus anholonomic objects of each connection defined in (4.90) and (4.97) are equal:

$$\bar{\kappa}^{\cdot\cdot\chi}_{\beta\alpha} = -\bar{\Gamma}^{\cdot\cdot\chi}_{[\beta\alpha]} = -\tilde{\Gamma}^{\cdot\cdot\chi}_{[\beta\alpha]} = \tilde{\kappa}^{\cdot\cdot\chi}_{\beta\alpha}. \tag{4.99}$$

Skew partial derivatives of convected basis vectors are computed as

$$\partial_{[\alpha}\tilde{\boldsymbol{g}}'_{\beta]} = \tilde{\Gamma}^{\cdot\cdot\chi}_{[\alpha\beta]}\tilde{\boldsymbol{g}}'_{\chi} = \tilde{\kappa}^{\cdot\cdot\chi}_{\beta\alpha}\tilde{\boldsymbol{g}}'_{\chi}, \qquad \partial_{[\alpha}\bar{\boldsymbol{g}}'_{\beta]} = \bar{\Gamma}^{\cdot\cdot\chi}_{[\alpha\beta]}\bar{\boldsymbol{g}}'_{\chi} = \bar{\kappa}^{\cdot\cdot\chi}_{\beta\alpha}\bar{\boldsymbol{g}}'_{\chi}. \tag{4.100}$$

Because $\det\tilde{\boldsymbol{F}} > 0$, the $\tilde{\boldsymbol{g}}'_{\alpha}$ form a (right-handed, but generally not orthonormal) basis; similarly, because $\det\bar{\boldsymbol{F}} > 0$, the $\bar{\boldsymbol{g}}'_{\alpha}$ are linearly independent. It follows that

$$\partial_{[\alpha}\tilde{\boldsymbol{g}}'_{\beta]} = 0 \Leftrightarrow \tilde{\Gamma}^{\cdot\cdot\chi}_{[\alpha\beta]} = 0, \qquad \partial_{[\alpha}\bar{\boldsymbol{g}}'_{\beta]} = 0 \Leftrightarrow \bar{\Gamma}^{\cdot\cdot\chi}_{[\alpha\beta]} = 0. \tag{4.101}$$

Therefore, the following local integrability conditions are equivalent:

$$\partial_{[\alpha}\tilde{\boldsymbol{g}}'_{\beta]} = 0 \Leftrightarrow \partial_{[\alpha}\bar{\boldsymbol{g}}'_{\beta]} = 0 \Leftrightarrow \tilde{\kappa}_{\beta\alpha}^{..\chi} = \bar{\kappa}_{\beta\alpha}^{..\chi} = 0$$
$$\Leftrightarrow \tilde{F}^{\alpha}_{.A} = \partial_A \tilde{x}^{\alpha} \Leftrightarrow \bar{F}^{-1\alpha}_{.a} = \partial_a \tilde{x}^{\alpha}. \quad (4.102)$$

In other words, if $\tilde{\boldsymbol{F}}(X,t)$ is integrable with respect to X^A, then $\bar{\boldsymbol{F}}^{-1}(x,t)$ is integrable with respect to $x^a(X,t)$ over the same simply connected neighborhood of X at the same instant of time t, *i.e.*, over the local neighborhood $x(X,t)$, and vice-versa. These conclusions may also be applied globally over B_0 presuming that the body of interest is simply connected and the quantities entering (4.100) are globally continuous over B_0. The existence of functions $X^A(\tilde{x}^{\alpha},t)$ and $x^a(\tilde{x}^{\alpha},t)$ is guaranteed over a local neighborhood of X by the inverse function theorem and the presumed invertibility of $\tilde{\boldsymbol{F}}$ and $\bar{\boldsymbol{F}}^{-1}$, but global continuity of such functions is not ensured even when (4.102) applies at each individual point $X \in B_0$ [25].

A necessary and sufficient condition for intermediate configuration $\tilde{B}$ to be labeled anholonomic is a nonzero anholonomic object. It follows from (4.99) that if $\tilde{\boldsymbol{F}}(X,t)$ is not integrable with respect to X^A over a given simply connected neighborhood of X, then $\bar{\boldsymbol{F}}^{-1}(x,t)$ is not integrable with respect to x^a over the corresponding neighborhood of $x(X,t)$, and vice-versa.

4.1.5 *Integrable connections*

Another set of connection coefficients that emerges frequently in differential geometry is of the following form:

$$\tilde{\Gamma}^{..A}_{BC} = \tilde{F}^{-1A}_{.\alpha}\partial_B \tilde{F}^{\alpha}_{.C} = -\tilde{F}^{\alpha}_{.C}\partial_B \tilde{F}^{-1A}_{.\alpha} = -\tilde{F}^{\alpha}_{.C}\tilde{F}^{\beta}_{.B}\partial_\beta \tilde{F}^{-1A}_{.\alpha}. \quad (4.103)$$

Pre-multiplication of (4.103) by $\tilde{\boldsymbol{F}}$ followed by partial differentiation with respect to X^D gives

$$\begin{aligned}\partial_D(\tilde{F}^{\alpha}_{.A}\tilde{\Gamma}^{..A}_{BC}) &= \partial_D(\tilde{F}^{\alpha}_{.A}\tilde{F}^{-1A}_{.\beta}\partial_B \tilde{F}^{\beta}_{.C}) \\ &= \partial_D(\partial_B \tilde{F}^{\alpha}_{.C}) \\ &= \partial_B(\partial_D \tilde{F}^{\alpha}_{.C}). \end{aligned} \quad (4.104)$$

Expanding the left of (4.104) with the product rule yields

$$\begin{aligned}\partial_D(\tilde{F}^{\alpha}_{.A}\tilde{\Gamma}^{..A}_{BC}) &= \tilde{\Gamma}^{..A}_{BC}\partial_D \tilde{F}^{\alpha}_{.A} + \tilde{F}^{\alpha}_{.A}\partial_D \tilde{\Gamma}^{..A}_{BC} \\ &= \tilde{F}^{\alpha}_{.A}(\tilde{\Gamma}^{..E}_{BC}\tilde{F}^{-1A}_{.\beta}\partial_D \tilde{F}^{\beta}_{.E} + \partial_D \tilde{\Gamma}^{..A}_{BC}) \\ &= \tilde{F}^{\alpha}_{.A}(\tilde{\Gamma}^{..E}_{BC}\tilde{\Gamma}^{..A}_{DE} + \partial_D \tilde{\Gamma}^{..A}_{BC}) \\ &= \tilde{F}^{\alpha}_{.A}(\partial_D \tilde{\Gamma}^{..A}_{BC} + \tilde{\Gamma}^{..A}_{DE}\tilde{\Gamma}^{..E}_{BC}). \end{aligned} \quad (4.105)$$

Combining skew parts of (4.104) and (4.105),

$$\begin{aligned}
0 &= 2\partial_{[B}\partial_{D]}\tilde{F}^{\alpha}_{\cdot C} \\
&= 2\partial_{[D}\partial_{B]}\tilde{F}^{\alpha}_{\cdot C} \\
&= 2\partial_{[D}(\tilde{F}^{\alpha}_{\cdot|A|}\tilde{\Gamma}^{\cdot\cdot A}_{B]C}) \\
&= 2\tilde{F}^{\alpha}_{\cdot A}(\partial_{[D}\tilde{\Gamma}^{\cdot\cdot A}_{B]C} + \tilde{\Gamma}^{\cdot\cdot A}_{[D|E|}\tilde{\Gamma}^{\cdot\cdot E}_{B]C}) \\
&= \tilde{F}^{\alpha}_{\cdot A}\tilde{R}^{\cdot\cdot\cdot A}_{DBC},
\end{aligned} \tag{4.106}$$

where the Riemann-Christoffel curvature tensor of the connection coefficients of (4.103) is defined as in (2.75):

$$\tilde{R}^{\cdot\cdot\cdot A}_{BCD} = 2\partial_{[B}\tilde{\Gamma}^{\cdot\cdot A}_{C]D} + 2\tilde{\Gamma}^{\cdot\cdot A}_{[B|E|}\tilde{\Gamma}^{\cdot\cdot E}_{C]D}. \tag{4.107}$$

Thus, local integrability conditions for the existence of an invertible differentiable map $\tilde{F}^{\alpha}_{\cdot A}(X,t)$ yielding $\tilde{\Gamma}^{\cdot\cdot A}_{BC}$ can be stated as [28]

$$\tilde{R}^{\cdot\cdot\cdot A}_{BCD} = 0 \Leftrightarrow \tilde{\Gamma}^{\cdot\cdot A}_{BC} = \tilde{F}^{-1A}_{\cdot\alpha}\partial_B\tilde{F}^{\alpha}_{\cdot C}. \tag{4.108}$$

A connection with the property of vanishing curvature as in (4.108) is said to be integrable or flat. Furthermore, when the torsion tensor of (4.103) defined as in (2.73) and (2.74),

$$\tilde{T}^{\cdot\cdot A}_{BC} = \tilde{\Gamma}^{\cdot\cdot A}_{[BC]} = \tilde{F}^{-1A}_{\cdot\alpha}\partial_{[B}\tilde{F}^{\alpha}_{\cdot C]}, \tag{4.109}$$

vanishes, it follows from (4.13) that deformation map $\tilde{\boldsymbol{F}}(X,t)$ itself is an integrable function of X^A, *i.e.*,

$$\tilde{T}^{\cdot\cdot A}_{BC} = 0 \Leftrightarrow \tilde{F}^{\alpha}_{\cdot A} = \partial_A\tilde{x}^{\alpha}. \tag{4.110}$$

The torsion tensor of (4.109) is related to the anholonomic object of (4.90) via

$$\tilde{T}^{\cdot\cdot A}_{BC} = \tilde{F}^{-1A}_{\cdot\alpha}\tilde{F}^{\beta}_{\cdot B}\tilde{F}^{\chi}_{\cdot C}\tilde{\kappa}^{\cdot\cdot\alpha}_{\beta\chi}. \tag{4.111}$$

Torsion $\tilde{T}^{\cdot\cdot A}_{BC}$ has physical meaning in the context of dislocation theories of crystalline solids [7].

Analogously to (4.103), consider now the connection coefficients referred to the spatial frame

$$\bar{\Gamma}^{\cdot\cdot a}_{bc} = \bar{F}^{a}_{\cdot\alpha}\partial_b\bar{F}^{-1\alpha}_{\cdot c} = -\bar{F}^{-1\alpha}_{\cdot c}\partial_b\bar{F}^{a}_{\cdot\alpha} - -\bar{F}^{-1\alpha}_{\cdot c}\bar{F}^{-1\beta}_{\cdot b}\partial_\beta\bar{F}^{a}_{\cdot\alpha}. \tag{4.112}$$

Pre-multiplication of (4.112) by $\bar{\boldsymbol{F}}^{-1}$ followed by partial differentiation with respect to x^d gives

$$\begin{aligned}
\partial_d(\bar{F}^{-1\alpha}_{\cdot a}\bar{\Gamma}^{\cdot\cdot a}_{bc}) &= \partial_d(\bar{F}^{-1\alpha}_{\cdot a}\bar{F}^{a}_{\cdot\beta}\partial_b\bar{F}^{-1\beta}_{\cdot c}) \\
&= \partial_d(\partial_b\bar{F}^{-1\alpha}_{\cdot c}) \\
&= \partial_b(\partial_d\bar{F}^{-1\alpha}_{\cdot c}).
\end{aligned} \tag{4.113}$$

Expanding the left of (4.113) with the product rule,

$$\begin{aligned}\partial_d(\bar{F}^{-1\alpha}_{\cdot a}\bar{\Gamma}^{\cdot\cdot a}_{bc}) &= \bar{\Gamma}^{\cdot\cdot a}_{bc}\partial_d\bar{F}^{-1\alpha}_{\cdot a} + \bar{F}^{-1\alpha}_{\cdot a}\partial_d\bar{\Gamma}^{\cdot\cdot a}_{bc}\\ &= \bar{F}^{-1\alpha}_{\cdot a}(\bar{\Gamma}^{\cdot\cdot e}_{bc}\bar{F}^{a}_{\cdot\beta}\partial_d\bar{F}^{-1\beta}_{\cdot e} + \partial_d\bar{\Gamma}^{\cdot\cdot a}_{bc})\\ &= \bar{F}^{-1\alpha}_{\cdot a}(\bar{\Gamma}^{\cdot\cdot e}_{bc}\bar{\Gamma}^{\cdot\cdot a}_{de} + \partial_d\bar{\Gamma}^{\cdot\cdot a}_{bc})\\ &= \bar{F}^{-1\alpha}_{\cdot a}(\partial_d\bar{\Gamma}^{\cdot\cdot a}_{bc} + \bar{\Gamma}^{\cdot\cdot a}_{de}\bar{\Gamma}^{\cdot\cdot e}_{bc}).\end{aligned} \tag{4.114}$$

Combining skew parts of (4.113) and (4.114),

$$\begin{aligned}0 &= 2\partial_{[b}\partial_{d]}\bar{F}^{-1\alpha}_{\cdot c}\\ &= 2\partial_{[d}\partial_{b]}\bar{F}^{-1\alpha}_{\cdot c}\\ &= 2\partial_{[d}(\bar{F}^{-1\alpha}_{\cdot |a|}\bar{\Gamma}^{\cdot\cdot a}_{b]c})\\ &= 2\bar{F}^{-1\alpha}_{\cdot a}(\partial_{[d}\bar{\Gamma}^{\cdot\cdot a}_{b]c} + \bar{\Gamma}^{\cdot\cdot a}_{[d|e|}\bar{\Gamma}^{\cdot\cdot e}_{b]c})\\ &= \bar{F}^{-1\alpha}_{\cdot a}\bar{R}^{\cdot\cdot\cdot a}_{dbc},\end{aligned} \tag{4.115}$$

where the Riemann-Christoffel curvature tensor of the connection coefficients defined in (4.112) is

$$\bar{R}^{\cdot\cdot\cdot a}_{bcd} = 2\partial_{[b}\bar{\Gamma}^{\cdot\cdot a}_{c]d} + 2\bar{\Gamma}^{\cdot\cdot a}_{[b|e|}\bar{\Gamma}^{\cdot\cdot e}_{c]d}. \tag{4.116}$$

Thus, local integrability conditions for the existence of an invertible differentiable map $\bar{F}^{a}_{\cdot\alpha}(x,t)$ yielding connection coefficients $\bar{\Gamma}^{\cdot\cdot a}_{bc}$ are [28]

$$\bar{R}^{\cdot\cdot\cdot a}_{bcd} = 0 \Leftrightarrow \bar{\Gamma}^{\cdot\cdot a}_{bc} = \bar{F}^{a}_{\cdot\alpha}\partial_b\bar{F}^{-1\alpha}_{\cdot c}. \tag{4.117}$$

When the torsion tensor of (4.112) defined as in (2.73) and (2.74),

$$\bar{T}^{\cdot\cdot a}_{bc} = \bar{\Gamma}^{\cdot\cdot a}_{[bc]} = \bar{F}^{a}_{\cdot\alpha}\partial_{[b}\bar{F}^{-1\alpha}_{\cdot c]}, \tag{4.118}$$

vanishes, it follows from (4.17) that deformation map $\bar{\boldsymbol{F}}^{-1}(x,t)$ itself is an integrable function of x^a, *i.e.*,

$$\bar{T}^{\cdot\cdot a}_{bc} = 0 \Leftrightarrow \bar{F}^{-1\alpha}_{\cdot a} = \partial_a\tilde{x}^{\alpha}. \tag{4.119}$$

Torsion tensor (4.118) is related to the anholonomic object of (4.97) by

$$\bar{T}^{\cdot\cdot a}_{bc} = \bar{F}^{a}_{\cdot\alpha}\bar{F}^{-1\beta}_{\cdot b}\bar{F}^{-1\chi}_{\cdot c}\bar{\kappa}^{\cdot\cdot\alpha}_{\beta\chi}. \tag{4.120}$$

Torsion $\bar{T}^{\cdot\cdot a}_{bc}$ is also important in dislocation theories of crystals [7].

4.1.6 *Contortion*

Let $\boldsymbol{Q}(X,t) = Q^{\alpha}_{.A}\tilde{\boldsymbol{g}}_{\alpha} \otimes \boldsymbol{G}^{A}$ be an invertible second-order, two-point tensor field continuously differentiable with respect to X^A. Furthermore, assume that $\boldsymbol{Q}$ obeys the orthonormality conditions

$$\boldsymbol{Q}^{\mathrm{T}} = \boldsymbol{Q}^{-1}, \qquad Q^{\mathrm{T}\,A}_{\ \ .\alpha} = Q^{\beta}_{.B} G^{AB} \tilde{g}_{\alpha\beta} = Q^{-1A}_{\ \ .\alpha}, \tag{4.121}$$

where the T superscript denotes transposition. The contortion field $K^{..A}_{BC}(X,t)$ associated with $\boldsymbol{Q}$ is defined [25] as the following connection coefficients, analogous to the integrable coefficients in the first two equalities of (4.103):

$$K^{..A}_{BC} = Q^{-1A}_{\ \ .\alpha} \partial_B Q^{\alpha}_{.C} = -Q^{\alpha}_{.C} \partial_B Q^{-1A}_{\ \ .\alpha}. \tag{4.122}$$

The torsion $S^{..A}_{BC}$ associated with the contortion is defined in the usual way for a linear connection referred to holonomic reference coordinates:

$$S^{..A}_{BC} = K^{..A}_{[BC]}. \tag{4.123}$$

From (4.121), contortion obeys the identities

$$\begin{aligned} K^{..A}_{BC} &= Q^{\mathrm{T}\,A}_{\ \ .\alpha} \partial_B Q^{\alpha}_{.C} \\ &= \tilde{g}_{\alpha\beta} G^{AD} Q^{\beta}_{.D} \partial_B Q^{\alpha}_{.C} \\ &= -Q^{\alpha}_{.C} \partial_B (\tilde{g}_{\alpha\beta} G^{AD} Q^{\beta}_{.D}) \\ &= -Q^{\alpha}_{.C} \partial_B Q^{\mathrm{T}\,A}_{\ \ .\alpha}. \end{aligned} \tag{4.124}$$

When spatially constant basis vectors, and hence constant metric tensors G_{AB} and $\tilde{g}_{\alpha\beta}$, are chosen for respective configurations B_0 and $\tilde{B}$, (4.124) can be written

$$\begin{aligned} K^{..A}_{BC} &= \tilde{g}_{\alpha\beta} G^{AD} Q^{\beta}_{.D} \partial_B Q^{\alpha}_{.C} \\ &= -\tilde{g}_{\alpha\beta} G^{AD} Q^{\alpha}_{.C} \partial_B Q^{\beta}_{.D} \\ &= -\tilde{g}_{\alpha\beta} G^{AD} Q^{\beta}_{.C} \partial_B Q^{\alpha}_{.D}. \end{aligned} \tag{4.125}$$

Defining a fully covariant version of the contortion as

$$K_{BCD} = K^{..A}_{BC} G_{AD}, \tag{4.126}$$

identity (4.125) becomes

$$K_{BCD} = Q_{\alpha D} \partial_B Q^{\alpha}_{.C} = -Q_{\alpha C} \partial_B Q^{\alpha}_{.D} = -K_{BDC} = K_{B[CD]}. \tag{4.127}$$

Defining the covariant torsion with (4.123) as

$$S_{BCD} = S^{..A}_{BC} G_{AD} = S_{[BC]D} = K_{[BC]D}, \tag{4.128}$$

verification is straightforward that torsion and contortion are related by [25]

$$\begin{aligned} K_{BCD} &= S_{BCD} - S_{CDB} - S_{BDC} \\ &= S_{BCD} - S_{CDB} + S_{DBC} \\ &= S_{\{BCD\}}, \end{aligned} \tag{4.129}$$

where notation in the final equality is introduced in (2.92). Thus when torsion vanishes then contortion must also vanish.

Consider the special case when $\boldsymbol{Q}(X,t)$ is an integrable field with respect to reference coordinates X^A. Then

$$Q^{\alpha}_{.A} = \partial_A \tilde{r}^{\alpha}, \tag{4.130}$$

where $\tilde{r}^{\alpha}(X,t)$ is a twice differentiable motion. In this case, again assuming constant metric tensors, the covariant contortion of (4.126) obeys

$$\begin{aligned} K_{BCD} &= Q_{\alpha D} \partial_B Q^{\alpha}_{.C} \\ &= (\partial_D \tilde{r}_{\alpha})(\partial_B \partial_C \tilde{r}^{\alpha}) \\ &= (\partial_D \tilde{r}_{\alpha})(\partial_C \partial_B \tilde{r}^{\alpha}) \\ &= K_{CBD}. \end{aligned} \tag{4.131}$$

It follows from (4.128) that torsion vanishes in this case, and hence from (4.129), contortion itself must also vanish. Since $\boldsymbol{Q}$ is invertible, it follows that $\boldsymbol{Q}$ must also be spatially constant, *e.g.*, a rigid body motion:

$$Q^{\alpha}_{.A} = \partial_A \tilde{r}^{\alpha} \Rightarrow S_{BCD} = 0 \Rightarrow K_{BCD} = 0 \Rightarrow \partial_B Q^{\alpha}_{.C} = 0. \tag{4.132}$$

Therefore, if an integrable deformation gradient field is a pure rotation [*i.e.*, $\boldsymbol{F} = \boldsymbol{R}$ in the polar decomposition $\boldsymbol{F} = \boldsymbol{RU}$ of (3.74)], then this rotation field must be homogeneous (*i.e.*, spatially constant). Furthermore, if $\boldsymbol{U}(X,t)$ is integrable with respect to X^A, then $\boldsymbol{R}$ is also integrable in the same local neighborhood by arguments analogous to (4.102). Thus, if stretch field $\boldsymbol{U}$ from an integrable deformation gradient field $F^a_{.A} = \partial_A x^a = R^a_{.B} U^B_{.A}$ is spatially constant (and thus necessarily integrable), then the corresponding rotation field $\boldsymbol{R}$ must also be spatially constant.

Now let $T_{BC}^{..A} = \Gamma_{[BC]}^{..A}$ be the torsion tensor of an arbitrary linear connection referred to holonomic reference coordinates, and define a general contortion tensor as

$$\begin{aligned} K_{ABC} &= T_{\{ABC\}} \\ &= T_{ABC} - T_{BCA} + T_{CAB} \\ &= -(-T_{CAB} + T_{BCA} - T_{ABC}) \\ &= -(T_{ACB} - T_{CBA} + T_{BAC}) \\ &= -T_{\{ACB\}} \\ &= -K_{ACB}, \end{aligned} \tag{4.133}$$

$$\begin{aligned}
K_{[AB]C} &= T_{[AB]C} - T_{[B|C|A]} + T_{C[AB]} \\
&= T_{[AB]C} + T_{C[BA]} + T_{C[AB]} \\
&= T_{[AB]C} + T_{C[BA]} - T_{C[BA]} \\
&= T_{[AB]C} \\
&= T_{ABC}.
\end{aligned} \tag{4.134}$$

Because torsion transforms under a change of coordinates as a tensor, contortion is also a tensor. Contortion can be substituted for terms involving the torsion in (2.94). The resulting connection becomes, when metric ($M_{BDC} = 0$), the sum of the Levi-Civita coefficients $\{{}^{..A}_{BC}\} = \{{}^{..A}_{CB}\}$ and the contortion:

$$\begin{aligned}
\Gamma^{..A}_{BC} &= \frac{1}{2} G^{AD}(2\chi_{\{BDC\}} - 2T_{\{BDC\}}) \\
&= \{{}^{..A}_{BC}\} - G^{AD} K_{BDC} \\
&= \{{}^{..A}_{BC}\} + G^{AD} K_{BCD} \\
&= \{{}^{..A}_{BC}\} + K^{..A}_{BC}.
\end{aligned} \tag{4.135}$$

Symmetric and skew parts of an otherwise arbitrary metric connection are then, respectively,

$$\Gamma^{..A}_{(BC)} = \{{}^{..A}_{BC}\} + K^{..A}_{(BC)}, \qquad \Gamma^{..A}_{[BC]} = K^{..A}_{[BC]}. \tag{4.136}$$

From (2.103) and (2.105), the Riemann-Christoffel curvature tensor of the total connection $\Gamma^{..A}_{BC}$ obeys

$$\begin{aligned}
R^{...A}_{BCD} &= 2\partial_{[B}\Gamma^{..A}_{C]D} + 2\Gamma^{..A}_{[B|E|}\Gamma^{..E}_{C]D} \\
&= \bar{R}^{...A}_{BCD} + 2\nabla_{[B}K^{..A}_{C]D} + 2T^{..E}_{BC}K^{..A}_{ED} + 2K^{..E}_{[B|D|}K^{..A}_{C]E} \\
&= \bar{R}^{...A}_{BCD} + 2\nabla_{[B}K^{..A}_{C]D} + 2K^{..E}_{[BC]}K^{..A}_{ED} + 2K^{..E}_{[B|D|}K^{..A}_{C]E} \\
&= \bar{R}^{...A}_{BCD} + 2\bar{\nabla}_{[B}K^{..A}_{C]D} + 2\{{}^{..E}_{[BC]}\}K^{..A}_{ED} - 2K^{..E}_{[B|D|}K^{..A}_{C]E} \\
&= \bar{R}^{...A}_{BCD} + 2\bar{\nabla}_{[B}K^{..A}_{C]D} - 2K^{..E}_{[B|D|}K^{..A}_{C]E},
\end{aligned} \tag{4.137}$$

where $\bar{R}^{...A}_{BCD}$ is the curvature tensor of the Levi-Civita connection $\{{}^{..A}_{BC}\}$, $\nabla(\cdot)$ is the covariant derivative with respect to total connection $\Gamma^{..A}_{BC}$, and $\bar{\nabla}(\cdot)$ is the covariant derivative with respect to $\{{}^{..A}_{BC}\}$. Therefore, if the Levi-Civita connection is flat ($\bar{R}^{...A}_{BCD} = 0$), the curvature of $\Gamma^{..A}_{BC}$ results only from the contortion and its covariant derivative. If torsion vanishes ($T^{..A}_{BC} = \Gamma^{..A}_{[BC]} = 0$), then by (4.133), contortion also vanishes ($K^{..A}_{BC} = 0$), so in that case the total curvature reduces to the contribution from the Levi-Civita connection, *i.e.*, $R^{...A}_{BCD} = \bar{R}^{...A}_{BCD}$.

4.2 Anholonomic Covariant Derivatives

Differentiation with respect to general anholonomic coordinates is developed. Covariant derivatives and the corresponding connection coefficients are defined. Various choices of anholonomic connection coefficients corresponding to different basis vectors in the intermediate configuration are examined. Total covariant derivatives of two- (and three-) point tensors with one or more indices referred to anholonomic space are defined. Gradient, divergence, curl, and Laplacian operations and the corresponding identities are presented.

4.2.1 *Differentiation*

Covariant differentiation with respect to anholonomic coordinates is defined similarly to (2.64). Let $\boldsymbol{A}$ be a vector or higher-order tensor field with components $A^{\alpha\ldots\phi}_{\gamma\ldots\mu}$. The covariant derivative of $\boldsymbol{A}$ is computed as

$$\begin{aligned}\nabla_\nu A^{\alpha\ldots\phi}_{\gamma\ldots\mu} = \partial_\nu A^{\alpha\ldots\phi}_{\gamma\ldots\mu} &+ \Gamma^{\cdot\cdot\alpha}_{\nu\rho} A^{\rho\ldots\phi}_{\gamma\ldots\mu} + \cdots + \Gamma^{\cdot\cdot\phi}_{\nu\rho} A^{\alpha\ldots\rho}_{\gamma\ldots\mu} \\ &- \Gamma^{\cdot\cdot\rho}_{\nu\gamma} A^{\alpha\ldots\phi}_{\rho\ldots\mu} - \cdots - \Gamma^{\cdot\cdot\rho}_{\nu\mu} A^{\alpha\ldots\phi}_{\gamma\ldots\rho}.\end{aligned} \tag{4.138}$$

Partial differentiation obeys (4.20), *i.e.*,

$$\partial_\nu(\cdot) = \partial_A(\cdot)\tilde{F}^{-1A}_{\ \ \cdot\nu} = \partial_a(\cdot)\bar{F}^{a}_{\cdot\nu}. \tag{4.139}$$

Connection coefficients referred to intermediate space $\tilde{B}$, written as $\Gamma^{\cdot\cdot\alpha}_{\nu\rho}$, in general consist of up to n^3 arbitrary entries, where n is the dimensionality of Euclidean spaces B_0 and B.

The most general form of anholonomic connection coefficients is given in (4.83). Using (2.58) and modern notation of (4.64), connection coefficients associated with a general connection ∇ can be defined as the following gradients of basis vectors:

$$\Gamma^{\cdot\cdot\alpha}_{\beta\chi}\boldsymbol{\partial}_\alpha = \nabla_{\boldsymbol{\partial}_\beta}\boldsymbol{\partial}_\chi. \tag{4.140}$$

Application of this definition to (4.65) leads to

$$\begin{aligned}0 &= \nabla_{\boldsymbol{\partial}_\alpha}\delta^\beta_\chi \\ &= \nabla_{\boldsymbol{\partial}_\alpha}\langle\boldsymbol{\partial}_\chi, \boldsymbol{d}\tilde{x}^\beta\rangle \\ &= \langle\nabla_{\boldsymbol{\partial}_\alpha}\boldsymbol{\partial}_\chi, \boldsymbol{d}\tilde{x}^\beta\rangle + \langle\boldsymbol{\partial}_\chi, \nabla_{\boldsymbol{\partial}_\alpha}\boldsymbol{d}\tilde{x}^\beta\rangle \\ &= \Gamma^{\cdot\cdot\delta}_{\alpha\chi}\langle\boldsymbol{\partial}_\delta, \boldsymbol{d}\tilde{x}^\beta\rangle + \langle\boldsymbol{\partial}_\chi, \nabla_{\boldsymbol{\partial}_\alpha}\boldsymbol{d}\tilde{x}^\beta\rangle \\ &= \Gamma^{\cdot\cdot\beta}_{\alpha\chi} + \langle\boldsymbol{\partial}_\chi, \nabla_{\boldsymbol{\partial}_\alpha}\boldsymbol{d}\tilde{x}^\beta\rangle,\end{aligned} \tag{4.141}$$

such that gradients of reciprocal basis vectors obey

$$\nabla_{\boldsymbol{\partial}_\alpha} \boldsymbol{d}\tilde{x}^\beta = -\Gamma^{\cdot\cdot\beta}_{\alpha\chi} \boldsymbol{d}\tilde{x}^\chi. \tag{4.142}$$

The covariant derivative of a differentiable vector field $\boldsymbol{A} = A^\alpha \boldsymbol{\partial}_\alpha$ can then be computed as

$$\begin{aligned} \nabla \boldsymbol{A} &= \nabla_{\boldsymbol{\partial}_\beta} \boldsymbol{A} \otimes \boldsymbol{d}\tilde{x}^\beta \\ &= \nabla_{\boldsymbol{\partial}_\beta} (A^\alpha \boldsymbol{\partial}_\alpha) \otimes \boldsymbol{d}\tilde{x}^\beta \\ &= \nabla_{\boldsymbol{\partial}_\beta} (A^\alpha) \boldsymbol{\partial}_\alpha \otimes \boldsymbol{d}\tilde{x}^\beta + A^\alpha \nabla_{\boldsymbol{\partial}_\beta} (\boldsymbol{\partial}_\alpha) \otimes \boldsymbol{d}\tilde{x}^\beta \\ &= \partial_\beta A^\alpha \boldsymbol{\partial}_\alpha \otimes \boldsymbol{d}\tilde{x}^\beta + A^\alpha \Gamma^{\cdot\cdot\chi}_{\beta\alpha} \boldsymbol{\partial}_\chi \otimes \boldsymbol{d}\tilde{x}^\beta \\ &= (\partial_\beta A^\alpha + A^\chi \Gamma^{\cdot\cdot\alpha}_{\beta\chi}) \boldsymbol{\partial}_\alpha \otimes \boldsymbol{d}\tilde{x}^\beta, \end{aligned} \tag{4.143}$$

where the notation $\nabla_{\boldsymbol{\partial}_\alpha}(\cdot) = \partial_\alpha(\cdot)$ when the argument is a scalar or scalar component of a vector or tensor. Outcomes of (4.138) and (4.143) are fully consistent.

Henceforward, coefficients $\Gamma^{\cdot\cdot\alpha}_{\beta\chi}$ in (4.138) are restricted, by definition, to obey the following relationships that result in a metric connection. Partial derivatives of intermediate basis vectors and their reciprocals are, by definition,

$$\partial_\beta \tilde{\boldsymbol{g}}_\alpha = \Gamma^{\cdot\cdot\chi}_{\beta\alpha} \tilde{\boldsymbol{g}}_\chi, \qquad \partial_\beta \tilde{\boldsymbol{g}}^\alpha = -\Gamma^{\cdot\cdot\alpha}_{\beta\chi} \tilde{\boldsymbol{g}}^\chi. \tag{4.144}$$

Therefore, symbolic covariant derivatives of intermediate basis vectors vanish identically according to the following notational scheme, analogous to that used in (2.125) and (2.126):

$$\nabla_\beta \tilde{\boldsymbol{g}}_\alpha = \partial_\beta \tilde{\boldsymbol{g}}_\alpha - \Gamma^{\cdot\cdot\chi}_{\beta\alpha} \tilde{\boldsymbol{g}}_\chi = 0, \qquad \nabla_\beta \tilde{\boldsymbol{g}}^\alpha = \partial_\beta \tilde{\boldsymbol{g}}^\alpha + \Gamma^{\cdot\cdot\alpha}_{\beta\chi} \tilde{\boldsymbol{g}}^\chi = 0. \tag{4.145}$$

For example, applying (4.138)-(4.144) to a generic contravariant vector field $\boldsymbol{A} = A^\alpha \tilde{\boldsymbol{g}}_\alpha$ gives the gradient of $\boldsymbol{A}$:

$$\begin{aligned} \partial_\beta \boldsymbol{A} \otimes \tilde{\boldsymbol{g}}^\beta &= \partial_\beta (A^\alpha \tilde{\boldsymbol{g}}_\alpha) \otimes \tilde{\boldsymbol{g}}^\beta \\ &= (\partial_\beta A^\alpha \tilde{\boldsymbol{g}}_\alpha + A^\alpha \partial_\beta \tilde{\boldsymbol{g}}_\alpha) \otimes \tilde{\boldsymbol{g}}^\beta \\ &- (\partial_\beta A^\alpha \tilde{\boldsymbol{g}}_\alpha + A^\alpha \Gamma^{\cdot\cdot\chi}_{\beta\alpha} \tilde{\boldsymbol{g}}_\chi) \otimes \tilde{\boldsymbol{g}}^\beta \\ &= (\partial_\beta A^\alpha + A^\chi \Gamma^{\cdot\cdot\alpha}_{\beta\chi}) \tilde{\boldsymbol{g}}_\alpha \otimes \tilde{\boldsymbol{g}}^\beta \\ &= \nabla_\beta A^\alpha \tilde{\boldsymbol{g}}_\alpha \otimes \tilde{\boldsymbol{g}}^\beta. \end{aligned} \tag{4.146}$$

By definition, the covariant derivative (*i.e.*, gradient) of a scalar field is the same as its partial derivative, *e.g.*,

$$\nabla_\alpha A = \partial_\alpha A = \tilde{F}^{-1B}_{\cdot\alpha} \partial_B A = \bar{F}^{b}_{\cdot\alpha} \partial_b A. \tag{4.147}$$

The following identity also applies analogously to (2.67):

$$\begin{aligned}\nabla_\alpha \delta^\beta_\chi &= \partial_\alpha \delta^\beta_\chi + \Gamma^{\cdot\cdot\beta}_{\alpha\delta}\delta^\delta_\chi - \Gamma^{\cdot\cdot\delta}_{\alpha\chi}\delta^\beta_\delta \\ &= \tilde{F}^{-1A}_{\cdot\alpha}\partial_A \delta^\beta_\chi + \Gamma^{\cdot\cdot\beta}_{\alpha\chi} - \Gamma^{\cdot\cdot\beta}_{\alpha\chi} \\ &= 0.\end{aligned} \tag{4.148}$$

From definitions in (4.144), the partial anholonomic derivative of the corresponding metric tensor of (4.27) is

$$\begin{aligned}\partial_\chi \tilde{g}_{\alpha\beta} &= \partial_\chi(\tilde{\boldsymbol{g}}_\alpha \cdot \tilde{\boldsymbol{g}}_\beta) \\ &= \tilde{\boldsymbol{g}}_\alpha \cdot \partial_\chi \tilde{\boldsymbol{g}}_\beta + \tilde{\boldsymbol{g}}_\beta \cdot \partial_\chi \tilde{\boldsymbol{g}}_\alpha \\ &= \tilde{\boldsymbol{g}}_\alpha \cdot \Gamma^{\cdot\cdot\delta}_{\chi\beta}\tilde{\boldsymbol{g}}_\delta + \tilde{\boldsymbol{g}}_\beta \cdot \Gamma^{\cdot\cdot\delta}_{\chi\alpha}\tilde{\boldsymbol{g}}_\delta \\ &= \Gamma^{\cdot\cdot\delta}_{\chi\beta}\tilde{g}_{\alpha\delta} + \Gamma^{\cdot\cdot\delta}_{\chi\alpha}\tilde{g}_{\beta\delta} \\ &= 2\,\Gamma^{\cdot\cdot\delta}_{\chi(\alpha}\tilde{g}_{\beta)\delta}.\end{aligned} \tag{4.149}$$

The partial derivative of its determinant is then, from (2.142),

$$\begin{aligned}\partial_\alpha \tilde{g} &= \partial_\alpha \det \tilde{\boldsymbol{g}} \\ &= \tilde{g}\,\tilde{g}^{\beta\chi}\partial_\alpha \tilde{g}_{\beta\chi} \\ &= \tilde{g}\tilde{g}^{\beta\chi}(\Gamma^{\cdot\cdot\delta}_{\alpha\beta}\tilde{g}_{\chi\delta} + \Gamma^{\cdot\cdot\delta}_{\alpha\chi}\tilde{g}_{\beta\delta}) \\ &= 2\tilde{g}\,\Gamma^{\cdot\cdot\delta}_{\alpha\delta}.\end{aligned} \tag{4.150}$$

Related identities are

$$\partial_\alpha \sqrt{\tilde{g}} = \sqrt{\tilde{g}}\,\Gamma^{\cdot\cdot\beta}_{\alpha\beta}, \qquad \partial_\alpha(\ln\sqrt{\tilde{g}}) = \Gamma^{\cdot\cdot\beta}_{\alpha\beta}. \tag{4.151}$$

As

$$\Gamma^{\cdot\cdot\beta}_{\alpha\beta} = \Gamma^{\cdot\cdot\beta}_{\alpha\beta} - \Gamma^{\cdot\cdot\beta}_{\beta\alpha} + \Gamma^{\cdot\cdot\beta}_{\beta\alpha} = 2\Gamma^{\cdot\cdot\beta}_{[\alpha\beta]} + \Gamma^{\cdot\cdot\beta}_{\beta\alpha}, \tag{4.152}$$

an identity like the last of (2.144) does not apply unless $\Gamma^{\cdot\cdot\beta}_{[\alpha\beta]} = 0$.

Connection coefficients of (4.144) are metric since the covariant derivative of the metric tensor vanishes by (4.149):

$$\begin{aligned}\nabla_\chi \tilde{g}_{\alpha\beta} &= \partial_\chi \tilde{g}_{\alpha\beta} - \Gamma^{\cdot\cdot\delta}_{\chi\alpha}\tilde{g}_{\delta\beta} - \Gamma^{\cdot\cdot\delta}_{\chi\beta}\tilde{g}_{\alpha\delta} \\ &= \partial_\chi \tilde{g}_{\alpha\beta} - 2\,\Gamma^{\cdot\cdot\delta}_{\chi(\alpha}\tilde{g}_{\beta)\delta} \\ &= 0.\end{aligned} \tag{4.153}$$

From similar arguments, for the inverse of the intermediate metric,

$$\partial_\chi \tilde{g}^{\alpha\beta} = -2\,\Gamma^{\cdot\cdot(\alpha}_{\chi\delta}\tilde{g}^{\beta)\delta}, \qquad \nabla_\chi \tilde{g}^{\alpha\beta} = 0. \tag{4.154}$$

Therefore, covariant differentiation via ∇_χ commutes with lowering (raising) indices via $\tilde{g}_{\alpha\beta}$ ($\tilde{g}^{\alpha\beta}$). Conditions (4.153) and (4.154) are necessary but apparently not sufficient for (4.144) to always apply.

Because the connection is metric, identities for permutation tensors analogous to (2.146) and (2.147) apply. For the covariant permutation tensor,

$$\begin{aligned}\nabla_\delta \epsilon_{\alpha\beta\chi} &= \partial_\delta \epsilon_{\alpha\beta\chi} - \Gamma^{..\epsilon}_{\delta\alpha}\epsilon_{\epsilon\beta\chi} - \Gamma^{..\epsilon}_{\delta\beta}\epsilon_{\alpha\epsilon\chi} - \Gamma^{..\epsilon}_{\delta\chi}\epsilon_{\alpha\beta\epsilon} \\ &= \partial_\delta \epsilon_{\alpha\beta\chi} - \Gamma^{..\epsilon}_{\delta\epsilon}\epsilon_{\alpha\beta\chi} \\ &= \partial_\delta(\sqrt{\tilde{g}})e_{\alpha\beta\chi} - \partial_\delta(\ln\sqrt{\tilde{g}})\epsilon_{\alpha\beta\chi} \\ &= \partial_\delta(\ln\sqrt{\tilde{g}})\epsilon_{\alpha\beta\chi} - \partial_\delta(\ln\sqrt{\tilde{g}})\epsilon_{\alpha\beta\chi} \\ &= 0.\end{aligned} \tag{4.155}$$

Similarly, for the contravariant permutation tensor,

$$\begin{aligned}\nabla_\delta \epsilon^{\alpha\beta\chi} &= \partial_\delta \epsilon^{\alpha\beta\chi} + \Gamma^{..\alpha}_{\delta\epsilon}\epsilon^{\epsilon\beta\chi} + \Gamma^{..\beta}_{\delta\epsilon}\epsilon^{\alpha\epsilon\chi} + \Gamma^{..\chi}_{\delta\epsilon}\epsilon^{\alpha\beta\epsilon} \\ &= \partial_\delta \epsilon^{\alpha\beta\chi} + \Gamma^{..\epsilon}_{\delta\epsilon}\epsilon^{\alpha\beta\chi} \\ &= \partial_\delta(1/\sqrt{\tilde{g}})e^{\alpha\beta\chi} + \partial_\delta(\ln\sqrt{\tilde{g}})\epsilon^{\alpha\beta\chi} \\ &= -\partial_\delta(\ln\sqrt{\tilde{g}})\epsilon^{\alpha\beta\chi} + \partial_\delta(\ln\sqrt{\tilde{g}})\epsilon^{\alpha\beta\chi} \\ &= 0.\end{aligned} \tag{4.156}$$

If the covariant derivative of the determinant $\tilde{g} = \det(\tilde{g}_{\alpha\beta})$ is defined by application of the chain rule as [8]

$$\nabla_\alpha \tilde{g} = \frac{\partial \tilde{g}}{\partial \tilde{g}_{\beta\chi}} \nabla_\alpha \tilde{g}_{\beta\chi} = \tilde{g}\tilde{g}^{\chi\beta}\nabla_\alpha \tilde{g}_{\beta\chi}, \tag{4.157}$$

then $\nabla_\alpha \tilde{g} = 0$ follows from (4.153).

Recall from (4.99) that the anholonomic object, denoted now generically by $\kappa^{..\chi}_{\beta\alpha}$, obeys

$$\begin{aligned}\kappa^{..\chi}_{\beta\alpha} &= \tilde{F}^{-1B}_{\;\;.\beta}\tilde{F}^{-1A}_{\;\;.\alpha}\partial_{[B}\tilde{F}^{\chi}_{.A]} \\ &= \tilde{F}^{-1B}_{\;\;.[\beta}\tilde{F}^{-1A}_{\;\;.\alpha]}\partial_B \tilde{F}^{\chi}_{.A} \\ &= -\tilde{F}^{\chi}_{.A}\tilde{F}^{-1B}_{\;\;.[\beta}\partial_{|B|}\tilde{F}^{-1A}_{\;\;.\alpha]} \\ &= -\tilde{F}^{\chi}_{.A}\partial_{[\beta}\tilde{F}^{-1A}_{\;\;.\alpha]} \\ &= \bar{F}^{b}_{.\beta}\bar{F}^{a}_{.\alpha}\partial_{[b}\bar{F}^{-1\chi}_{\;\;.a]} \\ &= \bar{F}^{b}_{.[\beta}\bar{F}^{a}_{.\alpha]}\partial_b \bar{F}^{-1\chi}_{\;\;.a} \\ &= -\bar{F}^{-1\chi}_{\;\;.a}\bar{F}^{b}_{.[\beta}\partial_{|b|}\bar{F}^{a}_{.\alpha]} \\ &= -\bar{F}^{-1\chi}_{\;\;.a}\partial_{[\beta}\bar{F}^{a}_{.\alpha]}.\end{aligned} \tag{4.158}$$

The intermediate torsion, denoted now generically by $T^{..\chi}_{\beta\alpha}$, is defined using only the final two equalities in (4.68):

$$T^{..\alpha}_{\beta\chi} = \Gamma^{..\alpha}_{[\beta\chi]} + \kappa^{..\alpha}_{\beta\chi} = T^{..\alpha}_{[\beta\chi]}. \tag{4.159}$$

Here, the first equality in (4.68) is not required to hold for this generic definition of the torsion of an anholonomic space [*i.e.*, the torsion in (4.159) need not map between configurations as a true tensor]. Similarly, the intermediate curvature, denoted now generically by $R_{\beta\chi\delta}^{\cdot\cdot\cdot\alpha}$, is defined using only the final two equalities in (4.71):

$$R_{\beta\chi\delta}^{\cdot\cdot\cdot\alpha} = 2\partial_{[\beta}\Gamma_{\chi]\delta}^{\cdot\cdot\alpha} + 2\Gamma_{[\beta|\epsilon|}^{\cdot\cdot\alpha}\Gamma_{\chi]\delta}^{\cdot\cdot\epsilon} + 2\kappa_{\beta\chi}^{\cdot\cdot\epsilon}\Gamma_{\epsilon\delta}^{\cdot\cdot\alpha} = R_{[\beta\chi]\delta}^{\cdot\cdot\cdot\alpha}. \tag{4.160}$$

Here, the first equality in (4.71) is not required to hold for this generic definition of the curvature of an anholonomic space [*i.e.*, the curvature in (4.160) need not map between configurations as a true tensor].

With definitions (4.158), (4.159), and (4.160) now in place, skew gradients can be obtained. Specifically, from (4.76),

$$\partial_{[\alpha}[\partial_{\beta]}(\cdot)] = -\kappa_{\alpha\beta}^{\cdot\cdot\chi}\partial_{\chi}(\cdot). \tag{4.161}$$

From (4.77), for a twice differentiable scalar field A,

$$\nabla_{[\alpha}(\nabla_{\beta]}A) = -T_{\alpha\beta}^{\cdot\cdot\chi}\partial_{\chi}A. \tag{4.162}$$

From (4.73) and (4.74),

$$\nabla_{[\beta}\nabla_{\chi]}V^{\alpha} = \frac{1}{2}R_{\beta\chi\delta}^{\cdot\cdot\cdot\alpha}V^{\delta} - T_{\beta\chi}^{\cdot\cdot\delta}\nabla_{\delta}V^{\alpha}, \tag{4.163}$$

$$\nabla_{[\beta}\nabla_{\chi]}\alpha_{\delta} = -\frac{1}{2}R_{\beta\chi\delta}^{\cdot\cdot\cdot\alpha}\alpha_{\alpha} - T_{\beta\chi}^{\cdot\cdot\alpha}\nabla_{\alpha}\alpha_{\delta}, \tag{4.164}$$

where V^{α} and α_{δ} denote components of twice differentiable vector and covector fields, respectively, referred to intermediate configuration $\tilde{B}$.

4.2.2 *Anholonomic connection coefficients*

Four particular classes of choices for intermediate/anholonomic connection coefficients $\Gamma_{\beta\chi}^{\cdot\cdot\alpha}$ entering (4.138) and (4.144) are discussed in what follows.

First consider the case when (4.13) applies, such that $\tilde{B}$ can be regarded as a Euclidean space and a position vector $\tilde{\boldsymbol{x}}$ can be assigned to any point $\tilde{x}(X,t) \in \tilde{B}$. In that case, basis vectors and metric tensor components can be defined as in (4.47) and (4.48), and any time-independent Euclidean coordinate system (*e.g.*, fixed curvilinear or Cartesian coordinates) can be used for $\tilde{x}^{\alpha}$. Coefficients of the Levi-Civita connection on $\tilde{B}$ are, analogously to (2.122),

$$\Gamma_{\beta\chi}^{\cdot\cdot\alpha}(\tilde{x}) = \overset{\tilde{g}}{\Gamma}{}_{\beta\chi}^{\cdot\cdot\alpha} = \frac{1}{2}\tilde{g}^{\alpha\delta}(\partial_{\beta}\tilde{g}_{\chi\delta} + \partial_{\chi}\tilde{g}_{\beta\delta} - \partial_{\delta}\tilde{g}_{\beta\chi}) = \overset{\tilde{g}}{\Gamma}{}_{\chi\beta}^{\cdot\cdot\alpha}. \tag{4.165}$$

Since $\tilde{x}^\alpha$ are holonomic coordinates in Euclidean space, the anholonomic object, torsion, and curvature formed from connection coefficients (4.165) all vanish identically. According to (4.153) or to more specific definition (4.165), covariant derivative $\nabla_\chi \tilde{g}_{\alpha\beta}$ vanishes identically as well.

Next consider the case when (4.14) applies, such that single-valued coordinates $\tilde{x}^\alpha(X,t)$ continuously differentiable with respect to X^A do not exist in $\tilde{B}$. When, following (4.49), identical coordinate systems are used in configurations B_0 and $\tilde{B}$:

$$\tilde{\boldsymbol{g}}_\alpha(X) = \delta^A_\alpha \boldsymbol{G}_A, \tag{4.166}$$

$$\tilde{\boldsymbol{g}}^\alpha(X) = \tilde{g}^{\alpha\beta}\tilde{\boldsymbol{g}}_\beta = \delta^\alpha_C \delta^\beta_D G^{CD} \delta^A_\beta \boldsymbol{G}_A = \delta^\alpha_C G^{CA} \boldsymbol{G}_A = \delta^\alpha_A \boldsymbol{G}^A, \tag{4.167}$$

$$\tilde{g}_{\alpha\beta}(X) = \delta^A_\alpha \delta^B_\beta G_{AB}, \qquad \tilde{g}^{\alpha\beta}(X) = \delta^\alpha_A \delta^\beta_B G^{AB}. \tag{4.168}$$

Taking the partial derivative of (4.166) and applying (2.125) and (4.15) gives

$$\begin{aligned}
\partial_\beta \tilde{\boldsymbol{g}}_\alpha &= \partial_\beta(\delta^A_\alpha \boldsymbol{G}_A) \\
&= \delta^A_\alpha \partial_\beta \boldsymbol{G}_A \\
&= \delta^A_\alpha \tilde{F}^{-1B}_{\cdot\beta} \partial_B \boldsymbol{G}_A \\
&= \delta^A_\alpha \tilde{F}^{-1B}_{\cdot\beta} \overset{G}{\Gamma}{}^{\cdot\cdot C}_{BA} \boldsymbol{G}_C \\
&= \delta^A_\alpha \tilde{F}^{-1B}_{\cdot\beta} \overset{G}{\Gamma}{}^{\cdot\cdot C}_{BA} \delta^\chi_C \tilde{\boldsymbol{g}}_\chi \\
&= \Gamma^{\cdot\cdot\chi}_{\beta\alpha} \tilde{\boldsymbol{g}}_\chi.
\end{aligned} \tag{4.169}$$

Similarly, taking the partial derivative of (4.167) and applying (2.126) and (4.15) gives

$$\begin{aligned}
\partial_\beta \tilde{\boldsymbol{g}}^\alpha &= \partial_\beta(\delta^\alpha_A \boldsymbol{G}^A) \\
&= \delta^\alpha_A \partial_\beta \boldsymbol{G}^A \\
&= \delta^\alpha_A \tilde{F}^{-1B}_{\cdot\beta} \partial_B \boldsymbol{G}^A \\
&- -\delta^\alpha_A \tilde{F}^{-1B}_{\cdot\beta} \overset{G}{\Gamma}{}^{\cdot\cdot A}_{BC} \boldsymbol{G}^C \\
&= -\delta^\alpha_A \tilde{F}^{-1B}_{\cdot\beta} \overset{G}{\Gamma}{}^{\cdot\cdot A}_{BC} \delta^C_\chi \tilde{\boldsymbol{g}}^\chi \\
&= -\Gamma^{\cdot\cdot\alpha}_{\beta\chi} \tilde{\boldsymbol{g}}^\chi.
\end{aligned} \tag{4.170}$$

Comparing (4.144), (4.169), and (4.170), connection coefficients consistent with (4.49) are time dependent:

$$\Gamma^{\cdot\cdot\alpha}_{\beta\chi}(X,t) = \delta^\alpha_A \delta^C_\chi \overset{G}{\Gamma}{}^{\cdot\cdot A}_{BC}(X) \tilde{F}^{-1B}_{\cdot\beta}(X,t). \tag{4.171}$$

The covariant derivative (*i.e.*, gradient) of a differentiable vector field $\boldsymbol{V} = V^\alpha \tilde{\boldsymbol{g}}_\alpha$ is computed in this case as

$$\begin{aligned}
\nabla \boldsymbol{V} &= \partial_\beta \boldsymbol{V} \otimes \tilde{\boldsymbol{g}}^\beta \\
&= \partial_\beta (V^\alpha \tilde{\boldsymbol{g}}_\alpha) \otimes \tilde{\boldsymbol{g}}^\beta \\
&= (\partial_\beta V^\alpha + \Gamma^{\cdot\cdot\alpha}_{\beta\chi} V^\chi) \tilde{\boldsymbol{g}}_\alpha \otimes \tilde{\boldsymbol{g}}^\beta \\
&= (\partial_\beta V^\alpha + \tilde{F}^{-1B}_{\ \ \cdot\beta} \delta^\alpha_A \delta^C_\chi \overset{G}{\Gamma}{}^{\cdot\cdot A}_{BC} V^\chi) \tilde{\boldsymbol{g}}_\alpha \otimes \tilde{\boldsymbol{g}}^\beta \\
&= \tilde{F}^{-1B}_{\ \ \cdot\beta} (\partial_B V^\alpha + \delta^\alpha_A \delta^C_\chi \overset{G}{\Gamma}{}^{\cdot\cdot A}_{BC} V^\chi) \tilde{\boldsymbol{g}}_\alpha \otimes \tilde{\boldsymbol{g}}^\beta \\
&= \tilde{F}^{-1B}_{\ \ \cdot\beta} (\delta^\alpha_A \partial_B V^A + \delta^\alpha_A \delta^C_\chi \delta^\chi_D \overset{G}{\Gamma}{}^{\cdot\cdot A}_{BC} V^D) \tilde{\boldsymbol{g}}_\alpha \otimes \tilde{\boldsymbol{g}}^\beta \\
&= \tilde{F}^{-1B}_{\ \ \cdot\beta} \delta^\alpha_A (\partial_B V^A + \overset{G}{\Gamma}{}^{\cdot\cdot A}_{BC} V^C) \tilde{\boldsymbol{g}}_\alpha \otimes \tilde{\boldsymbol{g}}^\beta \\
&= V^A_{;B} \tilde{F}^{-1B}_{\ \ \cdot\beta} \boldsymbol{G}_A \otimes \tilde{\boldsymbol{g}}^\beta \\
&= V^A_{;B} \tilde{F}^{-1B}_{\ \ \cdot\beta} \delta^\beta_C \boldsymbol{G}_A \otimes \boldsymbol{G}^C,
\end{aligned} \tag{4.172}$$

where (4.53) has been used for the shifter such that $V^\alpha = V^A g^\alpha_A = V^A \delta^\alpha_A$. Similarly, the covariant derivative of a covector field $\boldsymbol{\alpha} = \alpha_\alpha \tilde{\boldsymbol{g}}^\alpha$ is computed as

$$\begin{aligned}
\nabla \boldsymbol{\alpha} &= \partial_\beta \boldsymbol{\alpha} \otimes \tilde{\boldsymbol{g}}^\beta \\
&= \partial_\beta (\alpha_\alpha \tilde{\boldsymbol{g}}^\alpha) \otimes \tilde{\boldsymbol{g}}^\beta \\
&= (\partial_\beta \alpha_\alpha - \Gamma^{\cdot\cdot\chi}_{\beta\alpha} \alpha_\chi) \tilde{\boldsymbol{g}}^\alpha \otimes \tilde{\boldsymbol{g}}^\beta \\
&= (\partial_\beta \alpha_\alpha - \tilde{F}^{-1B}_{\ \ \cdot\beta} \delta^A_\alpha \delta^\chi_C \overset{G}{\Gamma}{}^{\cdot\cdot C}_{BA} \alpha_\chi) \tilde{\boldsymbol{g}}^\alpha \otimes \tilde{\boldsymbol{g}}^\beta \\
&= \tilde{F}^{-1B}_{\ \ \cdot\beta} (\partial_B \alpha_\alpha - \delta^A_\alpha \delta^\chi_C \overset{G}{\Gamma}{}^{\cdot\cdot C}_{BA} \alpha_\chi) \tilde{\boldsymbol{g}}^\alpha \otimes \tilde{\boldsymbol{g}}^\beta \\
&= \tilde{F}^{-1B}_{\ \ \cdot\beta} (\delta^A_\alpha \partial_B \alpha_A - \delta^A_\alpha \delta^\chi_C \delta^D_\chi \overset{G}{\Gamma}{}^{\cdot\cdot C}_{BA} \alpha_D) \tilde{\boldsymbol{g}}^\alpha \otimes \tilde{\boldsymbol{g}}^\beta \\
&= \tilde{F}^{-1B}_{\ \ \cdot\beta} \delta^A_\alpha (\partial_B \alpha_A - \overset{G}{\Gamma}{}^{\cdot\cdot C}_{BA} \alpha_C) \tilde{\boldsymbol{g}}^\alpha \otimes \tilde{\boldsymbol{g}}^\beta \\
&= \alpha_{A;B} \tilde{F}^{-1B}_{\ \ \cdot\beta} \boldsymbol{G}^A \otimes \tilde{\boldsymbol{g}}^\beta \\
&= \alpha_{A;B} \tilde{F}^{-1B}_{\ \ \cdot\beta} \delta^\beta_C \boldsymbol{G}^A \otimes \boldsymbol{G}^C.
\end{aligned} \tag{4.173}$$

By extension, components of the covariant derivative of a tensor field of arbitrary order with components $A^{\alpha\ldots\phi}_{\gamma\ldots\mu}$ are found as

$$\nabla_\nu A^{\alpha\ldots\phi}_{\gamma\ldots\mu} = \tilde{F}^{-1N}_{\ \ \cdot\nu} (\delta^\alpha_A \cdots \delta^\phi_F)(\delta^G_\gamma \cdots \delta^M_\mu) A^{A\ldots F}_{G\ldots M;N}. \tag{4.174}$$

Since Levi-Civita connection coefficients $\overset{G}{\Gamma}{}^{\cdot\cdot A}_{BC}$ are symmetric in covariant components, it follows that $\Gamma^{\cdot\cdot\alpha}_{\beta\chi}$ defined in (4.171) obeys

$$\begin{aligned}
\Gamma^{\cdot\cdot\alpha}_{\beta\chi} &= \delta^\alpha_A \delta^C_\chi \overset{G}{\Gamma}{}^{\cdot\cdot A}_{BC} \tilde{F}^{-1B}_{\ \ \cdot\beta} \\
&= \delta^\alpha_A \delta^C_\chi \overset{G}{\Gamma}{}^{\cdot\cdot A}_{CB} \tilde{F}^{-1B}_{\ \ \cdot\beta} \\
&= \delta^\alpha_A \delta^B_\chi \overset{G}{\Gamma}{}^{\cdot\cdot A}_{CB} \tilde{F}^{-1C}_{\ \ \cdot\beta}.
\end{aligned} \tag{4.175}$$

Covariant indices of $\Gamma^{\cdot\cdot\alpha}_{\beta\chi}$ are generally not symmetric; the left covariant component corresponding to differentiation by $\nabla_{\beta}(\cdot)$ in (4.138) correlates with the covariant component of $\tilde{F}^{-1C}_{\ \ \cdot\beta}$.

The torsion of anholonomic coefficients of (4.171) is defined as in (4.159):

$$
\begin{aligned}
T^{\cdot\cdot\alpha}_{\beta\chi} &= \Gamma^{\cdot\cdot\alpha}_{[\beta\chi]} + \tilde{\kappa}^{\cdot\cdot\alpha}_{\beta\chi} \\
&= \delta^{\alpha}_{A}\overset{G}{\Gamma}{}^{\cdot\cdot A}_{BC}\tilde{F}^{-1B}_{\ \ \cdot[\beta}\delta^{C}_{\chi]} + \tilde{F}^{-1A}_{\ \ \cdot\beta}\tilde{F}^{-1B}_{\ \ \cdot\chi}\partial_{[A}\tilde{F}^{\alpha}_{\cdot B]} \\
&= \delta^{\alpha}_{A}\overset{G}{\Gamma}{}^{\cdot\cdot A}_{BC}\tilde{F}^{-1B}_{\ \ \cdot[\beta}\delta^{C}_{\chi]} + \tilde{F}^{-1A}_{\ \ \cdot[\beta}\tilde{F}^{-1B}_{\ \ \cdot\chi]}\partial_{A}\tilde{F}^{\alpha}_{\cdot B},
\end{aligned}
\qquad (4.176)
$$

where $\tilde{\kappa}^{\cdot\cdot\alpha}_{\beta\chi}$ is the anholonomic object associated with $\tilde{\boldsymbol{F}}$ and is identical to that of (4.90). When anholonomic object $\tilde{\kappa}^{\cdot\cdot\alpha}_{\beta\chi}$ is nonzero, skew partial derivatives as in (4.161) are generally nonzero. When torsion $T^{\cdot\cdot\alpha}_{\beta\chi}$ is nonzero, skew covariant derivatives of a scalar field as in (4.162) are generally nonzero.

The Riemann-Christoffel curvature associated with anholonomic coefficients of (4.171) is defined as in (4.160):

$$
\begin{aligned}
R^{\cdot\cdot\cdot\alpha}_{\beta\chi\delta} &= 2\partial_{[\beta}\Gamma^{\cdot\cdot\alpha}_{\chi]\delta} + 2\Gamma^{\cdot\cdot\alpha}_{[\beta|\epsilon|}\Gamma^{\cdot\cdot\epsilon}_{\chi]\delta} + 2\tilde{\kappa}^{\cdot\cdot\epsilon}_{\beta\chi}\Gamma^{\cdot\cdot\alpha}_{\epsilon\delta} \\
&= 2\delta^{\alpha}_{A}\delta^{D}_{\delta}\overset{G}{\Gamma}{}^{\cdot\cdot A}_{CD}\partial_{[\beta}\tilde{F}^{-1C}_{\ \ \cdot\chi]} + 2\delta^{\alpha}_{A}\delta^{D}_{\delta}\overset{G}{\Gamma}{}^{\cdot\cdot A}_{BE}\overset{G}{\Gamma}{}^{\cdot\cdot E}_{CD}\tilde{F}^{-1B}_{\ \ \cdot[\beta}\tilde{F}^{-1C}_{\ \ \cdot\chi]} \\
&\quad + 2\delta^{\alpha}_{A}\delta^{D}_{\delta}\overset{G}{\Gamma}{}^{\cdot\cdot A}_{BD}\tilde{F}^{-1B}_{\ \ \cdot\epsilon}\tilde{F}^{-1E}_{\ \ \cdot[\beta}\tilde{F}^{-1F}_{\ \ \cdot\chi]}\partial_{E}\tilde{F}^{\epsilon}_{\cdot F} \\
&= 2\delta^{\alpha}_{A}\delta^{D}_{\delta}\overset{G}{\Gamma}{}^{\cdot\cdot A}_{CD}\partial_{[\beta}\tilde{F}^{-1C}_{\ \ \cdot\chi]} + 2\delta^{\alpha}_{A}\delta^{D}_{\delta}\overset{G}{\Gamma}{}^{\cdot\cdot A}_{[B|E|}\overset{G}{\Gamma}{}^{\cdot\cdot E}_{C]D}\tilde{F}^{-1B}_{\ \ \cdot\beta}\tilde{F}^{-1C}_{\ \ \cdot\chi} \\
&\quad - 2\delta^{\alpha}_{A}\delta^{D}_{\delta}\overset{G}{\Gamma}{}^{\cdot\cdot A}_{BD}\tilde{F}^{-1B}_{\ \ \cdot\epsilon}\tilde{F}^{-1E}_{\ \ \cdot[\beta}\tilde{F}^{\epsilon}_{\cdot|F|}\partial_{|E|}\tilde{F}^{-1F}_{\ \ \cdot\chi]} \\
&= 2\delta^{\alpha}_{A}\delta^{D}_{\delta}\overset{G}{\Gamma}{}^{\cdot\cdot A}_{CD}\partial_{[\beta}\tilde{F}^{-1C}_{\ \ \cdot\chi]} + 2\delta^{\alpha}_{A}\delta^{D}_{\delta}\overset{G}{\Gamma}{}^{\cdot\cdot A}_{[B|E|}\overset{G}{\Gamma}{}^{\cdot\cdot E}_{C]D}\tilde{F}^{-1B}_{\ \ \cdot\beta}\tilde{F}^{-1C}_{\ \ \cdot\chi} \\
&\quad - 2\delta^{\alpha}_{A}\delta^{D}_{\delta}\overset{G}{\Gamma}{}^{\cdot\cdot A}_{BD}\tilde{F}^{-1E}_{\ \ \cdot[\beta}\partial_{|E|}\tilde{F}^{-1B}_{\ \ \cdot\chi]} \\
&= 2\delta^{\alpha}_{A}\delta^{D}_{\delta}\tilde{F}^{-1B}_{\ \ \cdot\beta}\tilde{F}^{-1C}_{\ \ \cdot\chi}\overset{G}{\Gamma}{}^{\cdot\cdot A}_{[B|E|}\overset{G}{\Gamma}{}^{\cdot\cdot E}_{C]D} \\
&= -2\delta^{\alpha}_{A}\delta^{D}_{\delta}\tilde{F}^{-1B}_{\ \ \cdot\beta}\tilde{F}^{-1C}_{\ \ \cdot\chi}\partial_{[B}\overset{G}{\Gamma}{}^{\cdot\cdot A}_{C]D} \\
&= -2\delta^{\alpha}_{A}\delta^{D}_{\delta}\tilde{F}^{-1B}_{\ \ \cdot[\beta}\tilde{F}^{-1C}_{\ \ \cdot\chi]}\partial_{B}\overset{G}{\Gamma}{}^{\cdot\cdot A}_{CD},
\end{aligned}
\qquad (4.177)
$$

where the vanishing of the curvature tensor of Euclidean reference space (2.123) has been used. When the curvature $R^{\cdot\cdot\cdot\alpha}_{\beta\chi\delta}$ of (4.177) and/or torsion $T^{\cdot\cdot\alpha}_{\beta\chi}$ of (4.176) do not vanish, skew covariant derivatives of vector and covector fields are generally nonzero, as computed in (4.163) and (4.164), respectively. It is noted that torsion and curvature defined in this way can each be nonzero even when configuration $\tilde{B}$ is holonomic, *i.e.*, even when $\tilde{F}^{\alpha}_{\cdot A} = \partial_{A}\tilde{x}^{\alpha}$ and $\tilde{\kappa}^{\cdot\cdot\alpha}_{\beta\chi} = 0$.

Verifying (4.153), the (negative) covariant derivative of the metric tensor is computed directly as

$$\begin{aligned}
M_{\alpha\beta\chi} &= -\nabla_\alpha \tilde{g}_{\beta\chi} \\
&= -\partial_\alpha \tilde{g}_{\beta\chi} + \Gamma^{..\delta}_{\alpha\beta}\tilde{g}_{\delta\chi} + \Gamma^{..\delta}_{\alpha\chi}\tilde{g}_{\beta\delta} \\
&= -\delta^B_\beta \delta^C_\chi \tilde{F}^{-1A}_{\ \ \ .\alpha}\partial_A G_{BC} + \delta^B_\beta \delta^C_\chi \tilde{F}^{-1A}_{\ \ \ .\alpha} \overset{G}{\Gamma}{}^{..D}_{AB} G_{DC} \\
&\quad + \delta^B_\beta \delta^C_\chi \tilde{F}^{-1A}_{\ \ \ .\alpha} \overset{G}{\Gamma}{}^{..D}_{AC} G_{BD} \\
&= -\delta^B_\beta \delta^C_\chi \tilde{F}^{-1A}_{\ \ \ .\alpha}(\partial_A G_{BC} - \overset{G}{\Gamma}{}^{..D}_{AB} G_{DC} - \overset{G}{\Gamma}{}^{..D}_{AC} G_{BD}) \\
&= -\delta^B_\beta \delta^C_\chi \tilde{F}^{-1A}_{\ \ \ .\alpha} G_{BC;A} \\
&= 0,
\end{aligned} \tag{4.178}$$

where (2.133) has been used for vanishing of the covariant derivative of components of the referential metric tensor with respect to its Levi-Civita connection. Therefore, anholonomic covariant differentiation commutes with lowering indices via the metric $\tilde{g}_{\alpha\beta} = \delta^A_\alpha \delta^B_\beta G_{AB}$. Since

$$0 = \nabla_\alpha(\delta^\beta_\chi) = \nabla_\alpha(\tilde{g}^{\beta\delta}\tilde{g}_{\delta\chi}) = \tilde{g}_{\delta\chi}\nabla_\alpha \tilde{g}^{\beta\delta}, \tag{4.179}$$

the covariant derivative of inverse metric $\tilde{g}^{\alpha\beta} = \delta^\alpha_A \delta^\beta_B G^{AB}$ also vanishes, and covariant differentiation commutes with raising indices via inverse metric components $\tilde{g}^{\alpha\beta}$.

Now consider the case when (4.18) applies, such that single-valued coordinates $\tilde{x}^\alpha(x,t)$ continuously differentiable with respect to x^a do not exist in $\tilde{B}$. Following (4.55), identical coordinate systems are used in configurations B and $\tilde{B}$:

$$\tilde{\boldsymbol{g}}_\alpha(x) = \delta^a_\alpha \boldsymbol{g}_a, \tag{4.180}$$

$$\tilde{\boldsymbol{g}}^\alpha(x) = \tilde{g}^{\alpha\beta}\tilde{\boldsymbol{g}}_\beta = \delta^\alpha_c \delta^\beta_d g^{cd} \delta^a_\beta \boldsymbol{g}_a = \delta^\alpha_c g^{ca}\boldsymbol{g}_a = \delta^\alpha_a \boldsymbol{g}^a, \tag{4.181}$$

$$\tilde{g}_{\alpha\beta}(x) = \delta^a_\alpha \delta^b_\beta g_{ab}, \qquad \tilde{g}^{\alpha\beta}(x) = \delta^\alpha_a \delta^\beta_b g^{ab}. \tag{4.182}$$

Taking the partial derivative of (4.180) and applying (2.153) and (4.19) gives

$$\begin{aligned}
\partial_\beta \tilde{\boldsymbol{g}}_\alpha &= \partial_\beta(\delta^a_\alpha \boldsymbol{g}_a) \\
&= \delta^a_\alpha \partial_\beta \boldsymbol{g}_a \\
&= \delta^a_\alpha \bar{F}^b_{.\beta}\partial_b \boldsymbol{g}_a \\
&= \delta^a_\alpha \bar{F}^b_{.\beta} \overset{g}{\Gamma}{}^{..c}_{ba} \boldsymbol{g}_c \\
&= \delta^a_\alpha \bar{F}^b_{.\beta} \overset{g}{\Gamma}{}^{..c}_{ba} \delta^\chi_c \tilde{\boldsymbol{g}}_\chi \\
&= \Gamma^{..\chi}_{\beta\alpha}\tilde{\boldsymbol{g}}_\chi.
\end{aligned} \tag{4.183}$$

Taking the partial derivative of (4.181) and applying (2.154) and (4.19),

$$\begin{aligned}
\partial_\beta \tilde{\boldsymbol{g}}^\alpha &= \partial_\beta(\delta^\alpha_a \boldsymbol{g}^a) \\
&= \delta^\alpha_a \partial_\beta \boldsymbol{g}^a \\
&= \delta^\alpha_a \bar{F}^b_{\boldsymbol{\cdot}\beta} \partial_b \boldsymbol{g}^a \\
&= -\delta^\alpha_a \bar{F}^b_{\boldsymbol{\cdot}\beta} \overset{g}{\Gamma}{}^{\cdot\cdot a}_{bc} \boldsymbol{g}^c \\
&= -\delta^\alpha_a \bar{F}^b_{\boldsymbol{\cdot}\beta} \overset{g}{\Gamma}{}^{\cdot\cdot a}_{bc} \delta^c_\chi \tilde{\boldsymbol{g}}^\chi \\
&= -\Gamma^{\cdot\cdot\alpha}_{\beta\chi} \tilde{\boldsymbol{g}}^\chi.
\end{aligned} \tag{4.184}$$

Comparing (4.144), (4.183), and (4.184), connection coefficients consistent with (4.55) are time dependent:

$$\Gamma^{\cdot\cdot\alpha}_{\beta\chi}(x,t) = \delta^\alpha_a \delta^c_\chi \overset{g}{\Gamma}{}^{\cdot\cdot a}_{bc}(x) \bar{F}^b_{\boldsymbol{\cdot}\beta}(x,t). \tag{4.185}$$

The covariant derivative (*i.e.*, gradient) of a differentiable vector field $\boldsymbol{V} = V^\alpha \tilde{\boldsymbol{g}}_\alpha$ and of a covector field $\boldsymbol{\alpha} = \alpha_\alpha \tilde{\boldsymbol{g}}^\alpha$ are computed in this case as

$$\begin{aligned}
\nabla \boldsymbol{V} &= \partial_\beta \boldsymbol{V} \otimes \tilde{\boldsymbol{g}}^\beta \\
&= \partial_\beta (V^\alpha \tilde{\boldsymbol{g}}_\alpha) \otimes \tilde{\boldsymbol{g}}^\beta \\
&= (\partial_\beta V^\alpha + \Gamma^{\cdot\cdot\alpha}_{\beta\chi} V^\chi) \tilde{\boldsymbol{g}}_\alpha \otimes \tilde{\boldsymbol{g}}^\beta \\
&= (\partial_\beta V^\alpha + \bar{F}^b_{\boldsymbol{\cdot}\beta} \delta^\alpha_a \delta^c_\chi \overset{g}{\Gamma}{}^{\cdot\cdot a}_{bc} V^\chi) \tilde{\boldsymbol{g}}_\alpha \otimes \tilde{\boldsymbol{g}}^\beta \\
&= \bar{F}^b_{\boldsymbol{\cdot}\beta} (\partial_b V^\alpha + \delta^\alpha_a \delta^c_\chi \overset{g}{\Gamma}{}^{\cdot\cdot a}_{bc} V^\chi) \tilde{\boldsymbol{g}}_\alpha \otimes \tilde{\boldsymbol{g}}^\beta \\
&= \bar{F}^b_{\boldsymbol{\cdot}\beta} (\delta^\alpha_a \partial_b V^a + \delta^\alpha_a \delta^c_\chi \delta^\chi_d \overset{g}{\Gamma}{}^{\cdot\cdot a}_{bc} V^d) \tilde{\boldsymbol{g}}_\alpha \otimes \tilde{\boldsymbol{g}}^\beta \\
&= \bar{F}^b_{\boldsymbol{\cdot}\beta} \delta^\alpha_a (\partial_b V^a + \overset{g}{\Gamma}{}^{\cdot\cdot a}_{bc} V^c) \tilde{\boldsymbol{g}}_\alpha \otimes \tilde{\boldsymbol{g}}^\beta \\
&= V^a_{;b} \bar{F}^b_{\boldsymbol{\cdot}\beta} \boldsymbol{g}_a \otimes \tilde{\boldsymbol{g}}^\beta \\
&= V^a_{;b} \bar{F}^b_{\boldsymbol{\cdot}\beta} \delta^\beta_c \boldsymbol{g}_a \otimes \boldsymbol{g}^c,
\end{aligned} \tag{4.186}$$

$$\begin{aligned}
\nabla \boldsymbol{\alpha} &= \partial_\beta \boldsymbol{\alpha} \otimes \tilde{\boldsymbol{g}}^\beta \\
&= \partial_\beta (\alpha_\alpha \tilde{\boldsymbol{g}}^\alpha) \otimes \tilde{\boldsymbol{g}}^\beta \\
&= (\partial_\beta \alpha_\alpha - \Gamma^{\cdot\cdot\chi}_{\beta\alpha} \alpha_\chi) \tilde{\boldsymbol{g}}^\alpha \otimes \tilde{\boldsymbol{g}}^\beta \\
&= (\partial_\beta \alpha_\alpha - \bar{F}^b_{\boldsymbol{\cdot}\beta} \delta^a_\alpha \delta^\chi_c \overset{g}{\Gamma}{}^{\cdot\cdot c}_{ba} \alpha_\chi) \tilde{\boldsymbol{g}}^\alpha \otimes \tilde{\boldsymbol{g}}^\beta \\
&= \bar{F}^b_{\boldsymbol{\cdot}\beta} (\partial_b \alpha_\alpha - \delta^a_\alpha \delta^\chi_c \overset{g}{\Gamma}{}^{\cdot\cdot c}_{ba} \alpha_\chi) \tilde{\boldsymbol{g}}^\alpha \otimes \tilde{\boldsymbol{g}}^\beta \\
&= \bar{F}^b_{\boldsymbol{\cdot}\beta} (\delta^a_\alpha \partial_b \alpha_a - \delta^a_\alpha \delta^\chi_c \delta^d_\chi \overset{g}{\Gamma}{}^{\cdot\cdot c}_{ba} \alpha_d) \tilde{\boldsymbol{g}}^\alpha \otimes \tilde{\boldsymbol{g}}^\beta \\
&= \bar{F}^b_{\boldsymbol{\cdot}\beta} \delta^a_\alpha (\partial_b \alpha_a - \overset{g}{\Gamma}{}^{\cdot\cdot c}_{ba} \alpha_c) \tilde{\boldsymbol{g}}^\alpha \otimes \tilde{\boldsymbol{g}}^\beta \\
&= \alpha_{a;b} \bar{F}^b_{\boldsymbol{\cdot}\beta} \boldsymbol{g}^a \otimes \tilde{\boldsymbol{g}}^\beta \\
&= \alpha_{a;b} \bar{F}^b_{\boldsymbol{\cdot}\beta} \delta^\beta_c \boldsymbol{g}^a \otimes \boldsymbol{g}^c,
\end{aligned} \tag{4.187}$$

where (4.60) has been used for the shifter such that $V^\alpha = V^a g_a^\alpha = V^a \delta_a^\alpha$. Components of the covariant derivative of a tensor field of arbitrary order with components $A_{\gamma\ldots\mu}^{\alpha\ldots\phi}$ are found as

$$\nabla_\nu A_{\gamma\ldots\mu}^{\alpha\ldots\phi} = \bar{F}_{.\nu}^n (\delta_a^\alpha \cdots \delta_f^\phi)(\delta_\gamma^g \cdots \delta_\mu^m) A_{g\ldots m;n}^{a\ldots f}. \qquad (4.188)$$

Since Levi-Civita connection $\overset{g}{\Gamma}{}_{bc}^{\cdot\cdot a}$ is symmetric in covariant indices, it follows that $\Gamma_{\beta\chi}^{\cdot\cdot\alpha}$ of (4.185) obeys

$$\begin{aligned}
\Gamma_{\beta\chi}^{\cdot\cdot\alpha} &= \delta_a^\alpha \delta_\chi^c \overset{g}{\Gamma}{}_{bc}^{\cdot\cdot a} \bar{F}_{.\beta}^b \\
&= \delta_a^\alpha \delta_\chi^c \overset{g}{\Gamma}{}_{cb}^{\cdot\cdot a} \bar{F}_{.\beta}^b \\
&= \delta_a^\alpha \delta_\chi^b \overset{g}{\Gamma}{}_{cb}^{\cdot\cdot a} \bar{F}_{.\beta}^c. \qquad (4.189)
\end{aligned}$$

Covariant components of $\Gamma_{\beta\chi}^{\cdot\cdot\alpha}$ are generally not symmetric; here the left covariant component corresponding to differentiation by $\nabla_\beta(\cdot)$ in (4.138) correlates with the covariant component of $\bar{F}_{.\beta}^c$.

The torsion tensor of anholonomic coefficients (4.185) is, from (4.159),

$$\begin{aligned}
T_{\beta\chi}^{\cdot\cdot\alpha} &= \Gamma_{[\beta\chi]}^{\cdot\cdot\alpha} + \bar{\kappa}_{\beta\chi}^{\cdot\cdot\alpha} \\
&= \delta_a^{\alpha g}\overset{}{\Gamma}{}_{bc}^{\cdot\cdot a} \bar{F}_{.[\beta}^b \delta_{\chi]}^c + \bar{F}_{.\beta}^a \bar{F}_{.\chi}^b \partial_{[a} \bar{F}_{.b]}^{-1\alpha} \\
&= \delta_a^{\alpha g}\overset{}{\Gamma}{}_{bc}^{\cdot\cdot a} \bar{F}_{.[\beta}^b \delta_{\chi]}^c + \bar{F}_{.[\beta}^a \bar{F}_{.\chi]}^b \partial_a \bar{F}_{.b}^{-1\alpha}, \qquad (4.190)
\end{aligned}$$

where $\bar{\kappa}_{\beta\chi}^{\cdot\cdot\alpha}$ is the anholonomic object associated with $\bar{F}$ and is identical to that of (4.97). When anholonomic object $\bar{\kappa}_{\beta\chi}^{\cdot\cdot\alpha}$ is nonzero, skew partial derivatives as in (4.161) are generally nonzero. When torsion $T_{\beta\chi}^{\cdot\cdot\alpha}$ is nonzero, skew covariant derivatives of a scalar field as in (4.162) are generally nonzero.

The Riemann-Christoffel curvature tensor of anholonomic coefficients (4.185) is defined as in (4.160):

$$\begin{aligned}
R_{\beta\chi\delta}^{\cdot\cdot\cdot\alpha} &= 2\partial_{[\beta}\Gamma_{\chi]\delta}^{\cdot\cdot\alpha} + 2\Gamma_{[\beta\epsilon}^{\cdot\cdot\alpha}\Gamma_{\chi]\delta}^{\cdot\cdot\epsilon} + 2\bar{\kappa}_{\beta\chi}^{\cdot\cdot\epsilon}\Gamma_{\epsilon\delta}^{\cdot\cdot\alpha} \\
&= 2\delta_a^\alpha \delta_\delta^d \overset{g}{\Gamma}{}_{cd}^{\cdot\cdot a} \partial_{[\beta}\bar{F}_{.\chi]}^c + 2\delta_a^\alpha \delta_\delta^d \overset{g}{\Gamma}{}_{be}^{\cdot\cdot a}\overset{g}{\Gamma}{}_{cd}^{\cdot\cdot e} \bar{F}_{.[\beta}^b \bar{F}_{.\chi]}^c \\
&\quad + 2\delta_a^\alpha \delta_\delta^d \overset{g}{\Gamma}{}_{bd}^{\cdot\cdot a} \bar{F}_{.\epsilon}^b \bar{F}_{.[\beta}^e \bar{F}_{.\chi]}^f \partial_e \bar{F}_{.f}^{-1\epsilon} \\
&= 2\delta_a^\alpha \delta_\delta^d \overset{g}{\Gamma}{}_{cd}^{\cdot\cdot a} \partial_{[\beta}\bar{F}_{.\chi]}^c + 2\delta_a^\alpha \delta_\delta^d \overset{g}{\Gamma}{}_{[b|e|}^{\cdot\cdot a}\overset{g}{\Gamma}{}_{c]d}^{\cdot\cdot e} \bar{F}_{.\beta}^b \bar{F}_{.\chi}^c \\
&\quad - 2\delta_a^\alpha \delta_\delta^d \overset{g}{\Gamma}{}_{bd}^{\cdot\cdot a} \bar{F}_{.\epsilon}^b \bar{F}_{.[\beta}^e \bar{F}_{.|f|}^{-1\epsilon} \partial_{|e|}\bar{F}_{.\chi]}^f \\
&= 2\delta_a^\alpha \delta_\delta^d \overset{g}{\Gamma}{}_{cd}^{\cdot\cdot a} \partial_{[\beta}\bar{F}_{.\chi]}^c + 2\delta_a^\alpha \delta_\delta^d \overset{g}{\Gamma}{}_{[b|e|}^{\cdot\cdot a}\overset{g}{\Gamma}{}_{c]d}^{\cdot\cdot e} \bar{F}_{.\beta}^b \bar{F}_{.\chi}^c \\
&\quad - 2\delta_a^\alpha \delta_\delta^d \overset{g}{\Gamma}{}_{bd}^{\cdot\cdot a} \bar{F}_{.[\beta}^e \partial_{|e|}\bar{F}_{.\chi]}^b \\
&= 2\delta_a^\alpha \delta_\delta^d \bar{F}_{.\beta}^b \bar{F}_{.\chi}^c \overset{g}{\Gamma}{}_{[b|e|}^{\cdot\cdot a}\overset{g}{\Gamma}{}_{c]d}^{\cdot\cdot e} \\
&= -2\delta_a^\alpha \delta_\delta^d \bar{F}_{.\beta}^b \bar{F}_{.\chi}^c \partial_{[b}\overset{g}{\Gamma}{}_{c]d}^{\cdot\cdot a} \\
&= -2\delta_a^\alpha \delta_\delta^d \bar{F}_{.[\beta}^b \bar{F}_{.\chi]}^c \partial_b \overset{g}{\Gamma}{}_{cd}^{\cdot\cdot a}, \qquad (4.191)
\end{aligned}$$

where the vanishing of the curvature tensor of Euclidean current space (2.151) has been used. When the curvature $R_{\beta\chi\delta}^{\cdot\cdot\cdot\alpha}$ of (4.191) and/or torsion $T_{\beta\chi}^{\cdot\cdot\alpha}$ of (4.190) do not vanish, skew covariant derivatives of vector and covector fields are generally nonzero, as found via (4.163) and (4.164), respectively. Note that the curvature and torsion defined in this manner need not vanish even if configuration $\tilde{B}$ is holonomic, *i.e.*, even if $\bar{F}^{-1\alpha}_{\cdot a} = \partial_a \tilde{x}^\alpha$ and anholonomic object $\bar{\kappa}_{\beta\chi}^{\cdot\cdot\alpha} = 0$.

Verifying (4.153) for this generic choice of anholonomic coordinates, the (negative) covariant derivative of the metric tensor is

$$\begin{aligned} M_{\alpha\beta\chi} &= -\nabla_\alpha \tilde{g}_{\beta\chi} \\ &= -\partial_\alpha \tilde{g}_{\beta\chi} + \Gamma_{\alpha\beta}^{\cdot\cdot\delta} \tilde{g}_{\delta\chi} + \Gamma_{\alpha\chi}^{\cdot\cdot\delta} \tilde{g}_{\beta\delta} \\ &= -\delta^b_\beta \delta^c_\chi \bar{F}^a_{\cdot\alpha} \partial_a g_{bc} + \delta^b_\beta \delta^c_\chi \bar{F}^a_{\cdot\alpha} \overset{g}{\Gamma}{}_{ab}^{\cdot\cdot d} g_{dc} \\ &\quad + \delta^b_\beta \delta^c_\chi \bar{F}^a_{\cdot\alpha} \overset{g}{\Gamma}{}_{ac}^{\cdot\cdot d} g_{bd} \\ &= -\delta^b_\beta \delta^c_\chi \bar{F}^a_{\cdot\alpha} (\partial_a g_{bc} - \overset{g}{\Gamma}{}_{ab}^{\cdot\cdot d} g_{dc} - \overset{g}{\Gamma}{}_{ac}^{\cdot\cdot d} g_{bd}) \\ &= -\delta^b_\beta \delta^c_\chi \bar{F}^a_{\cdot\alpha} g_{bc;a} \\ &= 0, \end{aligned} \tag{4.192}$$

where (2.160) has been used for vanishing of the covariant derivative of the spatial metric tensor with respect to its Levi-Civita connection. Therefore, covariant differentiation commutes with lowering indices via the metric $\tilde{g}_{\alpha\beta} = \delta^a_\alpha \delta^b_\beta g_{ab}$. Since

$$0 = \nabla_\alpha (\delta^\beta_\chi) = \nabla_\alpha (\tilde{g}^{\beta\delta} \tilde{g}_{\delta\chi}) = \tilde{g}_{\delta\chi} \nabla_\alpha \tilde{g}^{\beta\delta}, \tag{4.193}$$

the covariant derivative of inverse metric $\tilde{g}^{\alpha\beta} = \delta^\alpha_A \delta^\beta_B g^{ab}$ likewise vanishes, and covariant differentiation commutes with raising indices via $\tilde{g}^{\alpha\beta}$.

Finally consider the simplest case whereby Cartesian bases are used for $\tilde{B}$, as in (4.62):

$$\tilde{\boldsymbol{g}}_\alpha = \boldsymbol{e}_\alpha, \qquad \tilde{\boldsymbol{g}}^\alpha = \boldsymbol{e}^\alpha; \tag{4.194}$$

$$\tilde{g}_{\alpha\beta} = \boldsymbol{e}_\alpha \cdot \boldsymbol{e}_\beta = \delta_{\alpha\beta}, \qquad \tilde{g}^{\alpha\beta} = \boldsymbol{e}^\alpha \cdot \boldsymbol{e}^\beta = \delta^{\alpha\beta}. \tag{4.195}$$

In this case, partial derivatives of basis vectors vanish identically:

$$\partial_\beta \tilde{\boldsymbol{g}}_\alpha = \partial_\beta \boldsymbol{e}_\alpha = \tilde{F}^{-1B}_{\cdot\beta} \partial_B \boldsymbol{e}_\alpha = \bar{F}^b_{\cdot\beta} \partial_b \boldsymbol{e}_\alpha = \Gamma_{\beta\alpha}^{\cdot\cdot\chi} \tilde{\boldsymbol{g}}_\chi = 0, \tag{4.196}$$

as do partial derivatives of their reciprocals:

$$\partial_\beta \tilde{\boldsymbol{g}}^\alpha = \partial_\beta \boldsymbol{e}^\alpha = \tilde{F}^{-1B}_{\cdot\beta} \partial_B \boldsymbol{e}^\alpha = \bar{F}^b_{\cdot\beta} \partial_b \boldsymbol{e}^\alpha = -\Gamma_{\beta\chi}^{\cdot\cdot\alpha} \tilde{\boldsymbol{g}}^\chi = 0. \tag{4.197}$$

It follows that connection coefficients on $\tilde{B}$ must also vanish in this case:

$$\Gamma_{\beta\chi}^{\cdot\cdot\alpha} = 0. \tag{4.198}$$

Accordingly, generic anholonomic connection coefficients of (4.83) degenerate progressively to

$$\begin{aligned}\Gamma^{..\alpha}_{\beta\chi} &= \frac{1}{2}\tilde{g}^{\alpha\delta}(2\chi_{\{\beta\delta\chi\}} - 2T_{\{\beta\delta\chi\}} + 2\kappa_{\{\beta\delta\chi\}} + M_{\{\beta\delta\chi\}}) \\ &= \frac{1}{2}\delta^{\alpha\delta}(\partial_{\{\beta}\delta_{\delta\chi\}} - 2T_{\{\beta\delta\chi\}} + 2\kappa_{\{\beta\delta\chi\}} - \nabla_{\{\beta}\delta_{\delta\chi\}}) \\ &= -\delta^{\alpha\delta}(T_{\{\beta\delta\chi\}} - \kappa_{\{\beta\delta\chi\}}) \\ &= 0. \end{aligned} \tag{4.199}$$

The covariant derivative (*i.e.*, gradient) of a differentiable vector field $\boldsymbol{V} = V^\alpha \tilde{\boldsymbol{g}}_\alpha$ is computed in this case as

$$\begin{aligned}\nabla \boldsymbol{V} &= \partial_\beta \boldsymbol{V} \otimes \tilde{\boldsymbol{g}}^\beta \\ &= \partial_\beta V^\alpha \boldsymbol{e}_\alpha \otimes \boldsymbol{e}^\beta \\ &= \tilde{F}^{-1A}_{\ \ .\beta}\partial_A V^\alpha \boldsymbol{e}_\alpha \otimes \boldsymbol{e}^\beta \\ &= \bar{F}^a_{.\beta}\partial_a V^\alpha \boldsymbol{e}_\alpha \otimes \boldsymbol{e}^\beta. \end{aligned} \tag{4.200}$$

Similarly, the covariant derivative of a covector field $\boldsymbol{\alpha} = \alpha_\alpha \tilde{\boldsymbol{g}}^\alpha$ is computed as

$$\begin{aligned}\nabla \boldsymbol{\alpha} &= \partial_\beta \boldsymbol{\alpha} \otimes \tilde{\boldsymbol{g}}^\beta \\ &= \partial_\beta \alpha_\alpha \boldsymbol{e}^\alpha \otimes \boldsymbol{e}^\beta \\ &= \tilde{F}^{-1A}_{\ \ .\beta}\partial_A \alpha_\alpha \boldsymbol{e}^\alpha \otimes \boldsymbol{e}^\beta \\ &= \bar{F}^a_{.\beta}\partial_a \alpha_\alpha \boldsymbol{e}^\alpha \otimes \boldsymbol{e}^\beta. \end{aligned} \tag{4.201}$$

Components of the covariant derivative of a tensor field of arbitrary order with components $A^{\alpha\ldots\phi}_{\gamma\ldots\mu}$ are found as

$$\nabla_\nu A^{\alpha\ldots\phi}_{\gamma\ldots\mu} = \partial_\nu A^{\alpha\ldots\phi}_{\gamma\ldots\mu} = \tilde{F}^{-1N}_{\ \ .\nu}\partial_N A^{\alpha\ldots\phi}_{\gamma\ldots\mu} = \bar{F}^n_{.\nu}\partial_n A^{\alpha\ldots\phi}_{\gamma\ldots\mu}. \tag{4.202}$$

Note, however, that the anholonomic object need not vanish; hence, the torsion of (4.159) is equal to the anholonomic object:

$$T^{..\alpha}_{\beta\chi} = \kappa^{..\alpha}_{\beta\chi}. \tag{4.203}$$

This confirms (4.199). The curvature tensor of (4.160) vanishes identically in this case since $\Gamma^{..\alpha}_{\beta\chi} = 0$. The covariant derivative of the metric also vanishes identically:

$$M_{\alpha\beta\chi} = -\nabla_\alpha \tilde{g}_{\beta\chi} = -\partial_\alpha \delta_{\beta\chi} = -\tilde{F}^{-1A}_{\ \ .\alpha}\partial_A \delta_{\beta\chi} = -\bar{F}^a_{.\alpha}\partial_a \delta_{\beta\chi} = 0. \tag{4.204}$$

Consider now skew partial derivatives of intermediate basis vectors which can be found using the first of (4.144):

$$\partial_{[\beta}\tilde{\boldsymbol{g}}_{\alpha]} = \Gamma^{..\chi}_{[\beta\alpha]}\tilde{\boldsymbol{g}}_\chi. \tag{4.205}$$

When configuration $\tilde{B}$ is holonomic and (4.165) applies, then necessary conditions are $\partial_{[\beta}\tilde{\boldsymbol{g}}_{\alpha]} = \partial_{[\beta}\partial_{\alpha]}\tilde{\boldsymbol{x}} = 0$. However, for alternative prescriptions (4.171) or (4.185), $\partial_{[\beta}\tilde{\boldsymbol{g}}_{\alpha]}$ may not vanish even when $\tilde{B}$ is holonomic. And for choice (4.194), $\partial_{[\beta}\tilde{\boldsymbol{g}}_{\alpha]}$ always vanishes even when $\tilde{B}$ is anholonomic. Thus, for all choices of intermediate basis vectors addressed herein, conditions $\partial_{[\beta}\tilde{\boldsymbol{g}}_{\alpha]} \neq 0$ are neither necessary nor sufficient to render configuration $\tilde{B}$ anholonomic.

4.2.3 *Total covariant derivatives*

Covariant differentiation of two-point tensor fields with one or more components referred to anholonomic coordinates (*i.e.*, one or more indices referred to intermediate configuration $\tilde{B}$) is defined following arguments similar to those of Section 2.3.2. First consider a generic two-point tensor of the form $\boldsymbol{A}(X,t)$, with components $A^{\alpha\ldots\phi\;A\ldots F}_{\gamma\ldots\mu\;G\ldots M}$. The total covariant derivative of $\boldsymbol{A}$ is computed as

$$
\begin{aligned}
(A^{\alpha\ldots\phi\;A\ldots F}_{\gamma\ldots\mu\;G\ldots M})_{:\nu} &= \partial_\nu(A^{\alpha\ldots\phi\;A\ldots F}_{\gamma\ldots\mu\;G\ldots M}) \\
&\quad + \Gamma^{\cdot\cdot\alpha}_{\nu\rho}A^{\rho\ldots\phi\;A\ldots F}_{\gamma\ldots\mu\;G\ldots M} + \cdots + \Gamma^{\cdot\cdot\phi}_{\nu\rho}A^{\alpha\ldots\rho\;A\ldots F}_{\gamma\ldots\mu\;G\ldots M} \\
&\quad - \Gamma^{\cdot\cdot\rho}_{\nu\gamma}A^{\alpha\ldots\phi\;A\ldots F}_{\rho\ldots\mu\;G\ldots M} - \cdots - \Gamma^{\cdot\cdot\rho}_{\nu\mu}A^{\alpha\ldots\phi\;A\ldots F}_{\gamma\ldots\rho\;G\ldots M} \\
&\quad + \tilde{F}^{-1N}_{\cdot\nu}\,{}^{G}_{\Gamma}{}^{\cdot\cdot A}_{NR}A^{\alpha\ldots\phi\;R\ldots F}_{\gamma\ldots\mu\;G\ldots M} + \cdots \\
&\quad + \tilde{F}^{-1N}_{\cdot\nu}\,{}^{G}_{\Gamma}{}^{\cdot\cdot F}_{NR}A^{\alpha\ldots\phi\;A\ldots R}_{\gamma\ldots\mu\;G\ldots M} \\
&\quad - \tilde{F}^{-1N}_{\cdot\nu}\,{}^{G}_{\Gamma}{}^{\cdot\cdot R}_{NG}A^{\alpha\ldots\phi\;A\ldots F}_{\gamma\ldots\mu\;R\ldots M} - \cdots \\
&\quad - \tilde{F}^{-1N}_{\cdot\nu}\,{}^{G}_{\Gamma}{}^{\cdot\cdot R}_{NM}A^{\alpha\ldots\phi\;A\ldots F}_{\gamma\ldots\mu\;G\ldots R} \\
&= \big[\partial_N(A^{\alpha\ldots\phi\;A\ldots F}_{\gamma\ldots\mu\;G\ldots M}) \\
&\quad + {}^{G}_{\Gamma}{}^{\cdot\cdot A}_{NR}A^{\alpha\ldots\phi\;R\ldots F}_{\gamma\ldots\mu\;G\ldots M} + \cdots + {}^{G}_{\Gamma}{}^{\cdot\cdot F}_{NR}A^{\alpha\ldots\phi\;A\ldots R}_{\gamma\ldots\mu\;G\ldots M} \\
&\quad - {}^{G}_{\Gamma}{}^{\cdot\cdot R}_{NG}A^{\alpha\ldots\phi\;A\ldots F}_{\gamma\ldots\mu\;R\ldots M} - \cdots \\
&\quad - {}^{G}_{\Gamma}{}^{\cdot\cdot R}_{NM}A^{\alpha\ldots\phi\;A\ldots F}_{\gamma\ldots\mu\;G\ldots R}\big]\tilde{F}^{-1N}_{\cdot\nu} \\
&\quad + \Gamma^{\cdot\cdot\alpha}_{\nu\rho}A^{\rho\ldots\phi\;A\ldots F}_{\gamma\ldots\mu\;G\ldots M} + \cdots + \Gamma^{\cdot\cdot\phi}_{\nu\rho}A^{\alpha\ldots\rho\;A\ldots F}_{\gamma\ldots\mu\;G\ldots M} \\
&\quad - \Gamma^{\cdot\cdot\rho}_{\nu\gamma}A^{\alpha\ldots\phi\;A\ldots F}_{\rho\ldots\mu\;G\ldots M} - \cdots - \Gamma^{\cdot\cdot\rho}_{\nu\mu}A^{\alpha\ldots\phi\;A\ldots F}_{\gamma\ldots\rho\;G\ldots M} \\
&= (A^{\alpha\ldots\phi\;A\ldots F}_{\gamma\ldots\mu\;G\ldots M})_{;N}\tilde{F}^{-1N}_{\cdot\nu} \\
&\quad + \Gamma^{\cdot\cdot\alpha}_{\nu\rho}A^{\rho\ldots\phi\;A\ldots F}_{\gamma\ldots\mu\;G\ldots M} + \cdots + \Gamma^{\cdot\cdot\phi}_{\nu\rho}A^{\alpha\ldots\rho\;A\ldots F}_{\gamma\ldots\mu\;G\ldots M} \\
&\quad - \Gamma^{\cdot\cdot\rho}_{\nu\gamma}A^{\alpha\ldots\phi\;A\ldots F}_{\rho\ldots\mu\;G\ldots M} - \cdots - \Gamma^{\cdot\cdot\rho}_{\nu\mu}A^{\alpha\ldots\phi\;A\ldots F}_{\gamma\ldots\rho\;G\ldots M} \\
&= (A^{\alpha\ldots\phi\;A\ldots F}_{\gamma\ldots\mu\;G\ldots M})_{:N}\tilde{F}^{-1N}_{\cdot\nu}.
\end{aligned} \tag{4.206}
$$

For example, letting $\boldsymbol{A}(X,t) = \tilde{\boldsymbol{F}}(X,t) = \tilde{F}^{\alpha}_{.A}\tilde{\boldsymbol{g}}_{\alpha} \otimes \boldsymbol{G}^A$, the material gradient is computed as

$$\begin{aligned}
\overset{G}{\nabla}\tilde{\boldsymbol{F}} &= \partial_B \tilde{\boldsymbol{F}} \otimes \boldsymbol{G}^B \\
&= \partial_B(\tilde{F}^{\alpha}_{.A}\tilde{\boldsymbol{g}}_{\alpha} \otimes \boldsymbol{G}^A) \otimes \boldsymbol{G}^B \\
&= \partial_B \tilde{F}^{\alpha}_{.A}\tilde{\boldsymbol{g}}_{\alpha} \otimes \boldsymbol{G}^A \otimes \boldsymbol{G}^B \\
&\quad + \tilde{F}^{\chi}_{.A}\tilde{F}^{\beta}_{.B}(\partial_\beta \tilde{\boldsymbol{g}}_{\chi}) \otimes \boldsymbol{G}^A \otimes \boldsymbol{G}^B \\
&\quad + \tilde{F}^{\alpha}_{.C}\tilde{\boldsymbol{g}}_{\alpha} \otimes (\partial_B \boldsymbol{G}^C) \otimes \boldsymbol{G}^B \\
&= \partial_B \tilde{F}^{\alpha}_{.A}\tilde{\boldsymbol{g}}_{\alpha} \otimes \boldsymbol{G}^A \otimes \boldsymbol{G}^B \\
&\quad + \tilde{F}^{\chi}_{.A}\tilde{F}^{\beta}_{.B}(\Gamma^{..\alpha}_{\beta\chi}\tilde{\boldsymbol{g}}_{\alpha}) \otimes \boldsymbol{G}^A \otimes \boldsymbol{G}^B \\
&\quad - \tilde{F}^{\alpha}_{.C}\tilde{\boldsymbol{g}}_{\alpha} \otimes (\overset{G}{\Gamma}{}^{..C}_{BA}\boldsymbol{G}^A) \otimes \boldsymbol{G}^B \\
&= \left(\partial_B \tilde{F}^{\alpha}_{.A} + \Gamma^{..\alpha}_{\beta\chi}\tilde{F}^{\chi}_{.A}\tilde{F}^{\beta}_{.B} - \overset{G}{\Gamma}{}^{..C}_{BA}\tilde{F}^{\alpha}_{.C}\right)\tilde{\boldsymbol{g}}_{\alpha} \otimes \boldsymbol{G}^A \otimes \boldsymbol{G}^B \\
&= \tilde{F}^{\alpha}_{.A:B}\tilde{\boldsymbol{g}}_{\alpha} \otimes \boldsymbol{G}^A \otimes \boldsymbol{G}^B.
\end{aligned} \tag{4.207}$$

Notice that unlike $\overset{G}{\nabla}\boldsymbol{F}$ in the first of (3.25), skew covariant components of $\overset{G}{\nabla}\tilde{\boldsymbol{F}}$ do not necessarily vanish:

$$\begin{aligned}
\tilde{F}^{\alpha}_{.[A:B]} &= \partial_{[B}\tilde{F}^{\alpha}_{.A]} + \Gamma^{..\alpha}_{\beta\chi}\tilde{F}^{\beta}_{.[B}\tilde{F}^{\chi}_{.A]} - \overset{G}{\Gamma}{}^{..C}_{[BA]}\tilde{F}^{\alpha}_{.C} \\
&= \partial_{[B}\tilde{F}^{\alpha}_{.A]} + \Gamma^{..\alpha}_{[\beta\chi]}\tilde{F}^{\beta}_{.B}\tilde{F}^{\chi}_{.A}.
\end{aligned} \tag{4.208}$$

From (4.208), the material gradient $\overset{G}{\nabla}\tilde{\boldsymbol{F}}$ is generally symmetric in covariant indices only when both (4.13) holds and symmetric connection coefficients $\Gamma^{..\alpha}_{\beta\chi} = \Gamma^{..\alpha}_{(\beta\chi)}$ are prescribed on $\tilde{B}$.

Applying the total covariant derivative operation twice in succession with respect to reference coordinates X^A to $\tilde{F}^{\alpha}_{.A}(X,t)$ gives the following complete 15-term expression:

$$\begin{aligned}
\tilde{F}^{\alpha}_{.A:BC} &= \partial_C(\tilde{F}^{\alpha}_{.A:B}) - \overset{G}{\Gamma}{}^{..D}_{CA}\tilde{F}^{\alpha}_{.D:B} - \overset{G}{\Gamma}{}^{..D}_{CB}\tilde{F}^{\alpha}_{.A:D} + \Gamma^{..\alpha}_{\chi\delta}\tilde{F}^{\delta}_{.A:B}\tilde{F}^{\chi}_{.C} \\
&= \partial_C\partial_B\tilde{F}^{\alpha}_{.A} - \tilde{F}^{\alpha}_{.D}\partial_C\overset{G}{\Gamma}{}^{..D}_{BA} + \tilde{F}^{\epsilon}_{.C}\tilde{F}^{\delta}_{.A}\tilde{F}^{\beta}_{.B}\partial_\epsilon\Gamma^{..\alpha}_{\beta\delta} \\
&\quad - \overset{G}{\Gamma}{}^{..D}_{BA}\partial_C\tilde{F}^{\alpha}_{.D} - \overset{G}{\Gamma}{}^{..D}_{CA}\partial_B\tilde{F}^{\alpha}_{.D} \\
&\quad - \overset{G}{\Gamma}{}^{..D}_{CB}\partial_D\tilde{F}^{\alpha}_{.A} + \Gamma^{..\alpha}_{\beta\delta}\tilde{F}^{\beta}_{.B}\partial_C\tilde{F}^{\delta}_{.A} \\
&\quad + \Gamma^{..\alpha}_{\beta\delta}\tilde{F}^{\delta}_{.A}\partial_C\tilde{F}^{\beta}_{.B} + \Gamma^{..\alpha}_{\chi\delta}\tilde{F}^{\chi}_{.C}\partial_B\tilde{F}^{\delta}_{.A} \\
&\quad - \overset{G}{\Gamma}{}^{..D}_{CA}\Gamma^{..\alpha}_{\beta\delta}\tilde{F}^{\delta}_{.D}\tilde{F}^{\beta}_{.B} - \overset{G}{\Gamma}{}^{..D}_{CB}\Gamma^{..\alpha}_{\delta\beta}\tilde{F}^{\beta}_{.A}\tilde{F}^{\delta}_{.D} \\
&\quad - \overset{G}{\Gamma}{}^{..E}_{BA}\Gamma^{..\alpha}_{\chi\delta}\tilde{F}^{\chi}_{.C}\tilde{F}^{\delta}_{.E} + \overset{G}{\Gamma}{}^{..D}_{CA}\overset{G}{\Gamma}{}^{..E}_{BD}\tilde{F}^{\alpha}_{.E} \\
&\quad + \overset{G}{\Gamma}{}^{..D}_{CB}\overset{G}{\Gamma}{}^{..E}_{DA}\tilde{F}^{\alpha}_{.E} + \Gamma^{..\alpha}_{\chi\delta}\Gamma^{..\delta}_{\beta\epsilon}\tilde{F}^{\chi}_{.C}\tilde{F}^{\epsilon}_{.A}\tilde{F}^{\beta}_{.B}.
\end{aligned} \tag{4.209}$$

Generally, covariant entries of $\tilde{F}^{\alpha}_{.A:BC}$ in (4.209) do not demonstrate the same symmetries as $F^{a}_{.A:BC} = F^{a}_{(.A:BC)} = x^{a}_{:(ABC)}$ of (3.22) unless the intermediate configuration is a Euclidean space (*i.e.*, holonomic) and intermediate connection coefficients correspond to a symmetric (*i.e.*, Levi-Civita) connection. Analogously to (3.27), the second total covariant derivative of $\tilde{\boldsymbol{F}}$ maps between configurations as

$$\begin{aligned}
\tilde{F}^{\alpha}_{.A:\beta\chi} &= (\tilde{F}^{\alpha}_{.A:\beta})_{:C}\tilde{F}^{-1C}_{\ \ .\chi} \\
&= (\tilde{F}^{\alpha}_{.A:B}\tilde{F}^{-1B}_{\ \ .\beta})_{:C}\tilde{F}^{-1C}_{\ \ .\chi} \\
&= \tilde{F}^{\alpha}_{.A:BC}\tilde{F}^{-1B}_{\ \ .\beta}\tilde{F}^{-1C}_{\ \ .\chi} + \tilde{F}^{\alpha}_{.A:B}\tilde{F}^{-1B}_{\ \ .\beta:\chi},
\end{aligned} \tag{4.210}$$

where, similarly to (3.19),

$$\tilde{F}^{-1A}_{\ \ .\alpha:\beta} = -\tilde{F}^{-1A}_{\ \ .\chi}\tilde{F}^{-1B}_{\ \ .\alpha}\tilde{F}^{-1C}_{\ \ .\beta}\tilde{F}^{\chi}_{.B:C}. \tag{4.211}$$

Directly, the total covariant derivative of $\tilde{\boldsymbol{F}}^{-1}$ also obeys, from (4.206),

$$\tilde{F}^{-1A}_{\ \ .\alpha:\beta} = \partial_{\beta}\tilde{F}^{-1A}_{\ \ .\alpha} - \Gamma^{..\chi}_{\beta\alpha}\tilde{F}^{-1A}_{\ \ .\chi} + \overset{G}{\Gamma}{}^{..A}_{BC}\tilde{F}^{-1C}_{\ \ .\alpha}\tilde{F}^{-1B}_{\ \ .\beta}. \tag{4.212}$$

Next consider a generic two-point tensor of the form $\boldsymbol{A}(x,t)$, with components $A^{\alpha...\phi\ a...f}_{\gamma...\mu\ g...m}$. The total covariant derivative of $\boldsymbol{A}$ is computed completely analogously to (4.206), replacing referential coordinates X^N with spatial coordinates x^n and $\tilde{F}^{-1N}_{\ \ .\nu}$ with $\bar{F}^{n}_{.\nu}$:

$$\begin{aligned}
(A^{\alpha...\phi\ a...f}_{\gamma...\mu\ g...m})_{:\nu} &= \partial_{\nu}(A^{\alpha...\phi\ a...f}_{\gamma...\mu\ g...m}) \\
&\quad + \Gamma^{..\alpha}_{\nu\rho}A^{\rho...\phi\ a...f}_{\gamma...\mu\ g...m} + \cdots + \Gamma^{..\phi}_{\nu\rho}A^{\alpha...\rho\ a...f}_{\gamma...\mu\ g...m} \\
&\quad - \Gamma^{..\rho}_{\nu\gamma}A^{\alpha...\phi\ a...f}_{\rho...\mu\ g...m} - \cdots - \Gamma^{..\rho}_{\nu\mu}A^{\alpha...\phi\ a...f}_{\gamma...\rho\ g...m} \\
&\quad + \bar{F}^{n}_{.\nu}\overset{g}{\Gamma}{}^{..a}_{nr}A^{\alpha...\phi\ r...f}_{\gamma...\mu\ g...m} + \cdots + \bar{F}^{n}_{.\nu}\overset{g}{\Gamma}{}^{..f}_{nr}A^{\alpha...\phi\ a...r}_{\gamma...\mu\ g...m} \\
&\quad - \bar{F}^{n}_{.\nu}\overset{g}{\Gamma}{}^{..r}_{ng}A^{\alpha...\phi\ a...f}_{\gamma...\mu\ r...m} - \cdots - \bar{F}^{n}_{.\nu}\overset{g}{\Gamma}{}^{..r}_{nm}A^{\alpha...\phi\ a...f}_{\gamma...\mu\ g...r} \\
&= \Big[\partial_{n}(A^{\alpha...\phi\ a...f}_{\gamma...\mu\ g...m}) \\
&\quad + \overset{g}{\Gamma}{}^{..a}_{nr}A^{\alpha...\phi\ r...f}_{\gamma...\mu\ g...m} + \cdots + \overset{g}{\Gamma}{}^{..f}_{nr}A^{\alpha...\phi\ a...r}_{\gamma...\mu\ g...m} \\
&\quad - \overset{g}{\Gamma}{}^{..r}_{ng}A^{\alpha...\phi\ a...f}_{\gamma...\mu\ r...m} - \cdots - \overset{g}{\Gamma}{}^{..r}_{nm}A^{\alpha...\phi\ a...g}_{\gamma...\mu\ g...r}\Big]\bar{F}^{n}_{.\nu} \\
&\quad + \Gamma^{..\alpha}_{\nu\rho}A^{\rho...\psi\ u...f}_{\gamma...\mu\ g...m} + \cdots + \Gamma^{..\phi}_{\nu\rho}A^{\alpha...\rho\ a...f}_{\gamma...\mu\ g...m} \\
&\quad - \Gamma^{..\rho}_{\nu\gamma}A^{\alpha...\phi\ a...f}_{\rho...\mu\ g...m} - \cdots - \Gamma^{..\rho}_{\nu\mu}A^{\alpha...\phi\ a...f}_{\gamma...\rho\ g...m} \\
&= (A^{\alpha...\phi\ a...f}_{\gamma...\mu\ g...m})_{;n}\bar{F}^{n}_{.\nu} \\
&\quad + \Gamma^{..\alpha}_{\nu\rho}A^{\rho...\phi\ a...f}_{\gamma...\mu\ g...m} + \cdots + \Gamma^{..\phi}_{\nu\rho}A^{\alpha...\rho\ a...f}_{\gamma...\mu\ g...m} \\
&\quad - \Gamma^{..\rho}_{\nu\gamma}A^{\alpha...\phi\ a...f}_{\rho...\mu\ g...m} - \cdots - \Gamma^{..\rho}_{\nu\mu}A^{\alpha...\phi\ a...f}_{\gamma...\rho\ g...m} \\
&= (A^{\alpha...\phi\ a...f}_{\gamma...\mu\ g...m})_{:n}\bar{F}^{n}_{.\nu}.
\end{aligned} \tag{4.213}$$

For example, letting $\boldsymbol{A}(x,t) = \bar{\boldsymbol{F}}^{-1}(x,t) = \bar{F}^{-1\alpha}_{\ \ \bullet a}\tilde{\boldsymbol{g}}_\alpha \otimes \boldsymbol{g}^a$, the spatial gradient is computed as

$$\begin{aligned}
\overset{g}{\nabla}(\bar{\boldsymbol{F}}^{-1}) &= \partial_b \bar{\boldsymbol{F}}^{-1} \otimes \boldsymbol{g}^b \\
&= \partial_b(\bar{F}^{-1\alpha}_{\ \ \bullet a}\tilde{\boldsymbol{g}}_\alpha \otimes \boldsymbol{g}^a) \otimes \boldsymbol{g}^b \\
&= \partial_b \bar{F}^{-1\alpha}_{\ \ \bullet a}\tilde{\boldsymbol{g}}_\alpha \otimes \boldsymbol{g}^a \otimes \boldsymbol{g}^b \\
&\quad + \bar{F}^{-1\chi}_{\ \ \bullet a}\bar{F}^{-1\beta}_{\ \ \bullet b}(\partial_\beta \tilde{\boldsymbol{g}}_\chi) \otimes \boldsymbol{g}^a \otimes \boldsymbol{g}^b \\
&\quad + \bar{F}^{-1\alpha}_{\ \ \bullet c}\tilde{\boldsymbol{g}}_\alpha \otimes (\partial_b \boldsymbol{g}^c) \otimes \boldsymbol{g}^b \\
&= \partial_b \bar{F}^{-1\alpha}_{\ \ \bullet a}\tilde{\boldsymbol{g}}_\alpha \otimes \boldsymbol{g}^a \otimes \boldsymbol{g}^b \\
&\quad + \bar{F}^{-1\chi}_{\ \ \bullet a}\bar{F}^{-1\beta}_{\ \ \bullet b}(\Gamma^{\bullet\bullet\alpha}_{\beta\chi}\tilde{\boldsymbol{g}}_\alpha) \otimes \boldsymbol{g}^a \otimes \boldsymbol{g}^b \\
&\quad - \bar{F}^{-1\alpha}_{\ \ \bullet c}\tilde{\boldsymbol{g}}_\alpha \otimes (\overset{g}{\Gamma}{}^{\bullet\bullet c}_{ba}\boldsymbol{g}^a) \otimes \boldsymbol{g}^b \\
&= \left(\partial_b \bar{F}^{-1\alpha}_{\ \ \bullet a} + \Gamma^{\bullet\bullet\alpha}_{\beta\chi}\bar{F}^{-1\chi}_{\ \ \bullet a}\bar{F}^{-1\beta}_{\ \ \bullet b} - \overset{g}{\Gamma}{}^{\bullet\bullet c}_{ba}\bar{F}^{-1\alpha}_{\ \ \bullet c}\right)\tilde{\boldsymbol{g}}_\alpha \otimes \boldsymbol{g}^a \otimes \boldsymbol{g}^b \\
&= \bar{F}^{-1\alpha}_{\ \ \bullet a:b}\tilde{\boldsymbol{g}}_\alpha \otimes \boldsymbol{g}^a \otimes \boldsymbol{g}^b. \qquad (4.214)
\end{aligned}$$

Skew covariant components of $\overset{g}{\nabla}(\bar{\boldsymbol{F}}^{-1})$ do not necessarily vanish:

$$\begin{aligned}
\bar{F}^{-1\alpha}_{\ \ \bullet[a:b]} &= \partial_{[b}\bar{F}^{-1\alpha}_{\ \ \bullet a]} + \Gamma^{\bullet\bullet\alpha}_{\beta\chi}\bar{F}^{-1\beta}_{\ \ \bullet[b}\bar{F}^{-1\chi}_{\ \ \bullet a]} - \overset{g}{\Gamma}{}^{\bullet\bullet c}_{[ba]}\bar{F}^{-1\alpha}_{\ \ \bullet c} \\
&= \partial_{[b}\bar{F}^{-1\alpha}_{\ \ \bullet a]} + \Gamma^{\bullet\bullet\alpha}_{[\beta\chi]}\bar{F}^{-1\beta}_{\ \ \bullet b}\bar{F}^{-1\chi}_{\ \ \bullet a}. \qquad (4.215)
\end{aligned}$$

From (4.215), spatial gradient $\overset{g}{\nabla}(\bar{\boldsymbol{F}}^{-1})$ is generally symmetric in covariant indices only when both (4.17) holds and symmetric connection coefficients $\Gamma^{\bullet\bullet\alpha}_{\beta\chi} = \Gamma^{\bullet\bullet\alpha}_{(\beta\chi)}$ are prescribed on $\tilde{B}$.

Finally, consider a differentiable tensor field of order three or higher, with components referred to all three configurations $\tilde{B}$, B, and B_0, written as $A^{\alpha\ldots\phi\ a\ldots f\ A\ldots F}_{\gamma\ldots\mu\ g\ldots m\ G\ldots M}$. By extension, the total covariant derivative of field $\boldsymbol{A}$ is computed as

$$\begin{aligned}
(A^{\alpha\ldots\phi\ a\ldots f\ A\ldots F}_{\gamma\ldots\mu\ g\ldots m\ G\ldots M})_{:\nu} &= \partial_\nu(A^{\alpha\ldots\phi\ a\ldots f\ A\ldots F}_{\gamma\ldots\mu\ g\ldots m\ G\ldots M}) \\
&\quad + \Gamma^{\bullet\bullet\alpha}_{\nu\rho}A^{\rho\ldots\phi\ a\ldots f\ A\ldots F}_{\gamma\ldots\mu\ g\ldots m\ G\ldots M} + \cdots \\
&\quad + \Gamma^{\bullet\bullet\phi}_{\nu\rho}A^{\alpha\ldots\rho\ a\ldots f\ A\ldots F}_{\gamma\ldots\mu\ g\ldots m\ G\ldots M} \\
&\quad - \Gamma^{\bullet\bullet\rho}_{\nu\gamma}A^{\alpha\ldots\phi\ a\ldots f\ A\ldots F}_{\rho\ldots\mu\ g\ldots m\ G\ldots M} - \cdots \\
&\quad - \Gamma^{\bullet\bullet\rho}_{\nu\mu}A^{\alpha\ldots\phi\ a\ldots f\ A\ldots F}_{\gamma\ldots\rho\ g\ldots m\ G\ldots M} \\
&\quad + \bar{F}^{n}_{\bullet\nu}\overset{g}{\Gamma}{}^{\bullet\bullet a}_{nr}A^{\alpha\ldots\phi\ r\ldots f\ A\ldots F}_{\gamma\ldots\mu\ g\ldots m\ G\ldots M} + \cdots \\
&\quad + \bar{F}^{n}_{\bullet\nu}\overset{g}{\Gamma}{}^{\bullet\bullet f}_{nr}A^{\alpha\ldots\phi\ A\ldots r\ A\ldots F}_{\gamma\ldots\mu\ g\ldots m\ G\ldots M} \\
&\quad - \bar{F}^{n}_{\bullet\nu}\overset{g}{\Gamma}{}^{\bullet\bullet r}_{ng}A^{\alpha\ldots\phi\ a\ldots f\ A\ldots F}_{\gamma\ldots\mu\ r\ldots m\ G\ldots M} - \cdots \\
&\quad - \bar{F}^{n}_{\bullet\nu}\overset{g}{\Gamma}{}^{\bullet\bullet r}_{nm}A^{\alpha\ldots\phi\ a\ldots f\ A\ldots F}_{\gamma\ldots\mu\ g\ldots r\ G\ldots M}
\end{aligned}$$

$$
\begin{aligned}
&+ \tilde{F}^{-1N}_{\bullet\nu} \overset{G}{\Gamma}{}^{\bullet\bullet A}_{NR} A^{\alpha\dots\phi\, a\dots f\,\, R\dots F}_{\gamma\dots\mu\, g\dots m\, G\dots M} + \cdots \\
&+ \tilde{F}^{-1N}_{\bullet\nu} \overset{G}{\Gamma}{}^{\bullet\bullet F}_{NR} A^{\alpha\dots\phi\, a\dots f\,\, A\dots R}_{\gamma\dots\mu\, g\dots m\, G\dots M} \\
&- \tilde{F}^{-1N}_{\bullet\nu} \overset{G}{\Gamma}{}^{\bullet\bullet R}_{NG} A^{\alpha\dots\phi\, a\dots f\,\, A\dots F}_{\gamma\dots\mu\, g\dots m\, R\dots M} - \cdots \\
&- \tilde{F}^{-1N}_{\bullet\nu} \overset{G}{\Gamma}{}^{\bullet\bullet R}_{NM} A^{\alpha\dots\phi\, a\dots f\,\, A\dots F}_{\gamma\dots\mu\, g\dots m\, G\dots R} \\
&= (A^{\alpha\dots\phi\, a\dots f\,\, A\dots F}_{\gamma\dots\mu\, g\dots m\, G\dots M})_{:n} \bar{F}^{n}_{\bullet\nu} \\
&= (A^{\alpha\dots\phi\, a\dots f\,\, A\dots F}_{\gamma\dots\mu\, g\dots m\, G\dots M})_{:N} \tilde{F}^{-1N}_{\bullet\nu} \,. \qquad (4.216)
\end{aligned}
$$

Consider a differential line element $\mathrm{d}\boldsymbol{X}$ in the reference configuration. Such an element is mapped to its representation in the intermediate configuration $\mathrm{d}\tilde{\boldsymbol{x}}$ analogously to (3.14) via the Taylor series [7]

$$
\begin{aligned}
\mathrm{d}\tilde{x}^{\alpha}(X) &= (\tilde{F}^{\alpha}_{\bullet A})\Big|_X \mathrm{d}X^A + \frac{1}{2!}(\tilde{F}^{\alpha}_{\bullet A:B})\Big|_X \mathrm{d}X^A\mathrm{d}X^B \\
&\quad + \frac{1}{3!}(\tilde{F}^{\alpha}_{\bullet A:BC})\Big|_X \mathrm{d}X^A\mathrm{d}X^B\mathrm{d}X^C + \cdots \\
&= (\tilde{F}^{\alpha}_{\bullet A})\Big|_X \mathrm{d}X^A + \frac{1}{2!}\tilde{F}^{\alpha}_{\bullet(A:B)}\Big|_X \mathrm{d}X^A\mathrm{d}X^B \\
&\quad + \frac{1}{3!}\tilde{F}^{\alpha}_{\bullet(A:BC)}\Big|_X \mathrm{d}X^A\mathrm{d}X^B\mathrm{d}X^C + \cdots \,, \qquad (4.217)
\end{aligned}
$$

where components of the total referential covariant derivative of $\tilde{\boldsymbol{F}}$ are given in (4.207), and components of $\tilde{F}^{\alpha}_{\bullet A:BC}$ are given in (4.209). Notice that only symmetric parts of gradients of $\tilde{\boldsymbol{F}}$ contribute to (4.217) because of the symmetries $\mathrm{d}X^A\mathrm{d}X^B = \mathrm{d}X^B\mathrm{d}X^A$ and $\mathrm{d}X^A\mathrm{d}X^B\mathrm{d}X^C = \mathrm{d}X^{(A}\mathrm{d}X^B\mathrm{d}X^{C)}$.

Similarly, consider a differential line element $\mathrm{d}\boldsymbol{x}$ in the current configuration. Such an element is mapped to its representation in the intermediate configuration $\mathrm{d}\tilde{\boldsymbol{x}}$ via the Taylor series [7]

$$
\begin{aligned}
\mathrm{d}\tilde{x}^{\alpha}(x) &= (\bar{F}^{-1\alpha}_{\bullet a})\Big|_x \mathrm{d}x^a + \frac{1}{2!}(\bar{F}^{-1\alpha}_{\bullet a:b})\Big|_x \mathrm{d}x^a\mathrm{d}x^b \\
&\quad + \frac{1}{3!}(\bar{F}^{-1\alpha}_{\bullet a:bc})\Big|_x \mathrm{d}x^a\mathrm{d}x^b\mathrm{d}x^c + \cdots \\
&= (\bar{F}^{-1\alpha}_{\bullet a})\Big|_x \mathrm{d}x^a + \frac{1}{2!}\bar{F}^{-1\alpha}_{\bullet(a:b)}\Big|_x \mathrm{d}x^a\mathrm{d}x^b \\
&\quad + \frac{1}{3!}\bar{F}^{-1\alpha}_{\bullet(a:bc)}\Big|_x \mathrm{d}x^a\mathrm{d}x^b\mathrm{d}x^c + \cdots \,, \qquad (4.218)
\end{aligned}
$$

where components of the total spatial covariant derivative of $\bar{\boldsymbol{F}}^{-1}$ are given in (4.214), and components of $\bar{F}^{-1\alpha}_{\bullet a:bc}$ can be obtained through iteration of (4.213). Notice that only symmetric parts of gradients of $\bar{\boldsymbol{F}}^{-1}$ contribute to (4.218) because of the symmetries $\mathrm{d}x^a\mathrm{d}x^b = \mathrm{d}x^b\mathrm{d}x^a$ and $\mathrm{d}x^a\mathrm{d}x^b\mathrm{d}x^c = \mathrm{d}x^{(a}\mathrm{d}x^b\mathrm{d}x^{c)}$.

To first order in $\mathrm{d}\boldsymbol{X}$ and $\mathrm{d}\boldsymbol{x}$, and using (3.28) and (4.2),

$$\begin{aligned}\mathrm{d}\tilde{x}^{\alpha}(X) &= (\tilde{F}^{\alpha}_{\cdot A})\Big|_{X}\mathrm{d}X^{A}\\ &= (\bar{F}^{-1\alpha}_{\cdot a}F^{a}_{\cdot A})\Big|_{X}\mathrm{d}X^{A}\\ &= (\bar{F}^{-1\alpha}_{\cdot a})\Big|_{X}\mathrm{d}x^{a}(X),\end{aligned} \tag{4.219}$$

or in direct notation, using (4.1),

$$\mathrm{d}\tilde{\boldsymbol{x}} = \tilde{\boldsymbol{F}}\,\mathrm{d}\boldsymbol{X} = \bar{\boldsymbol{F}}^{-1}\boldsymbol{F}\,\mathrm{d}\boldsymbol{X} = \bar{\boldsymbol{F}}^{-1}\,\mathrm{d}\boldsymbol{x}. \tag{4.220}$$

Relationship (4.220) is a standard proposition that will be used often in subsequent derivations.

4.2.4 *Divergence, curl, and Laplacian*

Definitions that follow correspond to anholonomic space $\tilde{B}$, for which (4.138)–(4.147) apply for covariant and partial differentiation. The divergence of a contravariant vector field $\boldsymbol{V} = V^{\alpha}\tilde{\boldsymbol{g}}_{\alpha}$ is

$$\begin{aligned}\langle\nabla,\boldsymbol{V}\rangle &= \mathrm{tr}\,(\nabla\boldsymbol{V})\\ &= \langle\partial_{\alpha}\boldsymbol{V},\tilde{\boldsymbol{g}}^{\alpha}\rangle\\ &= \langle\partial_{\alpha}(V^{\beta}\tilde{\boldsymbol{g}}_{\beta}),\tilde{\boldsymbol{g}}^{\alpha}\rangle\\ &= \langle\nabla_{\alpha}V^{\beta}\tilde{\boldsymbol{g}}_{\beta},\tilde{\boldsymbol{g}}^{\alpha}\rangle\\ &= \nabla_{\alpha}V^{\alpha}\\ &= \partial_{\alpha}V^{\alpha} + \Gamma^{\cdot\cdot\alpha}_{\alpha\beta}V^{\beta}.\end{aligned} \tag{4.221}$$

Notice that the analog of the final equality in (2.172) does not necessarily apply here because the final equality in (2.144) may not hold for connection coefficients on $\tilde{B}$ [*i.e.*, $\Gamma^{\cdot\cdot\alpha}_{\beta\chi}$ are metric coefficients but are not necessarily symmetric (Levi-Civita) connection coefficients]. Explicitly, invoking (4.151),

$$\begin{aligned}\frac{1}{\sqrt{\tilde{g}}}\partial_{\alpha}(\sqrt{\tilde{g}}V^{\alpha}) &= \partial_{\alpha}V^{\alpha} + \Gamma^{\cdot\cdot\beta}_{\alpha\beta}V^{\alpha}\\ &= \nabla_{\alpha}V^{\alpha} + 2\,\Gamma^{\cdot\cdot\beta}_{[\alpha\beta]}V^{\beta}.\end{aligned} \tag{4.222}$$

The vector cross product $\times$ obeys, for two vectors $\boldsymbol{V}$ and $\boldsymbol{W}$ and two covectors $\boldsymbol{\alpha}$ and $\boldsymbol{\beta}$,

$$\boldsymbol{V}\times\boldsymbol{W} = \epsilon_{\alpha\beta\chi}V^{\beta}W^{\chi}\tilde{\boldsymbol{g}}^{\alpha}, \tag{4.223}$$

$$\boldsymbol{\alpha}\times\boldsymbol{\beta} = \epsilon^{\alpha\beta\chi}\alpha_{\beta}\beta_{\chi}\tilde{\boldsymbol{g}}_{\alpha}. \tag{4.224}$$

The curl of a covariant vector field $\boldsymbol{\alpha} = \alpha_\alpha \tilde{\boldsymbol{g}}^\alpha$ is then defined as

$$\begin{aligned}\nabla \times \boldsymbol{\alpha} &= \tilde{\boldsymbol{g}}^\alpha \times \partial_\alpha(\alpha_\beta \tilde{\boldsymbol{g}}^\beta) \\ &= \tilde{\boldsymbol{g}}^\alpha \times \tilde{\boldsymbol{g}}^\beta \nabla_\alpha \alpha_\beta \\ &= \epsilon^{\alpha\beta\chi} \nabla_\alpha \alpha_\beta \tilde{\boldsymbol{g}}_\chi \\ &= \epsilon^{\alpha\beta\chi} \nabla_\beta \alpha_\chi \tilde{\boldsymbol{g}}_\alpha. \end{aligned} \tag{4.225}$$

A relationship like the final equality in (2.175) does not necessarily hold here since the analog of (2.176) need not apply for possibly non-symmetric coefficients $\Gamma^{\cdot\cdot\alpha}_{\beta\chi}$.

The Laplacian of a scalar field f is, from the symmetry of inverse metric $\tilde{g}^{\alpha\beta}$ and the vanishing of the covariant derivative of the inverse metric tensor in (4.154),

$$\begin{aligned}\nabla^2 f &= \tilde{g}^{\alpha\beta} \nabla_\alpha \nabla_\beta f \\ &= \tilde{g}^{\alpha\beta} \nabla_\beta \nabla_\alpha f \\ &= \nabla_\alpha (\tilde{g}^{\alpha\beta} \partial_\beta f) \\ &= \nabla_\beta (\tilde{g}^{\alpha\beta} \partial_\alpha f) \\ &= \nabla_\alpha \nabla_\beta (\tilde{g}^{\alpha\beta} f) \\ &= \nabla_\beta \nabla_\alpha (\tilde{g}^{\alpha\beta} f) \\ &= \tilde{g}^{\alpha\beta} \nabla_\alpha (\partial_\beta f) \\ &= \tilde{g}^{\alpha\beta} \nabla_\beta (\partial_\alpha f) \\ &= \tilde{g}^{\alpha\beta} (\partial_\beta \partial_\alpha f - \Gamma^{\cdot\cdot\chi}_{\beta\alpha} \partial_\chi f). \end{aligned} \tag{4.226}$$

Because the anholonomic connection coefficients need not be symmetric, an identity like the third equality in (2.177) does not necessarily apply here:

$$\begin{aligned}\frac{1}{\sqrt{\tilde{g}}} \partial_\beta (\sqrt{\tilde{g}} \tilde{g}^{\alpha\beta} \partial_\alpha f) &= \tilde{g}^{\alpha\beta} \partial_\beta \partial_\alpha f + \frac{\partial_\alpha f}{\sqrt{\tilde{g}}} \partial_\beta (\sqrt{\tilde{g}} \tilde{g}^{\alpha\beta}) \\ &= \tilde{g}^{\alpha\beta} \partial_\beta \partial_\alpha f \\ &\quad + \frac{\partial_\alpha f}{\sqrt{\tilde{g}}} (\tilde{g}^{\alpha\beta} \partial_\beta \sqrt{\tilde{g}} + \sqrt{\tilde{g}} \partial_\beta \tilde{g}^{\alpha\beta}) \\ &= \tilde{g}^{\alpha\beta} \partial_\beta \partial_\alpha f \\ &\quad + \partial_\alpha f (\tilde{g}^{\alpha\beta} \Gamma^{\cdot\cdot\delta}_{\beta\delta} - \tilde{g}^{\delta\alpha} \Gamma^{\cdot\cdot\beta}_{\beta\delta} - \tilde{g}^{\delta\beta} \Gamma^{\cdot\cdot\alpha}_{\beta\delta}) \\ &= \tilde{g}^{\alpha\beta} \partial_\beta \partial_\alpha f \\ &\quad + \partial_\alpha f (\tilde{g}^{\alpha\beta} \Gamma^{\cdot\cdot\delta}_{\beta\delta} - \tilde{g}^{\alpha\beta} \Gamma^{\cdot\cdot\delta}_{\delta\beta} - \tilde{g}^{\delta\beta} \Gamma^{\cdot\cdot\alpha}_{\beta\delta}) \\ &= \tilde{g}^{\alpha\beta} \partial_\beta \partial_\alpha f - \tilde{g}^{\alpha\beta} \Gamma^{\cdot\cdot\chi}_{\beta\alpha} \partial_\alpha f + 2 \partial_\alpha f (\tilde{g}^{\alpha\beta} \Gamma^{\cdot\cdot\delta}_{[\beta\delta]}) \\ &= \nabla^2 f + 2 \tilde{g}^{\alpha\beta} \Gamma^{\cdot\cdot\delta}_{[\beta\delta]} \partial_\alpha f. \end{aligned} \tag{4.227}$$

Unlike (2.178), the divergence of the curl of a (co)vector field does not necessarily vanish in an anholonomic space:

$$\begin{aligned}\langle\nabla,\nabla\times\boldsymbol{\alpha}\rangle &= \nabla_{\alpha}(\epsilon^{\alpha\beta\chi}\nabla_{\beta}\alpha_{\chi})\\ &= \nabla_{\alpha}[(1/\sqrt{\tilde{g}})e^{\alpha\beta\chi}\nabla_{\beta}\alpha_{\chi}]\\ &= \nabla_{\alpha}[(1/\sqrt{\tilde{g}})e^{\alpha\beta\chi}]\nabla_{\beta}\alpha_{\chi} + [(1/\sqrt{\tilde{g}})e^{\alpha\beta\chi}]\nabla_{[\alpha}\nabla_{\beta]}\alpha_{\chi}\\ &= \nabla_{\alpha}[(1/\sqrt{\tilde{g}})e^{\alpha\beta\chi}]\nabla_{\beta}\alpha_{\chi}\\ &\quad - \frac{1}{2\sqrt{\tilde{g}}}e^{\alpha\beta\chi}[R^{\cdot\cdot\cdot\delta}_{\alpha\beta\chi}\alpha_{\delta} + 2T^{\cdot\cdot\delta}_{\alpha\beta}\nabla_{\delta}\alpha_{\chi}]\\ &= -\frac{1}{2}\epsilon^{\alpha\beta\chi}[R^{\cdot\cdot\cdot\delta}_{\alpha\beta\chi}\alpha_{\delta} + 2T^{\cdot\cdot\delta}_{\alpha\beta}\nabla_{\delta}\alpha_{\chi}].\end{aligned} \tag{4.228}$$

Identity (4.164) has been used with the vanishing property of the covariant derivative of the permutation tensor (4.156). Consulting (4.162), the curl of the gradient of a scalar field may be nonzero in anholonomic coordinates:

$$\begin{aligned}\nabla\times\nabla f &= \epsilon^{\alpha\beta\chi}(\nabla_{\beta}\nabla_{\chi}f)\tilde{\boldsymbol{g}}_{\alpha}\\ &= \epsilon^{\alpha\beta\chi}[\nabla_{[\beta}(\partial_{\chi]}f)]\tilde{\boldsymbol{g}}_{\alpha}\\ &= \epsilon^{\alpha\beta\chi}(\partial_{[\beta}\partial_{\chi]}f - \Gamma^{\cdot\cdot\gamma}_{[\beta\chi]}\partial_{\gamma}f)\tilde{\boldsymbol{g}}_{\alpha}\\ &= \epsilon^{\alpha\beta\chi}(T^{\cdot\cdot\gamma}_{\chi\beta}\partial_{\gamma}f)\tilde{\boldsymbol{g}}_{\alpha}.\end{aligned} \tag{4.229}$$

Identities (4.228) and (4.229) are now examined for the four particular classes of connection coefficients $\Gamma^{\cdot\cdot\alpha}_{\beta\chi}$ defined in Section 4.2.2.

If configuration $\tilde{B}$ is a Euclidean space and (4.165) defines the connection coefficients, then equalities (4.228) and (4.229) are identically zero.

When (4.171) defines the connection coefficients, *i.e.*, when identical (possibly curvilinear) coordinate systems are used in reference and intermediate configurations, then (4.228) and (4.229) become, respectively,

$$\begin{aligned}\langle\nabla,\nabla\times\boldsymbol{\alpha}\rangle &= \nabla_{\alpha}[(1/\sqrt{\tilde{g}})e^{\alpha\beta\chi}]\nabla_{\beta}\alpha_{\chi}\\ &\quad - \frac{1}{2\sqrt{\tilde{g}}}e^{\alpha\beta\chi}[R^{\cdot\cdot\cdot\delta}_{\alpha\beta\chi}\alpha_{\delta} + 2T^{\cdot\cdot\delta}_{\alpha\beta}\nabla_{\delta}\alpha_{\chi}]\\ &= \frac{1}{\sqrt{G}}e^{\alpha\beta\chi}[\delta^{C}_{\chi}\delta^{\delta}_{D}\tilde{F}^{-1A}_{\cdot[\alpha}\tilde{F}^{-1B}_{\cdot\beta]}\partial_{A}(\overset{G}{\Gamma}{}^{\cdot\cdot D}_{BC})\alpha_{\delta}\\ &\quad - (\delta^{\delta}_{D}\overset{G}{\Gamma}{}^{\cdot\cdot D}_{AB}\tilde{F}^{-1A}_{\cdot[\alpha}\delta^{B}_{\beta]} + \tilde{F}^{-1A}_{\cdot[\alpha}\tilde{F}^{-1B}_{\cdot\beta]}\partial_{A}\tilde{F}^{\delta}_{\cdot B})\\ &\quad \times(\partial_{\delta}\alpha_{\chi} - \Gamma^{\cdot\cdot\epsilon}_{\delta\chi}\alpha_{\epsilon})]\\ &= \frac{1}{\sqrt{G}}e^{\alpha\beta\chi}[\delta^{C}_{\chi}\delta^{\delta}_{D}\tilde{F}^{-1A}_{\cdot[\alpha}\tilde{F}^{-1B}_{\cdot\beta]}\partial_{A}(\overset{G}{\Gamma}{}^{\cdot\cdot D}_{BC})\alpha_{\delta}\\ &\quad - (\delta^{\delta}_{D}\overset{G}{\Gamma}{}^{\cdot\cdot D}_{AB}\tilde{F}^{-1A}_{\cdot[\alpha}\delta^{B}_{\beta]} + \tilde{F}^{-1A}_{\cdot[\alpha}\tilde{F}^{-1B}_{\cdot\beta]}\partial_{A}\tilde{F}^{\delta}_{\cdot B})\\ &\quad \times\tilde{F}^{-1D}_{\cdot\delta}(\partial_{D}\alpha_{\chi} - \delta^{\epsilon}_{E}\delta^{C}_{\chi}\overset{G}{\Gamma}{}^{\cdot\cdot E}_{DC}\alpha_{\epsilon})],\end{aligned} \tag{4.230}$$

$$\nabla \times \nabla f = -\epsilon^{\alpha\beta\chi}(T^{\cdot\cdot\gamma}_{\beta\chi}\partial_\gamma f)\tilde{\boldsymbol{g}}_\alpha$$
$$= -\frac{1}{\sqrt{G}}e^{\alpha\beta\chi}[(\delta^{\gamma}_{A}{}^{G}_{\Gamma}{}^{\cdot\cdot A}_{BC}\tilde{F}^{-1B}_{\cdot[\beta}\delta^{C}_{\chi]} + \tilde{F}^{-1A}_{\cdot[\beta}\tilde{F}^{-1B}_{\cdot\chi]}\partial_A\tilde{F}^{\gamma}_{\cdot B})$$
$$\times \tilde{F}^{-1E}_{\cdot\gamma}\partial_E f]\delta^F_\alpha \boldsymbol{G}_F. \tag{4.231}$$

Definitions (4.176) and (4.177) have been used for anholonomic torsion and curvature, respectively. Recall that neither tensor need be zero even if $\tilde{\boldsymbol{F}}$ is integrable. Recall also that vanishing of the covariant derivative of permutation tensor $\epsilon^{\alpha\beta\chi} = (1/\sqrt{\tilde{g}})e^{\alpha\beta\chi}$ in (4.230) is a consequence of the metric property of connection coefficients $\Gamma^{\cdot\cdot\alpha}_{\beta\chi}$.

Analogously, when (4.185) defines the connection coefficients, *i.e.*, when identical (possibly curvilinear) coordinate systems are used in current and intermediate configurations, then (4.228) and (4.229) become, respectively,

$$\langle\nabla, \nabla\times\boldsymbol{\alpha}\rangle = \nabla_\alpha[(1/\sqrt{\bar{g}})e^{\alpha\beta\chi}]\nabla_\beta\alpha_\chi$$
$$- \frac{1}{2\sqrt{\bar{g}}}e^{\alpha\beta\chi}[R^{\cdot\cdot\cdot\delta}_{\alpha\beta\chi}\alpha_\delta + 2T^{\cdot\cdot\delta}_{\alpha\beta}\nabla_\delta\alpha_\chi]$$
$$= \frac{1}{\sqrt{g}}e^{\alpha\beta\chi}[\delta^c_\chi\delta^\delta_d\bar{F}^a_{\cdot[\alpha}\bar{F}^b_{\cdot\beta]}\partial_a({}^{g}_{\Gamma}{}^{\cdot\cdot d}_{bc})\alpha_\delta$$
$$- (\delta^{\delta g}_{d\Gamma}{}^{\cdot\cdot d}_{ab}\bar{F}^a_{\cdot[\alpha}\delta^b_{\beta]} + \bar{F}^a_{\cdot[\alpha}\bar{F}^b_{\cdot\beta]}\partial_a\bar{F}^{-1\delta}_{\cdot b})$$
$$\times(\partial_\delta\alpha_\chi - \Gamma^{\cdot\cdot\epsilon}_{\delta\chi}\alpha_\epsilon)]$$
$$= \frac{1}{\sqrt{g}}e^{\alpha\beta\chi}[\delta^c_\chi\delta^\delta_d\bar{F}^a_{\cdot[\alpha}\bar{F}^b_{\cdot\beta]}\partial_a({}^{g}_{\Gamma}{}^{\cdot\cdot d}_{bc})\alpha_\delta$$
$$- (\delta^{\delta g}_{d\Gamma}{}^{\cdot\cdot d}_{ab}\bar{F}^a_{\cdot[\alpha}\delta^b_{\beta]} + \bar{F}^a_{\cdot[\alpha}\bar{F}^b_{\cdot\beta]}\partial_a\bar{F}^{-1\delta}_{\cdot b})$$
$$\times \bar{F}^d_{\cdot\delta}(\partial_d\alpha_\chi - \delta^\epsilon_e\delta^c_\chi{}^{g}_{\Gamma}{}^{\cdot\cdot e}_{dc}\alpha_\epsilon)], \tag{4.232}$$

$$\nabla \times \nabla f = -\epsilon^{\alpha\beta\chi}(T^{\cdot\cdot\gamma}_{\beta\chi}\partial_\gamma f)\tilde{\boldsymbol{g}}_\alpha$$
$$= -\frac{1}{\sqrt{g}}e^{\alpha\beta\chi}[(\delta^{\gamma g}_{a\Gamma}{}^{\cdot\cdot a}_{bc}\bar{F}^b_{\cdot[\beta}\delta^c_{\chi]} + \bar{F}^a_{\cdot[\beta}\bar{F}^b_{\cdot\chi]}\partial_a\bar{F}^{-1\gamma}_{\cdot b})$$
$$\times \bar{F}^e_{\cdot\gamma}\partial_e f]\delta^f_\alpha\boldsymbol{g}_f. \tag{4.233}$$

Definitions (4.190) and (4.191) have been used for anholonomic torsion and curvature, respectively. Recall that neither tensor need be zero even if $\bar{\boldsymbol{F}}^{-1}$ is integrable.

Finally, consider the case when (4.198) defines the (null) connection coefficients, *i.e.*, when a Cartesian system is used in the intermediate configuration, such that $\Gamma^{\cdot\cdot\alpha}_{\beta\chi} = 0$ identically. In that case, (4.228) and (4.229) become, respectively,

$$\begin{aligned}\langle \nabla, \nabla \times \boldsymbol{\alpha} \rangle &= \nabla_\alpha (\epsilon^{\alpha\beta\chi} \nabla_\beta \alpha_\chi) \\ &= e^{\alpha\beta\chi} \nabla_\alpha (\nabla_\beta \alpha_\chi) \\ &= e^{\alpha\beta\chi} \partial_{[\alpha} (\partial_{\beta]} \alpha_\chi) \\ &= e^{\alpha\beta\chi} \kappa_{\beta\alpha}^{\cdot\cdot\delta} \partial_\delta \alpha_\chi, \end{aligned} \tag{4.234}$$

$$\begin{aligned}\nabla \times \nabla f &= \epsilon^{\alpha\beta\chi} (\nabla_\beta \nabla_\chi f) \tilde{\boldsymbol{g}}_\alpha \\ &= e^{\alpha\beta\chi} [\partial_{[\beta} (\partial_{\chi]} f)] \boldsymbol{e}_\alpha \\ &= e^{\alpha\beta\chi} [\kappa_{\chi\beta}^{\cdot\cdot\delta} \partial_\delta f] \boldsymbol{e}_\alpha. \end{aligned} \tag{4.235}$$

Thus, even if intermediate connection coefficients vanish (*i.e.*, even if $\Gamma_{\beta\chi}^{\cdot\cdot\alpha} = 0$), (4.234) and (4.235) may be nonzero as a consequence of (4.21) since in general anholonomic coordinates, $\partial_\alpha[\partial_\beta(\cdot)] \neq \partial_\beta[\partial_\alpha(\cdot)]$; specifically, from (4.161), $\partial_{[\alpha}[\partial_{\beta]}(\cdot)] = \kappa_{\beta\alpha}^{\cdot\cdot\chi}\partial_\chi(\cdot)$, where $\kappa_{\beta\alpha}^{\cdot\cdot\chi}$ is the anholonomic object.

4.2.5 *Anholonomic cylindrical coordinates*

As an illustrative example of curvilinear intermediate (anholonomic) coordinates, consider cylindrical spatial coordinates of Section 3.4.2 mapped to the intermediate configuration $\tilde{B}$. Because cylindrical spatial coordinates exhibit relatively simple forms for metric tensors and connection coefficients, corresponding quantities mapped to the intermediate configuration can be derived explicitly by inspection.

In this example, (4.55)–(4.61) apply for basis vectors and metric tensors. Since the angular coordinate in current configuration B is labeled θ, overbars are used here to denote particular (anholonomic) coordinates referred to the intermediate configuration $\tilde{B}$. Specifically, holonomic coordinates on B are denoted by the usual (r, θ, z), while anholonomic coordinates on $\tilde{B}$ are denoted by $(\bar{r}, \bar{\theta}, \bar{z})$. In what follows, indices (r, θ, z) and $(\bar{r}, \bar{\theta}, \bar{z})$ refer to specific coordinates and are exempt from the summation convention. On the other hand, generic indices $(a, b, c, d \ldots)$ and $(\alpha, \beta, \chi, \delta \ldots)$ refer to free spatial and intermediate quantities, respectively, and are subject to summation over repeated indices.

Intermediate basis vectors, from (3.241) and (4.55), are

$$\tilde{\boldsymbol{g}}_{\bar{r}}(x) = \partial_r \boldsymbol{x}, \qquad \tilde{\boldsymbol{g}}_{\bar{\theta}}(x) = \partial_\theta \boldsymbol{x}, \qquad \tilde{\boldsymbol{g}}_{\bar{z}}(x) = \partial_z \boldsymbol{x}. \tag{4.236}$$

The metric tensor and its inverse, from (3.237) and (4.57), are

$$[\tilde{g}_{\alpha\beta}] = \begin{bmatrix} 1 & 0 & 0 \\ 0 & r^2 & 0 \\ 0 & 0 & 1 \end{bmatrix}, \qquad [\tilde{g}^{\alpha\beta}] = \begin{bmatrix} 1 & 0 & 0 \\ 0 & 1/r^2 & 0 \\ 0 & 0 & 1 \end{bmatrix}. \tag{4.237}$$

Determinants of the intermediate metric tensor and its inverse, from (3.238) and (4.59), are

$$\tilde{g} = r^2, \qquad \tilde{g}^{-1} = 1/r^2. \tag{4.238}$$

To second order in $\mathrm{d}\boldsymbol{x}$, from (4.220), the squared length of an intermediate line element is

$$\begin{aligned}
|\mathrm{d}\tilde{\boldsymbol{x}}|^2 &= \mathrm{d}\tilde{\boldsymbol{x}} \cdot \mathrm{d}\tilde{\boldsymbol{x}} \\
&= \bar{F}^{-1\alpha}_{\cdot a}\mathrm{d}x^a \tilde{g}_{\alpha\beta}\bar{F}^{-1\beta}_{\cdot b}\mathrm{d}x^b \\
&= (\bar{F}^{-1\alpha}_{\cdot r}\mathrm{d}r + \bar{F}^{-1\alpha}_{\cdot\theta}\mathrm{d}\theta + \bar{F}^{-1\alpha}_{\cdot z}\mathrm{d}z)\tilde{g}_{\alpha\beta} \\
&\quad \times (\bar{F}^{-1\beta}_{\cdot r}\mathrm{d}r + \bar{F}^{-1\beta}_{\cdot\theta}\mathrm{d}\theta + \bar{F}^{-1\beta}_{\cdot z}\mathrm{d}z).
\end{aligned} \tag{4.239}$$

As derived formally later in (5.34) and (5.35) for general coordinate systems, an intermediate volume element is, from (3.239),

$$\mathrm{d}\tilde{V} = (\det \bar{\boldsymbol{F}}^{-1})r \, \mathrm{d}r \, \mathrm{d}\theta \, \mathrm{d}z. \tag{4.240}$$

In the present example, definitions (4.185), (4.190), and (4.191) apply for respective connection coefficients, torsion, and curvature referred to the intermediate configuration. From (3.240) and (4.185), anholonomic connection coefficients consist of up to nine nonzero components:

$$\Gamma^{\cdot\cdot\bar{\theta}}_{\bar{r}\bar{\theta}} = \bar{F}^r_{\bar{r}}/r, \qquad \Gamma^{\cdot\cdot\bar{\theta}}_{\bar{\theta}\bar{\theta}} = \bar{F}^r_{\bar{\theta}}/r, \qquad \Gamma^{\cdot\cdot\bar{\theta}}_{\bar{z}\bar{\theta}} = \bar{F}^r_{\bar{z}}/r; \tag{4.241}$$

$$\Gamma^{\cdot\cdot\bar{\theta}}_{\bar{r}\bar{r}} = \bar{F}^\theta_{\bar{r}}/r, \qquad \Gamma^{\cdot\cdot\bar{\theta}}_{\bar{\theta}\bar{r}} = \bar{F}^\theta_{\bar{\theta}}/r, \qquad \Gamma^{\cdot\cdot\bar{\theta}}_{\bar{z}\bar{r}} = \bar{F}^\theta_{\bar{z}}/r; \tag{4.242}$$

$$\Gamma^{\cdot\cdot\bar{r}}_{\bar{r}\bar{\theta}} = -\bar{F}^\theta_{\bar{r}}r, \qquad \Gamma^{\cdot\cdot\bar{r}}_{\bar{\theta}\bar{\theta}} = -\bar{F}^\theta_{\bar{\theta}}r, \qquad \Gamma^{\cdot\cdot\bar{r}}_{\bar{z}\bar{\theta}} = -\bar{F}^\theta_{\bar{z}}r. \tag{4.243}$$

Possibly nonzero skew covariant components of these coefficients are

$$\Gamma^{\cdot\cdot\bar{\theta}}_{[\bar{r}\bar{\theta}]} = \frac{1}{2r}(\bar{F}^r_{\bar{r}} - \bar{F}^\theta_{\bar{\theta}}), \qquad \Gamma^{\cdot\cdot\bar{\theta}}_{[\bar{z}\bar{\theta}]} = \frac{1}{2r}\bar{F}^r_{\bar{z}}; \tag{4.244}$$

$$\Gamma^{\cdot\cdot\bar{\theta}}_{[\bar{z}\bar{r}]} = \frac{1}{2r}\bar{F}^\theta_{\bar{z}}; \tag{4.245}$$

$$\Gamma^{\cdot\cdot\bar{r}}_{[\bar{r}\bar{\theta}]} = -\frac{r}{2}\bar{F}^\theta_{\bar{r}}, \qquad \Gamma^{\cdot\cdot\bar{r}}_{[\bar{z}\bar{\theta}]} = -\frac{r}{2}\bar{F}^\theta_{\bar{z}}. \tag{4.246}$$

Torsion can then be computed from (4.190):

$$T^{\cdot\cdot\alpha}_{\beta\chi} = \Gamma^{\cdot\cdot\alpha}_{[\beta\chi]} + \bar{\kappa}^{\cdot\cdot\alpha}_{\beta\chi}, \tag{4.247}$$

noting that the anholonomic object here is

$$\bar{\kappa}^{\cdot\cdot\alpha}_{\beta\chi} = \bar{F}^a_{\cdot\beta}\bar{F}^b_{\cdot\chi}\partial_{[a}\bar{F}^{-1\alpha}_{\cdot b]}. \tag{4.248}$$

Spatial connection coefficients (3.240) have the following nonzero skew partial derivatives:

$$\partial_{[r} \overset{g}{\Gamma}{}^{\cdot\cdot\theta}_{\theta]r} = -\frac{1}{2r^2}, \qquad \partial_{[r} \overset{g}{\Gamma}{}^{\cdot\cdot r}_{\theta]\theta} = -\frac{1}{2}. \tag{4.249}$$

Therefore, possibly non-vanishing components of anholonomic curvature (4.191) follow as

$$R^{\cdot\cdot\cdot\bar{\theta}}_{\beta\chi\bar{r}} = \frac{2}{r^2}\bar{F}^{r}_{\cdot[\beta}\bar{F}^{\theta}_{\cdot\chi]}, \qquad R^{\cdot\cdot\cdot\bar{r}}_{\beta\chi\bar{\theta}} = 2\bar{F}^{r}_{\cdot[\beta}\bar{F}^{\theta}_{\cdot\chi]}. \tag{4.250}$$

Written out completely with no free indices, (4.250) yields the six possibly non-zero components

$$R^{\cdot\cdot\cdot\bar{\theta}}_{\bar{r}\bar{\theta}\bar{r}} = \frac{1}{r^2}(\bar{F}^{r}_{\cdot\bar{r}}\bar{F}^{\theta}_{\cdot\bar{\theta}} - \bar{F}^{r}_{\cdot\bar{\theta}}\bar{F}^{\theta}_{\cdot\bar{r}}), \tag{4.251}$$

$$R^{\cdot\cdot\cdot\bar{\theta}}_{\bar{\theta}\bar{z}\bar{r}} = \frac{1}{r^2}(\bar{F}^{r}_{\cdot\bar{\theta}}\bar{F}^{\theta}_{\cdot\bar{z}} - \bar{F}^{r}_{\cdot\bar{z}}\bar{F}^{\theta}_{\cdot\bar{\theta}}), \tag{4.252}$$

$$R^{\cdot\cdot\cdot\bar{\theta}}_{\bar{z}\bar{r}\bar{r}} = \frac{1}{r^2}(\bar{F}^{r}_{\cdot\bar{z}}\bar{F}^{\theta}_{\cdot\bar{r}} - \bar{F}^{r}_{\cdot\bar{r}}\bar{F}^{\theta}_{\cdot\bar{z}}); \tag{4.253}$$

$$R^{\cdot\cdot\cdot\bar{r}}_{\bar{r}\bar{\theta}\bar{\theta}} = \bar{F}^{r}_{\cdot\bar{r}}\bar{F}^{\theta}_{\cdot\bar{\theta}} - \bar{F}^{r}_{\cdot\bar{\theta}}\bar{F}^{\theta}_{\cdot\bar{r}}, \tag{4.254}$$

$$R^{\cdot\cdot\cdot\bar{r}}_{\bar{\theta}\bar{z}\bar{\theta}} = \bar{F}^{r}_{\cdot\bar{\theta}}\bar{F}^{\theta}_{\cdot\bar{z}} - \bar{F}^{r}_{\cdot\bar{z}}\bar{F}^{\theta}_{\cdot\bar{\theta}}, \tag{4.255}$$

$$R^{\cdot\cdot\cdot\bar{r}}_{\bar{z}\bar{r}\bar{\theta}} = \bar{F}^{r}_{\cdot\bar{z}}\bar{F}^{\theta}_{\cdot\bar{r}} - \bar{F}^{r}_{\cdot\bar{r}}\bar{F}^{\theta}_{\cdot\bar{z}}. \tag{4.256}$$

Clearly, skew components of the connection (and hence the torsion) as well as the curvature need not vanish even if the anholonomic object of (4.248) is identically zero.

Chapter 5

Kinematics of Anholonomic Deformation

In this chapter, a mathematical framework describing kinematics associated with non-integrable deformation mappings is given. The description includes a number of derived anholonomic quantities (*i.e.*, quantities referred to an intermediate configuration) and related tensor identities.

5.1 Multiplicative Kinematics

Jacobian determinants, Nanson's formula, stretch, rotation, and strain tensors, and rate equations associated with terms entering a multiplicative decomposition of the total deformation gradient are described.

5.1.1 *Jacobian determinants*

Jacobian determinant $\tilde{J}$ provides the relationship between differential volumes in reference and intermediate configurations. Differential volume element in the reference configuration $\mathrm{d}V$ is written symbolically as

$$\begin{aligned} \mathrm{d}V &= \mathrm{d}\boldsymbol{X}_1 \cdot (\mathrm{d}\boldsymbol{X}_2 \times \mathrm{d}\boldsymbol{X}_3) \\ &= \epsilon_{ABC}\mathrm{d}X^A\mathrm{d}X^B\mathrm{d}X^C \\ &= \sqrt{G}\,\mathrm{d}X^1\mathrm{d}X^2\mathrm{d}X^3 \subset B_0. \end{aligned} \tag{5.1}$$

Differential volume element in the intermediate configuration $\mathrm{d}\tilde{V}$ is written

$$\begin{aligned} \mathrm{d}\tilde{V} &= \mathrm{d}\tilde{\boldsymbol{x}}_1 \cdot (\mathrm{d}\tilde{\boldsymbol{x}}_2 \times \mathrm{d}\tilde{\boldsymbol{x}}_3) \\ &= \epsilon_{\alpha\beta\chi}\mathrm{d}\tilde{x}^\alpha\mathrm{d}\tilde{x}^\beta\mathrm{d}\tilde{x}^\chi \\ &= \sqrt{\tilde{g}}\,\mathrm{d}\tilde{x}^1\mathrm{d}\tilde{x}^2\mathrm{d}\tilde{x}^3 \subset \tilde{B}. \end{aligned} \tag{5.2}$$

It follows from (2.10) that

$$6\,\mathrm{d}X^A\mathrm{d}X^B\mathrm{d}X^C = \epsilon^{ABC}\mathrm{d}V. \tag{5.3}$$

Using linear transformation $\mathrm{d}\tilde{x}^\alpha = \tilde{F}^\alpha_{.A}\mathrm{d}X^A$ of (4.219),

$$\begin{aligned}\frac{\mathrm{d}\tilde{V}}{\mathrm{d}V} &= \frac{\epsilon_{\alpha\beta\chi}\mathrm{d}\tilde{x}^\alpha \mathrm{d}\tilde{x}^\beta \mathrm{d}\tilde{x}^\chi}{\mathrm{d}V}\\ &= \frac{\epsilon_{\alpha\beta\chi}\tilde{F}^\alpha_{.A}\mathrm{d}X^A\tilde{F}^\beta_{.B}\mathrm{d}X^B\tilde{F}^\chi_{.C}\mathrm{d}X^C}{\mathrm{d}V}\\ &= \epsilon_{\alpha\beta\chi}\tilde{F}^\alpha_{.A}\tilde{F}^\beta_{.B}\tilde{F}^\chi_{.C}\frac{\mathrm{d}X^A\mathrm{d}X^B\mathrm{d}X^C}{\mathrm{d}V}\\ &= \epsilon_{\alpha\beta\chi}\tilde{F}^\alpha_{.A}\tilde{F}^\beta_{.B}\tilde{F}^\chi_{.C}(\epsilon^{ABC}/6)\\ &= \tilde{J}. \end{aligned}\tag{5.4}$$

Thus the Jacobian determinant $\tilde{J}[\tilde{\boldsymbol{F}}(X,t),\tilde{g},G(X)]$ is given by

$$\begin{aligned}\tilde{J} &= 1/\tilde{J}^{-1}\\ &= \frac{1}{6}\epsilon^{ABC}\epsilon_{\alpha\beta\chi}\,\tilde{F}^\alpha_{.A}\tilde{F}^\beta_{.B}\tilde{F}^\chi_{.C}\\ &= \frac{1}{6}\sqrt{\tilde{g}/G}\;e^{ABC}e_{\alpha\beta\chi}\tilde{F}^\alpha_{.A}\tilde{F}^\beta_{.B}\tilde{F}^\chi_{.C}\\ &= \sqrt{\tilde{g}/G}\;\det\tilde{\boldsymbol{F}}\\ &= \sqrt{\det(\tilde{g}_{\alpha\beta})/\det(G_{AB})}\;\det(\tilde{F}^\alpha_{.A}),\end{aligned}\tag{5.5}$$

where, explicitly, the determinant of mixed-variant, two-point tensor $\tilde{\boldsymbol{F}}$ is

$$\begin{aligned}\det\tilde{\boldsymbol{F}} &= \det(\tilde{F}^\alpha_{.A})\\ &= \det\begin{bmatrix}\tilde{F}^1_{.1} & \tilde{F}^1_{.2} & \tilde{F}^1_{.3}\\ \tilde{F}^2_{.1} & \tilde{F}^2_{.2} & \tilde{F}^2_{.3}\\ \tilde{F}^3_{.1} & \tilde{F}^3_{.2} & \tilde{F}^3_{.3}\end{bmatrix}\\ &= \tilde{F}^1_{.1}(\tilde{F}^2_{.2}\tilde{F}^3_{.3} - \tilde{F}^3_{.2}\tilde{F}^2_{.3}) - \tilde{F}^1_{.2}(\tilde{F}^2_{.1}\tilde{F}^3_{.3} - \tilde{F}^3_{.1}\tilde{F}^2_{.3})\\ &\quad + \tilde{F}^1_{.3}(\tilde{F}^2_{.1}\tilde{F}^3_{.2} - \tilde{F}^3_{.1}\tilde{F}^2_{.2})\\ &= \tilde{F}^1_{.1}\tilde{F}^2_{.2}\tilde{F}^3_{.3} + \tilde{F}^1_{.2}\tilde{F}^2_{.3}\tilde{F}^3_{.1} + \tilde{F}^1_{.3}\tilde{F}^2_{.1}\tilde{F}^3_{.2}\\ &\quad - \tilde{F}^1_{.3}\tilde{F}^2_{.2}\tilde{F}^3_{.1} - \tilde{F}^1_{.2}\tilde{F}^2_{.1}\tilde{F}^3_{.3} - \tilde{F}^1_{.1}\tilde{F}^2_{.3}\tilde{F}^3_{.2}.\end{aligned}\tag{5.6}$$

Likewise, inverse Jacobian determinant $\tilde{J}^{-1}[\tilde{\boldsymbol{F}}^{-1}(x,t),G(X),\tilde{g}]$ is

$$\begin{aligned}\tilde{J}^{-1} &= 1/\tilde{J}\\ &= \frac{1}{6}\epsilon^{\alpha\beta\chi}\epsilon_{ABC}\,\tilde{F}^{-1A}_{\ \ .\alpha}\tilde{F}^{-1B}_{\ \ .\beta}\tilde{F}^{-1C}_{\ \ .\chi}\\ &= \frac{1}{6}\sqrt{G/\tilde{g}}\;e^{\alpha\beta\chi}e_{ABC}\tilde{F}^{-1A}_{\ \ .\alpha}\tilde{F}^{-1B}_{\ \ .\beta}\tilde{F}^{-1C}_{\ \ .\chi}\\ &= \sqrt{G/\tilde{g}}\;\det\tilde{\boldsymbol{F}}^{-1}\\ &= \sqrt{\det(G_{AB})/\det(\tilde{g}_{\alpha\beta})}\;\det(\tilde{F}^{-1A}_{\ \ .\alpha}).\end{aligned}\tag{5.7}$$

When the mapping is restricted to rigid translation (or to no motion at all), then $\tilde{F}^{\alpha}_{.A} = g^{\alpha}_{A}$ [the shifter of (4.37)], and

$$\tilde{\boldsymbol{F}} = g^{\alpha}_{A}\tilde{\boldsymbol{g}}_{\alpha} \otimes \boldsymbol{G}^{A} = \tilde{\boldsymbol{g}}_{\alpha} \otimes \tilde{\boldsymbol{g}}^{\alpha} = \boldsymbol{G}_{A} \otimes \boldsymbol{G}^{A}$$
$$\Rightarrow \tilde{J} = \sqrt{\tilde{g}/G}\,\det(g^{\alpha}_{A}) = 1 \quad \text{(rigid translation)} \tag{5.8}$$

follows from (4.41). Notice that these definitions of $\tilde{J}$ and $\tilde{J}^{-1}$ do not require explicit introduction of transformations to/from any Cartesian coordinate system. The physical requirement that referential and intermediate volume elements remain positive and bounded leads to the constraints

$$0 < \mathrm{d}\tilde{V} < \infty \Rightarrow 0 < \tilde{J} < \infty, \qquad 0 < \mathrm{d}V < \infty \Rightarrow 0 < \tilde{J}^{-1} < \infty. \tag{5.9}$$

From identity (2.142), it follows that

$$\begin{aligned}\frac{\partial \tilde{J}}{\partial \tilde{F}^{\alpha}_{.A}} &= \frac{\partial(\sqrt{\tilde{g}/G}\ \det \tilde{\boldsymbol{F}})}{\partial \tilde{F}^{\alpha}_{.A}} \\ &= \sqrt{\tilde{g}/G}\ \frac{\partial \det \tilde{\boldsymbol{F}}}{\partial \tilde{F}^{\alpha}_{.A}} \\ &= \sqrt{\tilde{g}/G}\ \det \tilde{\boldsymbol{F}}\ \tilde{F}^{-1A}_{.\alpha} \\ &= \tilde{J}\tilde{F}^{-1A}_{.\alpha},\end{aligned} \tag{5.10}$$

$$\begin{aligned}\frac{\partial \tilde{J}^{-1}}{\partial \tilde{F}^{-1A}_{.\alpha}} &= \frac{\partial(\sqrt{G/\tilde{g}}\ \det \tilde{\boldsymbol{F}}^{-1})}{\partial \tilde{F}^{-1A}_{.\alpha}} \\ &= \sqrt{G/\tilde{g}}\ \frac{\partial \det \tilde{\boldsymbol{F}}^{-1}}{\partial \tilde{F}^{-1A}_{.\alpha}} \\ &= \sqrt{G/\tilde{g}}\ \det \tilde{\boldsymbol{F}}^{-1}\ \tilde{F}^{\alpha}_{.A} \\ &= \tilde{J}^{-1}\tilde{F}^{\alpha}_{.A}.\end{aligned} \tag{5.11}$$

Therefore, inverses obey

$$\tilde{F}^{-1A}_{.\alpha} = \frac{\partial \ln \tilde{J}}{\partial \tilde{F}^{\alpha}_{.A}}, \tag{5.12}$$

$$\tilde{F}^{\alpha}_{.A} = \frac{\partial \ln(\tilde{J}^{-1})}{\partial \tilde{F}^{-1A}_{.\alpha}}. \tag{5.13}$$

Application of (2.143) leads to

$$\begin{aligned}\partial_A[\ln(\det \tilde{\boldsymbol{F}})] &= \frac{1}{\det \tilde{\boldsymbol{F}}}\partial_A \det \tilde{\boldsymbol{F}} \\ &= \frac{1}{\det \tilde{\boldsymbol{F}}}\frac{\partial \det \tilde{\boldsymbol{F}}}{\partial \tilde{F}^{\alpha}_{.B}}\partial_A \tilde{F}^{\alpha}_{.B} \\ &= \tilde{F}^{-1B}_{.\alpha}\partial_A \tilde{F}^{\alpha}_{.B} \\ &= \tilde{F}^{-1B}_{.\alpha}\tilde{F}^{\beta}_{.A}\partial_\beta \tilde{F}^{\alpha}_{.B},\end{aligned} \tag{5.14}$$

$$\begin{aligned}\partial_\alpha[\ln(\det \tilde{\boldsymbol{F}}^{-1})] &= \frac{1}{\det \tilde{\boldsymbol{F}}^{-1}}\partial_\alpha \det \tilde{\boldsymbol{F}}^{-1} \\ &= \frac{1}{\det \tilde{\boldsymbol{F}}^{-1}}\frac{\partial \det \tilde{\boldsymbol{F}}^{-1}}{\partial \tilde{F}^{-1B}_{\ \ \ \cdot\beta}}\partial_\alpha \tilde{F}^{-1B}_{\ \ \ \cdot\beta} \\ &= \tilde{F}^{\beta}_{\cdot B}\partial_\alpha \tilde{F}^{-1B}_{\ \ \ \cdot\beta} \\ &= \tilde{F}^{\beta}_{\cdot B}\tilde{F}^{-1A}_{\ \ \ \cdot\alpha}\partial_A \tilde{F}^{-1B}_{\ \ \ \cdot\beta}.\end{aligned} \tag{5.15}$$

The following additional identities can be derived:

$$\tilde{J}\tilde{F}^{-1A}_{\ \ \ \cdot\alpha} = \frac{1}{2}\epsilon_{\alpha\beta\chi}\epsilon^{ABC}\tilde{F}^{\beta}_{\cdot B}\tilde{F}^{\chi}_{\cdot C}, \tag{5.16}$$

$$\tilde{J}^{-1}\tilde{F}^{\alpha}_{\cdot A} = \frac{1}{2}\epsilon_{ABC}\epsilon^{\alpha\beta\chi}\tilde{F}^{-1B}_{\ \ \ \cdot\beta}\tilde{F}^{-1C}_{\ \ \ \cdot\chi}, \tag{5.17}$$

$$\tilde{J}\delta^{A}_{B} = \frac{1}{2}\epsilon_{\alpha\chi\delta}\epsilon^{ACD}\tilde{F}^{\chi}_{\cdot C}\tilde{F}^{\delta}_{\cdot D}\tilde{F}^{\alpha}_{\cdot B}, \tag{5.18}$$

$$\tilde{J}^{-1}\delta^{\alpha}_{\beta} = \frac{1}{2}\epsilon_{ACD}\epsilon^{\alpha\chi\delta}\tilde{F}^{-1C}_{\ \ \ \cdot\chi}\tilde{F}^{-1D}_{\ \ \ \cdot\delta}\tilde{F}^{-1A}_{\ \ \ \cdot\beta}. \tag{5.19}$$

Permutation tensors map between configurations via

$$\epsilon^{ABC} = \tilde{J}\epsilon^{\alpha\beta\chi}\tilde{F}^{-1A}_{\ \ \ \cdot\alpha}\tilde{F}^{-1B}_{\ \ \ \cdot\beta}\tilde{F}^{-1C}_{\ \ \ \cdot\chi}, \tag{5.20}$$

$$\epsilon_{\alpha\beta\chi} = \tilde{J}\epsilon_{ABC}\tilde{F}^{-1A}_{\ \ \ \cdot\alpha}\tilde{F}^{-1B}_{\ \ \ \cdot\beta}\tilde{F}^{-1C}_{\ \ \ \cdot\chi}, \tag{5.21}$$

$$\epsilon_{ABC} = \tilde{J}^{-1}\epsilon_{\alpha\beta\chi}\tilde{F}^{\alpha}_{\cdot A}\tilde{F}^{\beta}_{\cdot B}\tilde{F}^{\chi}_{\cdot C}, \tag{5.22}$$

$$\epsilon^{\alpha\beta\chi} = \tilde{J}^{-1}\epsilon^{ABC}\tilde{F}^{\alpha}_{\cdot A}\tilde{F}^{\beta}_{\cdot B}\tilde{F}^{\chi}_{\cdot C}. \tag{5.23}$$

The anholonomic gradient of the intermediate Jacobian determinant obeys, from the chain rule, (4.150), and (5.10),

$$\begin{aligned}\partial_\alpha \tilde{J} &= \frac{\partial \tilde{J}}{\partial \tilde{F}^{\beta}_{\cdot A}}\partial_\alpha \tilde{F}^{\beta}_{\cdot A} + \frac{\partial \tilde{J}}{\partial \tilde{g}}\partial_\alpha \tilde{g} + \frac{\partial \tilde{J}}{\partial G}\partial_\alpha G \\ &= \tilde{J}\tilde{F}^{-1A}_{\ \ \ \cdot\beta}\partial_\alpha \tilde{F}^{\beta}_{\cdot A} + 2\tilde{g}\,\Gamma^{\cdot\cdot\beta}_{\alpha\beta}\frac{\partial \tilde{J}}{\partial \tilde{g}} + 2G\,\overset{G}{\Gamma}{}^{\cdot\cdot B}_{AB}\tilde{F}^{-1A}_{\ \ \ \cdot\alpha}\frac{\partial \tilde{J}}{\partial G} \\ &= \tilde{J}\tilde{F}^{-1A}_{\ \ \ \cdot\beta}\partial_\alpha \tilde{F}^{\beta}_{\cdot A} + \tilde{J}\,\Gamma^{\cdot\cdot\beta}_{\alpha\beta} - \tilde{J}\tilde{F}^{-1A}_{\ \ \ \cdot\alpha}\overset{G}{\Gamma}{}^{\cdot\cdot B}_{AB} \\ &= \tilde{J}\tilde{F}^{-1A}_{\ \ \ \cdot\beta}(\partial_\alpha \tilde{F}^{\beta}_{\cdot A} + \Gamma^{\cdot\cdot\beta}_{\alpha\chi}\tilde{F}^{\chi}_{\cdot A} - \overset{G}{\Gamma}{}^{\cdot\cdot B}_{DA}\tilde{F}^{-1D}_{\ \ \ \cdot\alpha}\tilde{F}^{\beta}_{\cdot B}) \\ &= \tilde{J}\tilde{F}^{-1A}_{\ \ \ \cdot\beta}\tilde{F}^{\beta}_{\cdot A:\alpha} \\ &= \frac{\partial \tilde{J}}{\partial \tilde{F}^{\beta}_{\cdot A}}\tilde{F}^{\beta}_{\cdot A:\alpha}.\end{aligned} \tag{5.24}$$

It follows from (5.11) that for the inverse

$$\partial_\alpha(\tilde{J}^{-1}) = \tilde{J}^{-1}\tilde{F}^{\beta}_{\cdot A}\tilde{F}^{-1A}_{\cdot\beta:\alpha}$$
$$= -\tilde{J}^{-1}\tilde{F}^{-1A}_{\cdot\beta}\tilde{F}^{\beta}_{\cdot A:\alpha}$$
$$= \frac{\partial\tilde{J}^{-1}}{\partial\tilde{F}^{-1A}_{\cdot\beta}}\tilde{F}^{-1A}_{\cdot\beta:\alpha}. \tag{5.25}$$

Notice also the identities

$$\partial_\alpha\tilde{J} = \tilde{F}^{-1A}_{\cdot\alpha}\partial_A\tilde{J}, \qquad \partial_\alpha(\tilde{J}^{-1}) = \tilde{F}^{-1A}_{\cdot\alpha}\partial_A(\tilde{J}^{-1}). \tag{5.26}$$

Taking the divergence of (5.10) and using (5.16) results in [7]

$$(\tilde{J}\tilde{F}^{-1A}_{\cdot\alpha})_{:A} = \frac{1}{2}\epsilon_{\alpha\beta\chi}\epsilon^{ABC}\left(\tilde{F}^{\beta}_{\cdot B}\tilde{F}^{\chi}_{\cdot C:A} + \tilde{F}^{\beta}_{\cdot B:A}\tilde{F}^{\chi}_{\cdot C}\right)$$
$$= \frac{1}{2}\epsilon_{\alpha\beta\chi}\epsilon^{ABC}\left(\tilde{F}^{\beta}_{\cdot B}\tilde{F}^{\chi}_{\cdot[C:A]} + \tilde{F}^{\chi}_{\cdot C}\tilde{F}^{\beta}_{\cdot[B:A]}\right)$$
$$= \frac{1}{2}\epsilon_{\alpha\beta\chi}\epsilon^{ABC}\left(\tilde{F}^{\beta}_{\cdot[B}\tilde{F}^{\chi}_{\cdot C:A]} - \tilde{F}^{\beta}_{\cdot[C}\tilde{F}^{\chi}_{\cdot B:A]}\right)$$
$$= \frac{1}{2}\epsilon_{\alpha\beta\chi}\epsilon^{ABC}\left(\tilde{F}^{\beta}_{\cdot[B}\tilde{F}^{\chi}_{\cdot C:A]} + \tilde{F}^{\beta}_{\cdot[B}\tilde{F}^{\chi}_{\cdot C:A]}\right)$$
$$= \epsilon_{\alpha\beta\chi}\epsilon^{ABC}\tilde{F}^{\beta}_{\cdot A}\tilde{F}^{\chi}_{\cdot[B:C]}. \tag{5.27}$$

Recall from (4.208) that material gradient $\overset{G}{\nabla}\tilde{\boldsymbol{F}}$ is generally symmetric in covariant indices only when (4.13) holds and symmetric connection coefficients $\Gamma^{\cdot\cdot\alpha}_{\beta\chi} = \Gamma^{\cdot\cdot\alpha}_{(\beta\chi)}$ are prescribed on $\tilde{B}$. Thus, unless such symmetry conditions hold, the analog of Piola's identity (3.63) does not necessarily apply for the divergence of $\tilde{J}\tilde{F}^{-1A}_{\cdot\alpha}$. Notice that the vanishing property of the (total) covariant derivative of each permutation tensor with respect to a metric connection has been used in deriving (5.27), *i.e.*,

$$\epsilon^{BCD}_{\quad\ :A} = \epsilon^{BCD}_{\quad\ ;A} = 0, \qquad \epsilon_{\alpha\beta\chi:A} = \tilde{F}^{\delta}_{\cdot A}\nabla_\delta\epsilon_{\alpha\beta\chi} = 0. \tag{5.28}$$

Consider a first example. Let vector field $\boldsymbol{A} = A^A\boldsymbol{G}_A$ be the Piola transform of $\tilde{\boldsymbol{a}} = \tilde{a}^\alpha\tilde{\boldsymbol{g}}_\alpha$:

$$A^A = \tilde{J}\tilde{F}^{-1A}_{\cdot\alpha}\tilde{a}^\alpha. \tag{5.29}$$

Taking the divergence of (5.29) and applying the product rule for covariant differentiation with definition (4.206) results in

$$A^A_{;A} = A^A_{:A}$$
$$= (\tilde{J}\tilde{F}^{-1A}_{\cdot\alpha})_{:A}\tilde{a}^\alpha + \tilde{J}\tilde{F}^{-1A}_{\cdot\alpha}\tilde{a}^\alpha_{:A}$$
$$= (\tilde{J}\tilde{F}^{-1A}_{\cdot\alpha})_{:A}\tilde{a}^\alpha + \tilde{J}\tilde{a}^\alpha_{:\alpha}$$
$$= \tilde{J}[\tilde{J}^{-1}(\tilde{J}\tilde{F}^{-1A}_{\cdot\alpha})_{:A}\tilde{a}^\alpha + \tilde{a}^\alpha_{:\alpha}]$$
$$= \tilde{J}\,\hat{\nabla}_\alpha\tilde{a}^\alpha, \tag{5.30}$$

where the anholonomic covariant derivative operator $\hat{\nabla}(\cdot)$ is defined as [7]

$$\begin{aligned}\hat{\nabla}_\alpha(\cdot) &= (\cdot)_{:A}\tilde{F}^{-1A}_{\ \ .\alpha} + (\cdot)\tilde{J}^{-1}(\tilde{J}\tilde{F}^{-1A}_{\ \ .\alpha})_{:A} \\ &= (\cdot)_{:\alpha} + (\cdot)\tilde{J}^{-1}\epsilon_{\alpha\beta\chi}\epsilon^{ABC}\tilde{F}^{\beta}_{.A}\tilde{F}^{\chi}_{.B:C}. \end{aligned} \tag{5.31}$$

The second term on the right of (5.31) vanishes when the right of (5.27) vanishes, in which case (5.30) becomes similar to (3.72).

As a second example, consider the following Piola transformation between contravariant second-order tensor $\boldsymbol{A} = A^{AB}\boldsymbol{G}_A \otimes \boldsymbol{G}_B$ and its two-point counterpart $\tilde{\boldsymbol{a}} = \tilde{a}^{A\alpha}\boldsymbol{G}_A \otimes \tilde{\boldsymbol{g}}_\alpha$:

$$A^{AB} = \tilde{J}\tilde{F}^{-1B}_{\ \ .\alpha}\tilde{a}^{A\alpha}. \tag{5.32}$$

Divergence on the second leg of $\boldsymbol{A}$ is computed as

$$\begin{aligned} A^{AB}_{\ \ ;B} &= A^{AB}_{\ \ :B} \\ &= (\tilde{J}\tilde{F}^{-1B}_{\ \ .\alpha})_{:B}\tilde{a}^{A\alpha} + \tilde{J}\tilde{F}^{-1B}_{\ \ .\alpha}\tilde{a}^{A\alpha}_{\ \ :B} \\ &= (\tilde{J}\tilde{F}^{-1A}_{\ \ .\alpha})_{:B}\tilde{a}^{A\alpha} + \tilde{J}\tilde{a}^{A\alpha}_{\ \ :\alpha} \\ &= \tilde{J}[\tilde{J}^{-1}(\tilde{J}\tilde{F}^{-1B}_{\ \ .\alpha})_{:B}\tilde{a}^{A\alpha} + \tilde{a}^{A\alpha}_{\ \ :\alpha}] \\ &= \tilde{J}\,\hat{\nabla}_\alpha\tilde{a}^{A\alpha} \\ &= \tilde{J}(\partial_\alpha\tilde{a}^{A\alpha} + \Gamma^{..\alpha}_{\alpha\beta}\tilde{a}^{A\beta} + \overset{G}{\Gamma}{}^{..A}_{BC}\tilde{F}^{-1B}_{\ \ .\alpha}\tilde{a}^{C\alpha}) \\ &\quad + \epsilon_{\alpha\beta\chi}\epsilon^{DBC}\tilde{a}^{A\alpha}\tilde{F}^{\beta}_{.D}\tilde{F}^{\chi}_{.B:C} \\ &= \tilde{J}(\partial_\alpha\tilde{a}^{A\alpha} + \Gamma^{..\alpha}_{\alpha\beta}\tilde{a}^{A\beta} + \overset{G}{\Gamma}{}^{..A}_{BC}\tilde{F}^{-1B}_{\ \ .\alpha}\tilde{a}^{C\alpha}) \\ &\quad + \epsilon_{\alpha\beta\chi}\epsilon^{DBC}\tilde{a}^{A\alpha}\tilde{F}^{\beta}_{.D}(\partial_C\tilde{F}^{\chi}_{.B} + \Gamma^{..\chi}_{\delta\epsilon}\tilde{F}^{\delta}_{.C}\tilde{F}^{\epsilon}_{.B}). \end{aligned} \tag{5.33}$$

Definitions and arguments in relations (5.1)–(5.33) can be repeated for the Jacobian determinant of $\bar{\boldsymbol{F}}$. Key results are listed in what follows next. Differential volume elements in spatial and intermediate configurations are related by

$$\frac{\mathrm{d}v}{\mathrm{d}\tilde{V}} = \bar{J}, \tag{5.34}$$

where the Jacobian determinant $\bar{J}[\bar{\boldsymbol{F}}(X,t), g(x), \tilde{g}]$ is

$$\begin{aligned} \bar{J} &= 1/\bar{J}^{-1} \\ &= \frac{1}{6}\epsilon_{abc}\epsilon^{\alpha\beta\chi}\,\bar{F}^{a}_{.\alpha}\bar{F}^{b}_{.\beta}\bar{F}^{c}_{.\chi} \\ &= \frac{1}{6}\sqrt{g/\tilde{g}}\;e_{abc}e^{\alpha\beta\chi}\bar{F}^{a}_{.\alpha}\bar{F}^{b}_{.\beta}\bar{F}^{c}_{.\chi} \\ &= \sqrt{g/\tilde{g}}\;\det\bar{\boldsymbol{F}} \\ &= \sqrt{\det(g_{ab})/\det(\tilde{g}_{\alpha\beta})}\;\det(\bar{F}^{a}_{.\alpha}). \end{aligned} \tag{5.35}$$

Explicitly, the determinant of the mixed-variant, two-point tensor $\bar{\boldsymbol{F}}$ entering $\bar{J}$ is

$$\begin{aligned}
\det \bar{\boldsymbol{F}} &= \det(\bar{F}^{a}_{.\alpha}) \\
&= \det \begin{bmatrix} \bar{F}^{1}_{.1} & \bar{F}^{1}_{.2} & \bar{F}^{1}_{.3} \\ \bar{F}^{2}_{.1} & \bar{F}^{2}_{.2} & \bar{F}^{2}_{.3} \\ \bar{F}^{3}_{.1} & \bar{F}^{3}_{.2} & \bar{F}^{3}_{.3} \end{bmatrix} \\
&= \bar{F}^{1}_{.1}(\bar{F}^{2}_{.2}\bar{F}^{3}_{.3} - \bar{F}^{3}_{.2}\bar{F}^{2}_{.3}) - \bar{F}^{1}_{.2}(\bar{F}^{2}_{.1}\bar{F}^{3}_{.3} - \bar{F}^{3}_{.1}\bar{F}^{2}_{.3}) \\
&\quad + \bar{F}^{1}_{.3}(\bar{F}^{2}_{.1}\bar{F}^{3}_{.2} - \bar{F}^{3}_{.1}\bar{F}^{2}_{.2}) \\
&= \bar{F}^{1}_{.1}\bar{F}^{2}_{.2}\bar{F}^{3}_{.3} + \bar{F}^{1}_{.2}\bar{F}^{2}_{.3}\bar{F}^{3}_{.1} + \bar{F}^{1}_{.3}\bar{F}^{2}_{.1}\bar{F}^{3}_{.2} \\
&\quad - \bar{F}^{1}_{.3}\bar{F}^{2}_{.2}\bar{F}^{3}_{.1} - \bar{F}^{1}_{.2}\bar{F}^{2}_{.1}\bar{F}^{3}_{.3} - \bar{F}^{1}_{.1}\bar{F}^{2}_{.3}\bar{F}^{3}_{.2}.
\end{aligned} \tag{5.36}$$

From similar arguments, evaluation of the inverse Jacobian determinant $\bar{J}^{-1}[\bar{\boldsymbol{F}}^{-1}(x,t), \tilde{g}, g(x)]$ proceeds as

$$\begin{aligned}
\bar{J}^{-1} &= 1/\bar{J} \\
&= \frac{1}{6}\epsilon_{\alpha\beta\chi}\epsilon^{abc}\, \bar{F}^{-1\alpha}_{.a}\bar{F}^{-1\beta}_{.b}\bar{F}^{-1\chi}_{.c} \\
&= \frac{1}{6}\sqrt{\tilde{g}/g}\; e_{\alpha\beta\chi}e^{abc}\bar{F}^{-1\alpha}_{.a}\bar{F}^{-1\beta}_{.b}\bar{F}^{-1\chi}_{.c} \\
&= \sqrt{\tilde{g}/g}\;\det \bar{\boldsymbol{F}}^{-1} \\
&= \sqrt{\det(\tilde{g}_{\alpha\beta})/\det(g_{ab})}\;\det(\bar{F}^{-1\alpha}_{.a}).
\end{aligned} \tag{5.37}$$

The physical requirement that intermediate and spatial volume elements remain positive and bounded leads to the constraints

$$0 < \mathrm{d}\tilde{V} < \infty \Rightarrow 0 < \bar{J}^{-1} < \infty, \qquad 0 < \mathrm{d}v < \infty \Rightarrow 0 < \bar{J} < \infty. \tag{5.38}$$

From identity (2.142), it follows that, analogously to (5.10) and (5.11),

$$\frac{\partial \bar{J}}{\partial \bar{F}^{a}_{.\alpha}} = \bar{J}\bar{F}^{-1\alpha}_{.a}, \tag{5.39}$$

$$\frac{\partial \bar{J}^{-1}}{\partial \bar{F}^{-1\alpha}_{.a}} = \bar{J}^{-1}\bar{F}^{a}_{.\alpha}. \tag{5.40}$$

Therefore, inverses obey

$$\bar{F}^{-1\alpha}_{.a} = \frac{\partial \ln \bar{J}}{\partial \bar{F}^{a}_{.\alpha}}, \tag{5.41}$$

$$\bar{F}^{a}_{.\alpha} = \frac{\partial \ln(\bar{J}^{-1})}{\partial \bar{F}^{-1\alpha}_{.a}}. \tag{5.42}$$

Application of (2.143) leads to

$$\begin{aligned}\partial_\alpha[\ln(\det \bar{\boldsymbol{F}})] &= \frac{1}{\det \bar{\boldsymbol{F}}}\partial_\alpha \det \bar{\boldsymbol{F}} \\ &= \frac{1}{\det \bar{\boldsymbol{F}}}\frac{\partial \det \bar{\boldsymbol{F}}}{\partial \bar{F}^a_{.\beta}}\partial_\alpha \bar{F}^a_{.\beta} \\ &= \bar{F}^{-1\beta}_{.a}\partial_\alpha \bar{F}^a_{.\beta} \\ &= \bar{F}^{-1\beta}_{.a}\bar{F}^b_{.\alpha}\partial_b \bar{F}^a_{.\beta},\end{aligned} \tag{5.43}$$

$$\begin{aligned}\partial_a[\ln(\det \bar{\boldsymbol{F}}^{-1})] &= \frac{1}{\det \bar{\boldsymbol{F}}^{-1}}\partial_a \det \bar{\boldsymbol{F}}^{-1} \\ &= \frac{1}{\det \bar{\boldsymbol{F}}^{-1}}\frac{\partial \det \bar{\boldsymbol{F}}^{-1}}{\partial \bar{F}^{-1\beta}_{.b}}\partial_a \bar{F}^{-1\beta}_{.b} \\ &= \bar{F}^b_{.\beta}\partial_a \bar{F}^{-1\beta}_{.b} \\ &= \bar{F}^b_{.\beta}\bar{F}^{-1\alpha}_{.a}\partial_\alpha \bar{F}^{-1\beta}_{.b}.\end{aligned} \tag{5.44}$$

The following additional identities can be derived regarding the Jacobian determinant $\bar{J}$:

$$\bar{J}^{-1}\bar{F}^a_{.\alpha} = \frac{1}{2}\epsilon_{\alpha\beta\chi}\epsilon^{abc}\bar{F}^\beta_{.b}\bar{F}^\chi_{.c}, \tag{5.45}$$

$$\bar{J}\bar{F}^{-1\alpha}_{.a} = \frac{1}{2}\epsilon_{abc}\epsilon^{\alpha\beta\chi}\bar{F}^b_{.\beta}\bar{F}^c_{.\chi}, \tag{5.46}$$

$$\bar{J}^{-1}\delta^A_B = \frac{1}{2}\epsilon_{\alpha\chi\delta}\epsilon^{acd}\bar{F}^{-1\chi}_{.c}\bar{F}^{-1\delta}_{.d}\bar{F}^{-1\alpha}_{.b}, \tag{5.47}$$

$$\bar{J}\delta^\alpha_\beta = \frac{1}{2}\epsilon_{acd}\epsilon^{\alpha\chi\delta}\bar{F}^c_{.\chi}\bar{F}^d_{.\delta}\bar{F}^a_{.\beta}. \tag{5.48}$$

Permutation tensors map between spatial and intermediate configurations via the relationships

$$\epsilon^{abc} = \bar{J}^{-1}\epsilon^{\alpha\beta\chi}\bar{F}^a_{.\alpha}\bar{F}^b_{.\beta}\bar{F}^c_{.\chi}, \tag{5.49}$$

$$\epsilon_{\alpha\beta\chi} = \bar{J}^{-1}\epsilon_{abc}\bar{F}^a_{.\alpha}\bar{F}^b_{.\beta}\bar{F}^c_{.\chi}, \tag{5.50}$$

$$\epsilon_{abc} = \bar{J}\epsilon_{\alpha\beta\chi}\bar{F}^{-1\alpha}_{.a}\bar{F}^{-1\beta}_{.b}\bar{F}^{-1\chi}_{.c}, \tag{5.51}$$

$$\epsilon^{\alpha\beta\chi} = \bar{J}\epsilon^{abc}\bar{F}^{-1\alpha}_{.a}\bar{F}^{-1\beta}_{.b}\bar{F}^{-1\chi}_{.c}. \tag{5.52}$$

The anholonomic gradient of the intermediate Jacobian determinant obeys, from the chain rule, (4.150), and (5.39),

$$\begin{aligned}
\partial_\alpha \bar{J} &= \frac{\partial \bar{J}}{\partial \bar{F}^a_{.\beta}} \partial_\alpha \bar{F}^a_{.\beta} + \frac{\partial \bar{J}}{\partial \tilde{g}} \partial_\alpha \tilde{g} + \frac{\partial \bar{J}}{\partial g} \partial_\alpha g \\
&= \bar{J} \bar{F}^{-1\beta}_{.a} \partial_\alpha \bar{F}^a_{.\beta} + 2\tilde{g}\, \Gamma^{..\beta}_{\alpha\beta} \frac{\partial \bar{J}}{\partial \tilde{g}} + 2g \overset{g}{\Gamma}{}^{..b}_{ab} \bar{F}^a_{.\alpha} \frac{\partial \bar{J}}{\partial g} \\
&= \bar{J} \bar{F}^{-1\beta}_{.a} \partial_\alpha \bar{F}^a_{.\beta} - \bar{J}\, \Gamma^{..\beta}_{\alpha\beta} + \bar{J} \bar{F}^a_{.\alpha} \overset{g}{\Gamma}{}^{..b}_{ab} \\
&= \bar{J} \bar{F}^{-1\beta}_{.a} (\partial_\alpha \bar{F}^a_{.\beta} - \Gamma^{..\chi}_{\alpha\beta} \bar{F}^a_{.\chi} + \overset{g}{\Gamma}{}^{..a}_{dc} \bar{F}^d_{.\alpha} \bar{F}^c_{.\beta}) \\
&= \bar{J} \bar{F}^{-1\beta}_{.a} \bar{F}^a_{.\beta:\alpha} \\
&= \frac{\partial \bar{J}}{\partial \bar{F}^a_{.\beta}} \bar{F}^a_{.\beta:\alpha}.
\end{aligned} \tag{5.53}$$

It follows for the inverse that

$$\begin{aligned}
\partial_\alpha (\bar{J}^{-1}) &= \bar{J}^{-1} \bar{F}^a_{.\beta} \bar{F}^{-1\beta}_{.a:\alpha} \\
&= -\bar{J}^{-1} \bar{F}^{-1\beta}_{.a} \bar{F}^a_{.\beta:\alpha} \\
&= \frac{\partial \bar{J}^{-1}}{\partial \bar{F}^a_{.\beta}} \bar{F}^a_{.\beta:\alpha}.
\end{aligned} \tag{5.54}$$

Notice also the identities

$$\partial_\alpha \bar{J} = \bar{F}^a_{.\alpha} \partial_a \bar{J}, \qquad \partial_\alpha (\bar{J}^{-1}) = \bar{F}^a_{.\alpha} \partial_a (\bar{J}^{-1}). \tag{5.55}$$

Taking the divergence of (5.40), noting that covariant derivatives of permutation tensors vanish, and using (5.45), it follows that

$$\begin{aligned}
(\bar{J}^{-1} \bar{F}^a_{.\alpha})_{:a} &= \frac{1}{2} \epsilon_{\alpha\beta\chi} \epsilon^{abc} \left(\bar{F}^{-1\beta}_{.b} \bar{F}^{-1\chi}_{.c:a} + \bar{F}^{-1\beta}_{.b:a} \bar{F}^{-1\chi}_{.c} \right) \\
&= \frac{1}{2} \epsilon_{\alpha\beta\chi} \epsilon^{abc} \left(\bar{F}^{-1\beta}_{.b} \bar{F}^{-1\chi}_{.[c:a]} + \bar{F}^{-1\chi}_{.c} \bar{F}^{-1\beta}_{.[b:a]} \right) \\
&= \frac{1}{2} \epsilon_{\alpha\beta\chi} \epsilon^{abc} \left(\bar{F}^{-1\beta}_{.[b} \bar{F}^{-1\chi}_{.c:a]} - \bar{F}^{-1\beta}_{.[c} \bar{F}^{-1\chi}_{.b:a]} \right) \\
&= \frac{1}{2} \epsilon_{\alpha\beta\chi} \epsilon^{abc} \left(\bar{F}^{-1\beta}_{.[b} \bar{F}^{-1\chi}_{.c:a]} + \bar{F}^{-1\beta}_{.[b} \bar{F}^{-1\chi}_{.c:a]} \right) \\
&= \epsilon_{\alpha\beta\chi} \epsilon^{abc} \bar{F}^{-1\beta}_{.a} \bar{F}^{-1\chi}_{.[b:c]}.
\end{aligned} \tag{5.56}$$

Recall from (4.215) that spatial gradient $\overset{g}{\nabla}(\bar{\boldsymbol{F}}^{-1})$ is generally symmetric in covariant indices only when (4.17) holds and symmetric connection coefficients $\Gamma^{..\alpha}_{\beta\chi} = \Gamma^{..\alpha}_{(\beta\chi)}$ are prescribed on $\tilde{B}$. Only in such cases does the divergence of $\bar{J}^{-1} \bar{F}^a_{.\alpha}$ generally vanish so that an identity akin to that in (3.62) holds.

As a first example, let spatial vector field $\boldsymbol{A} = A^a \boldsymbol{g}_a$ be the Piola transform of $\tilde{\boldsymbol{a}} = \tilde{a}^\alpha \tilde{\boldsymbol{g}}_\alpha$:

$$A^a = \bar{J}^{-1} \bar{F}^a_{.\alpha} \tilde{a}^\alpha. \tag{5.57}$$

Taking the divergence of (5.57) and applying the product rule for covariant differentiation with definition (4.213) results in

$$\begin{aligned} A^a_{;\,a} &= A^a_{:a} \\ &= (\bar{J}^{-1} \bar{F}^a_{.\alpha})_{:a} \tilde{a}^\alpha + \bar{J}^{-1} \bar{F}^a_{.\alpha} \tilde{a}^\alpha_{:a} \\ &= (\bar{J}^{-1} \bar{F}^a_{.\alpha})_{:a} \tilde{a}^\alpha + \bar{J}^{-1} \tilde{a}^\alpha_{:\alpha} \\ &= \bar{J}^{-1} [\bar{J} (\bar{J}^{-1} \bar{F}^a_{.\alpha})_{:a} \tilde{a}^\alpha + \tilde{a}^\alpha_{:\alpha}] \\ &= \bar{J}^{-1}\, \hat{\nabla}_\alpha \tilde{a}^\alpha, \end{aligned} \tag{5.58}$$

where the anholonomic covariant derivative is defined analogously to that of (5.31) [7]:

$$\begin{aligned} \hat{\nabla}_\alpha (\cdot) &= (\cdot)_{:a} \bar{F}^a_{.\alpha} + (\cdot) \bar{J} (\bar{J}^{-1} \bar{F}^a_{.\alpha})_{:a} \\ &= (\cdot)_{:\alpha} + (\cdot) \bar{J} \epsilon_{\alpha\beta\chi} \epsilon^{abc} \bar{F}^{-1\beta}_{.a} \bar{F}^{-1\chi}_{.b:c}. \end{aligned} \tag{5.59}$$

The second term on the right of (5.59) vanishes when the right of (5.56) vanishes, in which case (5.58) becomes similar to (3.72).

As a second example, consider the following Piola transformation between contravariant second-order spatial tensor $\boldsymbol{A} = A^{ab} \boldsymbol{g}_a \otimes \boldsymbol{g}_b$ and two-point tensor $\tilde{\boldsymbol{a}} = \tilde{a}^{a\alpha} \boldsymbol{g}_a \otimes \tilde{\boldsymbol{g}}_\alpha$:

$$A^{ab} = \bar{J}^{-1} \bar{F}^b_{.\alpha} \tilde{a}^{a\alpha}. \tag{5.60}$$

The divergence on the second leg of $\boldsymbol{A}$ is computed similarly to (5.33):

$$\begin{aligned} A^{ab}_{;b} &= A^{ab}_{:b} \\ &= (\bar{J}^{-1} \bar{F}^b_{.\alpha})_{:b} \tilde{a}^{a\alpha} + \bar{J}^{-1} \bar{F}^b_{.\alpha} \tilde{a}^{a\alpha}_{:b} \\ &= (\bar{J}^{-1} \bar{F}^a_{.\alpha})_{:b} \tilde{a}^{a\alpha} + \bar{J}^{-1} \tilde{a}^{a\alpha}_{:\alpha} \\ &= \bar{J}^{-1} [\bar{J} (\bar{J}^{-1} \bar{F}^b_{.\alpha})_{:b} \tilde{a}^{a\alpha} + \tilde{a}^{a\alpha}_{:\alpha}] \\ &= \bar{J}^{-1}\, \hat{\nabla}_\alpha \tilde{a}^{a\alpha} \\ &= \bar{J}^{-1} (\partial_\alpha \tilde{a}^{a\alpha} + \Gamma^{..\alpha}_{\alpha\beta} \tilde{a}^{a\beta} + \overset{g}{\Gamma}{}^{..a}_{bc} \bar{F}^b_{.\alpha} \tilde{a}^{c\alpha}) \\ &\quad + \epsilon_{\alpha\beta\chi} \epsilon^{dbc} \tilde{a}^{a\alpha} \bar{F}^{-1\beta}_{.d} \bar{F}^{-1\chi}_{.b:c} \\ &= \bar{J}^{-1} (\partial_\alpha \tilde{a}^{a\alpha} + \Gamma^{..\alpha}_{\alpha\beta} \tilde{a}^{a\beta} + \overset{g}{\Gamma}{}^{..a}_{bc} \bar{F}^b_{.\alpha} \tilde{a}^{c\alpha}) \\ &\quad + \epsilon_{\alpha\beta\chi} \epsilon^{dbc} \tilde{a}^{a\alpha} \bar{F}^{-1\beta}_{.d} (\partial_c \bar{F}^{-1\chi}_{.b} + \Gamma^{..\chi}_{\delta\epsilon} \bar{F}^{-1\delta}_{.c} \bar{F}^{-1\epsilon}_{.b}). \end{aligned} \tag{5.61}$$

From (2.188) and (3.62), it follows that covariant derivative operations in (5.31) and (5.59) are equivalent:

$$\begin{aligned}
\hat{\nabla}_\alpha(\cdot) &= (\cdot)_{:a}\bar{F}^a_{\bullet\alpha} + (\cdot)\bar{J}(\bar{J}^{-1}\bar{F}^a_{\bullet\alpha})_{:a} \\
&= (\cdot)_{:A}\bar{F}^a_{\bullet\alpha}F^{-1A}_{\ \ \bullet a} + (\cdot)\bar{J}(J^{-1}\tilde{J}F^a_{\bullet A}\tilde{F}^{-1A}_{\ \ \bullet\alpha})_{:a} \\
&= (\cdot)_{:A}\tilde{F}^{-1A}_{\ \ \bullet\alpha} + (\cdot)\bar{J}J^{-1}F^a_{\bullet A}(\tilde{J}\tilde{F}^{-1A}_{\ \ \bullet\alpha})_{:a} \\
&= (\cdot)_{:A}\tilde{F}^{-1A}_{\ \ \bullet\alpha} + (\cdot)\tilde{J}^{-1}(\tilde{J}\tilde{F}^{-1A}_{\ \ \bullet\alpha})_{:A}.
\end{aligned} \tag{5.62}$$

Finally, the following basic identities are noted for products of Jacobian determinants:

$$\begin{aligned}
\frac{\mathrm{d}v}{\mathrm{d}V} &= \frac{\mathrm{d}v}{\mathrm{d}\tilde{V}}\frac{\mathrm{d}\tilde{V}}{\mathrm{d}V} \\
&= \bar{J}\tilde{J} \\
&= \sqrt{g/\tilde{g}}\ \det\bar{\boldsymbol{F}}\sqrt{\tilde{g}/G}\ \det\tilde{\boldsymbol{F}} \\
&= \sqrt{g/G}\det\boldsymbol{F} \\
&= J,
\end{aligned} \tag{5.63}$$

and similarly for products of their inverses:

$$\begin{aligned}
\frac{\mathrm{d}V}{\mathrm{d}v} &= \frac{\mathrm{d}V}{\mathrm{d}\tilde{V}}\frac{\mathrm{d}\tilde{V}}{\mathrm{d}v} \\
&= \tilde{J}^{-1}\bar{J}^{-1} \\
&= \sqrt{G/\tilde{g}}\ \det\tilde{\boldsymbol{F}}^{-1}\sqrt{\tilde{g}/g}\ \det\bar{\boldsymbol{F}}^{-1} \\
&= \sqrt{G/g}\det\boldsymbol{F}^{-1} \\
&= J^{-1}.
\end{aligned} \tag{5.64}$$

5.1.2 *Nanson's formula*

Oriented area elements in reference and intermediate configurations can be written respectively as

$$N_A\mathrm{d}S = \epsilon_{ABC}\mathrm{d}X^B\mathrm{d}X^C, \qquad \tilde{n}_\alpha\mathrm{d}\tilde{s} = \epsilon_{\alpha\beta\chi}\mathrm{d}\tilde{x}^\beta\mathrm{d}\tilde{x}^\chi; \tag{5.65}$$

where it is understood that $\mathrm{d}X^B$ and $\mathrm{d}X^C$ represent components of different line segments [hence $\mathrm{d}X^B\mathrm{d}X^C \neq \mathrm{d}X^{(B}\mathrm{d}X^{C)}$], and similarly for $\mathrm{d}\tilde{x}^\beta$ and $\mathrm{d}\tilde{x}^\chi$. In configuration B_0, $\boldsymbol{N}(X)$ is a unit normal covariant vector to scalar differential area element $\mathrm{d}S$. In configuration $\tilde{B}$, $\tilde{\boldsymbol{n}}(X,t)$ is a unit normal covariant vector to scalar differential area element $\mathrm{d}\tilde{s}$. Oriented area elements map between configurations via Nanson's formula, which can be derived directly using (4.219) and (5.21):

$$\begin{aligned}\tilde{n}_\alpha \mathrm{d}\tilde{s} &= (\epsilon_{\alpha\beta\chi})(\mathrm{d}\tilde{x}^\beta \mathrm{d}\tilde{x}^\chi) \\ &= (\tilde{J}\epsilon_{AEF}\tilde{F}^{-1A}_{\ \ \ \boldsymbol{.}\alpha}\tilde{F}^{-1E}_{\ \ \ \boldsymbol{.}\beta}\tilde{F}^{-1F}_{\ \ \ \boldsymbol{.}\chi})(\tilde{F}^{\beta}_{\boldsymbol{.}B}\mathrm{d}X^B\tilde{F}^{\chi}_{\boldsymbol{.}C}\mathrm{d}X^C) \\ &= \tilde{J}\tilde{F}^{-1A}_{\ \ \ \boldsymbol{.}\alpha}\epsilon_{ABC}\mathrm{d}X^B\mathrm{d}X^C \\ &= \tilde{J}\tilde{F}^{-1A}_{\ \ \ \boldsymbol{.}\alpha}N_A\mathrm{d}S. \end{aligned} \tag{5.66}$$

Squaring (5.66), noting $N_A N^A = \tilde{n}_\alpha\tilde{n}^\alpha = 1$, dividing by $\mathrm{d}S^2$, and taking the square root leads to the ratio

$$\frac{\mathrm{d}\tilde{s}}{\mathrm{d}S} = \tilde{J}(\tilde{F}^{-1A}_{\ \ \ \boldsymbol{.}\alpha}\tilde{g}^{\alpha\beta}\tilde{F}^{-1B}_{\ \ \ \boldsymbol{.}\beta}N_A N_B)^{1/2}. \tag{5.67}$$

Repeating the above derivations using (5.50) and spatial area element

$$n_a \mathrm{d}s = \epsilon_{abc}\mathrm{d}x^b\mathrm{d}x^c, \tag{5.68}$$

a local intermediate area element also obeys the following relationships

$$\tilde{n}_\alpha \mathrm{d}\tilde{s} = \bar{J}^{-1}\bar{F}^{a}_{\boldsymbol{.}\alpha}n_a\mathrm{d}s, \tag{5.69}$$

$$\frac{\mathrm{d}\tilde{s}}{\mathrm{d}s} = \bar{J}^{-1}(\bar{F}^{a}_{\boldsymbol{.}\alpha}\tilde{g}^{\alpha\beta}\bar{F}^{b}_{\boldsymbol{.}\beta}n_a n_b)^{1/2}. \tag{5.70}$$

When $\tilde{B}$ is anholonomic, $\mathrm{d}\tilde{s}$ and $\mathrm{d}\tilde{V}$ are strictly defined only in differential form. A global surface $\tilde{s}$ or volume $\tilde{V}$ need not exist in the intermediate configuration, even if the body in configurations B_0 and B is simply connected.

5.1.3 *Rotation, stretch, and strain*

First consider the mapping $\bar{\boldsymbol{F}}(X,t)$ entering (4.1). According to the polar decomposition theorem, because $\det\bar{\boldsymbol{F}} > 0$, this deformation mapping can be decomposed into a product of a rotation tensor $\bar{\boldsymbol{R}}$ and a positive definite stretch tensor $\bar{\boldsymbol{U}}$ or $\bar{\boldsymbol{V}}$:

$$\bar{\boldsymbol{F}} = \bar{\boldsymbol{R}}\bar{\boldsymbol{U}} = \bar{\boldsymbol{V}}\bar{\boldsymbol{R}}, \qquad \bar{F}^{a}_{\boldsymbol{.}\alpha} = \bar{R}^{a}_{\boldsymbol{.}\beta}\bar{U}^{\beta}_{\boldsymbol{.}\alpha} = \bar{V}^{a}_{\boldsymbol{.}b}\bar{R}^{b}_{\boldsymbol{.}\alpha}. \tag{5.71}$$

Rotation tensor $\bar{\boldsymbol{R}}(X,t)$ is a two-point tensor by convention, and is proper orthogonal:

$$\bar{\boldsymbol{R}}^{-1} = \bar{\boldsymbol{R}}^{\mathrm{T}}, \qquad \bar{R}^{a}_{\boldsymbol{.}\alpha}\bar{R}^{\mathrm{T}\alpha}_{\ \ \boldsymbol{.}b} = \delta^a_b, \qquad \bar{R}^{a}_{\boldsymbol{.}\beta}\bar{R}^{\mathrm{T}\alpha}_{\ \ \boldsymbol{.}a} = \delta^\alpha_\beta; \tag{5.72}$$

$$\frac{1}{6}\epsilon^{\alpha\beta\chi}\epsilon_{abc}\,\bar{R}^{a}_{\boldsymbol{.}\alpha}\bar{R}^{b}_{\boldsymbol{.}\beta}\bar{R}^{c}_{\boldsymbol{.}\chi} = \sqrt{g/\tilde{g}}\,\det\bar{\boldsymbol{R}} = 1. \tag{5.73}$$

Right stretch tensor $\bar{\boldsymbol{U}}$ and left stretch tensor $\bar{\boldsymbol{V}}$ are symmetric:

$$\tilde{g}_{\alpha\beta}\bar{U}^{\beta}_{.\chi} = \tilde{g}_{\chi\beta}\bar{U}^{\beta}_{.\alpha}, \qquad g_{ab}\bar{V}^{b}_{.c} = g_{cb}\bar{V}^{b}_{.a}. \tag{5.74}$$

From (5.35) and (5.73), it follows that the Jacobian $\bar{J}$ is fully determined by the stretch tensors:

$$\bar{J} = \sqrt{g/\tilde{g}}\ \det \bar{\boldsymbol{R}} \det \bar{\boldsymbol{U}} = \sqrt{g/\tilde{g}}\ \det \bar{\boldsymbol{V}} \det \bar{\boldsymbol{R}} = \det \bar{\boldsymbol{U}} = \det \bar{\boldsymbol{V}}. \tag{5.75}$$

Symmetric deformation tensors are introduced as

$$\bar{\boldsymbol{C}} = \bar{\boldsymbol{F}}^{\mathrm{T}}\bar{\boldsymbol{F}} = \bar{\boldsymbol{U}}^{2}, \qquad \bar{C}_{\alpha\beta} = \bar{F}^{a}_{.\alpha}g_{ab}\bar{F}^{b}_{.\beta}; \tag{5.76}$$

$$\bar{\boldsymbol{B}} = \bar{\boldsymbol{F}}\bar{\boldsymbol{F}}^{\mathrm{T}} = \bar{\boldsymbol{V}}^{2}, \qquad \bar{B}^{ab} = \bar{F}^{a}_{.\alpha}\tilde{g}^{\alpha\beta}\bar{F}^{b}_{.\beta}. \tag{5.77}$$

Deformation tensor $\bar{\boldsymbol{C}}(X,t)$ is given further attention. Inverting linear transformation $\mathrm{d}\tilde{x}^{\alpha} = \bar{F}^{-1\alpha}_{.a}\mathrm{d}x^{a}$ of (4.219), $\bar{\boldsymbol{C}}$ can be used to determine the squared length of a deformed line element:

$$\begin{aligned} |\mathrm{d}\boldsymbol{x}|^2 &= \mathrm{d}\boldsymbol{x}\cdot\mathrm{d}\boldsymbol{x} \\ &= \mathrm{d}x^{a}\,\mathrm{d}x_{a} \\ &= \bar{F}^{a}_{.\alpha}\,\mathrm{d}\tilde{x}^{\alpha}\,g_{ab}\,\bar{F}^{b}_{.\beta}\,\mathrm{d}\tilde{x}^{\beta} \\ &= \bar{C}_{\alpha\beta}\,\mathrm{d}\tilde{x}^{\alpha}\,\mathrm{d}\tilde{x}^{\beta} \\ &= \langle\bar{\boldsymbol{C}}\,\mathrm{d}\tilde{\boldsymbol{x}},\mathrm{d}\tilde{\boldsymbol{x}}\rangle. \end{aligned} \tag{5.78}$$

Similarly to (3.12) and (3.13), the following identities can be derived:

$$\frac{\partial \bar{F}^{a}_{.\alpha}}{\partial \bar{F}^{-1\chi}_{.c}} = -\bar{F}^{a}_{.\chi}\bar{F}^{c}_{.\alpha}, \qquad \frac{\partial \bar{F}^{-1\alpha}_{.a}}{\partial \bar{F}^{c}_{.\chi}} = -\bar{F}^{-1\alpha}_{.c}\bar{F}^{-1\chi}_{.a}. \tag{5.79}$$

It then follows that $\bar{\boldsymbol{C}}$ obeys

$$\bar{C}_{\alpha\beta} = \bar{F}^{a}_{.\alpha}g_{ab}\bar{F}^{b}_{.\beta} = g_{ab}\bar{F}^{a}_{.\beta}\bar{F}^{b}_{.\alpha} = -g_{ab}\frac{\partial \bar{F}^{a}_{.\alpha}}{\partial \bar{F}^{-1\beta}_{.b}}. \tag{5.80}$$

A covariant strain tensor $\boldsymbol{E}(X,t)$ can also be introduced:

$$\bar{\boldsymbol{E}} = \frac{1}{2}(\bar{\boldsymbol{F}}^{\mathrm{T}}\bar{\boldsymbol{F}} - \tilde{\boldsymbol{g}}), \qquad \bar{E}_{\alpha\beta} = \frac{1}{2}(\bar{F}^{a}_{.\alpha}g_{ab}\bar{F}^{b}_{.\beta} - \tilde{g}_{\alpha\beta}). \tag{5.81}$$

Notice that $\bar{\boldsymbol{E}}$ provides the difference in squared lengths of deformed and intermediate line elements:

$$\mathrm{d}\boldsymbol{x}\cdot\mathrm{d}\boldsymbol{x} - \mathrm{d}\tilde{\boldsymbol{x}}\cdot\mathrm{d}\tilde{\boldsymbol{x}} = 2\langle\bar{\boldsymbol{E}}\,\mathrm{d}\tilde{\boldsymbol{x}},\mathrm{d}\tilde{\boldsymbol{x}}\rangle. \tag{5.82}$$

Noting that

$$\begin{aligned}\det \bar{\boldsymbol{C}} &= \det(\bar{C}^{\alpha}_{.\beta})\\ &= \det(g_{ab}\tilde{g}^{\alpha\chi}\bar{F}^{a}_{.\chi}\bar{F}^{b}_{.\beta})\\ &= (g/\tilde{g})(\det \bar{\boldsymbol{F}})^2\\ &= \bar{J}^2,\end{aligned} \tag{5.83}$$

and using (2.142), the following identity is derived, holding $\tilde{g}_{\alpha\beta}$ fixed:

$$\begin{aligned}\frac{\partial \bar{J}}{\partial \bar{E}_{\alpha\beta}} &= 2\frac{\partial \bar{J}}{\partial \bar{C}_{\alpha\beta}}\\ &= \frac{1}{\bar{J}}\frac{\partial(\bar{J}^2)}{\partial \bar{C}_{\alpha\beta}}\\ &= \frac{1}{\bar{J}}\tilde{g}^{\alpha\chi}\frac{\partial(\bar{J}^2)}{\partial \bar{C}^{\chi}_{.\beta}}\\ &= \frac{1}{\bar{J}}\tilde{g}^{\alpha\chi}\frac{\partial \det \bar{\boldsymbol{C}}}{\partial \bar{C}^{\chi}_{.\beta}}\\ &= \frac{1}{\bar{J}}\tilde{g}^{\alpha\chi}(\det \bar{\boldsymbol{C}})\bar{C}^{-1\beta}_{\quad .\chi}\\ &= \bar{J}\bar{C}^{-1\beta\alpha}\\ &= \bar{J}\bar{C}^{-1\alpha\beta}\\ &= \bar{J}\bar{F}^{-1\alpha}_{\quad .a}g^{ab}\bar{F}^{-1\beta}_{\quad .b}.\end{aligned} \tag{5.84}$$

Now consider the mapping $\tilde{\boldsymbol{F}}(X,t)$ entering (4.1). According to the polar decomposition theorem, since $\det \tilde{\boldsymbol{F}} > 0$, this deformation mapping can be decomposed into a product of a rotation tensor $\tilde{\boldsymbol{R}}$ and a positive definite stretch tensor $\tilde{\boldsymbol{U}}$ or $\tilde{\boldsymbol{V}}$:

$$\tilde{\boldsymbol{F}} = \tilde{\boldsymbol{R}}\tilde{\boldsymbol{U}} = \tilde{\boldsymbol{V}}\tilde{\boldsymbol{R}}, \qquad \tilde{F}^{\alpha}_{.A} = \tilde{R}^{\alpha}_{.B}\tilde{U}^{B}_{.A} = \tilde{V}^{\alpha}_{.\beta}\tilde{R}^{\beta}_{.A}. \tag{5.85}$$

Rotation tensor $\tilde{\boldsymbol{R}}(X,t)$ is a two-point tensor by convention, and is proper orthogonal:

$$\tilde{\boldsymbol{R}}^{-1} = \tilde{\boldsymbol{R}}^{\mathrm{T}}, \qquad \tilde{R}^{\alpha}_{.A}\tilde{R}^{\mathrm{T}A}_{\quad .\beta} = \delta^{\alpha}_{\beta}, \qquad \tilde{R}^{\alpha}_{.B}\tilde{R}^{\mathrm{T}A}_{\quad .\alpha} = \delta^{A}_{B}; \tag{5.86}$$

$$\frac{1}{6}\epsilon_{\alpha\beta\chi}\epsilon^{ABC}\,\tilde{R}^{\alpha}_{.A}\tilde{R}^{\beta}_{.B}\tilde{R}^{\chi}_{.C} = \sqrt{\tilde{g}/G}\,\det \tilde{\boldsymbol{R}} = 1. \tag{5.87}$$

Right stretch tensor $\tilde{\boldsymbol{U}}$ and left stretch tensor $\tilde{\boldsymbol{V}}$ are symmetric:

$$G_{AB}\tilde{U}^{B}_{.C} = G_{CB}\tilde{U}^{B}_{.A}, \qquad \tilde{g}_{\alpha\beta}\tilde{V}^{\beta}_{.\chi} = \tilde{g}_{\chi\beta}\tilde{V}^{\beta}_{.\alpha}. \tag{5.88}$$

From (5.5) and (5.87), it follows that Jacobian $\tilde{J}$ is determined by either stretch tensor:

$$\tilde{J} = \sqrt{\tilde{g}/G}\ \det \tilde{\boldsymbol{R}} \det \tilde{\boldsymbol{U}} = \sqrt{\tilde{g}/G}\ \det \tilde{\boldsymbol{V}} \det \tilde{\boldsymbol{R}} = \det \tilde{\boldsymbol{U}} = \det \tilde{\boldsymbol{V}}. \quad (5.89)$$

Symmetric deformation tensors are introduced as

$$\tilde{\boldsymbol{C}} = \tilde{\boldsymbol{F}}^{\mathrm{T}} \tilde{\boldsymbol{F}} = \tilde{\boldsymbol{U}}^2, \qquad \tilde{C}_{AB} = \tilde{F}^{\alpha}_{.A} \tilde{g}_{\alpha\beta} \tilde{F}^{\beta}_{.B}; \quad (5.90)$$

$$\tilde{\boldsymbol{B}} = \tilde{\boldsymbol{F}} \tilde{\boldsymbol{F}}^{\mathrm{T}} = \tilde{\boldsymbol{V}}^2, \qquad \tilde{B}^{\alpha\beta} = \tilde{F}^{\alpha}_{.A} G^{AB} \tilde{F}^{\beta}_{.B}. \quad (5.91)$$

Deformation tensor $\tilde{\boldsymbol{C}}(X,t)$ is given further attention. Applying linear transformation $\mathrm{d}\tilde{x}^{\alpha} = \tilde{F}^{\alpha}_{.A} \mathrm{d}X^A$ of (4.219), $\tilde{\boldsymbol{C}}$ can be used to determine the squared length of an intermediate line element:

$$\begin{aligned} |\mathrm{d}\tilde{\boldsymbol{x}}|^2 &= \mathrm{d}\tilde{\boldsymbol{x}} \cdot \mathrm{d}\tilde{\boldsymbol{x}} \\ &= \mathrm{d}\tilde{x}^{\alpha}\ \mathrm{d}\tilde{x}_{\alpha} \\ &= \tilde{F}^{\alpha}_{.A}\ \mathrm{d}X^A\ \tilde{g}_{\alpha\beta}\ \tilde{F}^{\beta}_{.B}\ \mathrm{d}X^B \\ &= \tilde{C}_{AB} \mathrm{d}X^A \mathrm{d}X^B \\ &= \langle \tilde{\boldsymbol{C}}\ \mathrm{d}\boldsymbol{X}, \mathrm{d}\boldsymbol{X} \rangle, \end{aligned} \quad (5.92)$$

where covariant components of $\mathrm{d}\tilde{\boldsymbol{x}}$ are defined as $\mathrm{d}\tilde{x}_{\alpha} = \mathrm{d}\tilde{x}^{\beta} \tilde{g}_{\alpha\beta}$. Analogously to (3.12) and (3.13), differentiation of Kronecker's delta symbols produces

$$\frac{\partial \tilde{F}^{\alpha}_{.A}}{\partial \tilde{F}^{-1C}_{\ \ \ .\chi}} = -\tilde{F}^{\alpha}_{.C} \tilde{F}^{\chi}_{.A}, \qquad \frac{\partial \tilde{F}^{-1A}_{\ \ \ .\alpha}}{\partial \tilde{F}^{\chi}_{.C}} = -\tilde{F}^{-1A}_{\ \ \ .\chi} \tilde{F}^{-1C}_{\ \ \ .\alpha}. \quad (5.93)$$

Deformation tensor $\tilde{\boldsymbol{C}}$ therefore satisfies

$$\tilde{C}_{\alpha\beta} = \tilde{F}^{\alpha}_{.A} \tilde{g}_{\alpha\beta} \tilde{F}^{\beta}_{.B} = \tilde{g}_{\alpha\beta} \tilde{F}^{\alpha}_{.B} \tilde{F}^{\beta}_{.A} = -\tilde{g}_{\alpha\beta} \frac{\partial \tilde{F}^{\alpha}_{.A}}{\partial \tilde{F}^{-1B}_{\ \ \ .\beta}}. \quad (5.94)$$

A covariant strain tensor $\tilde{\boldsymbol{E}}(X,t)$ can also be introduced:

$$\tilde{\boldsymbol{E}} = \frac{1}{2}(\tilde{\boldsymbol{F}}^{\mathrm{T}} \tilde{\boldsymbol{F}} - \boldsymbol{G}), \qquad \tilde{E}_{AB} = \frac{1}{2}(\tilde{F}^{\alpha}_{.A} \tilde{g}_{\alpha\beta} \tilde{F}^{\beta}_{.B} - G_{AB}). \quad (5.95)$$

Notice that $\tilde{\boldsymbol{E}}$ provides the difference in squared lengths of intermediate and reference line elements:

$$\mathrm{d}\tilde{\boldsymbol{x}} \cdot \mathrm{d}\tilde{\boldsymbol{x}} - \mathrm{d}\boldsymbol{X} \cdot \mathrm{d}\boldsymbol{X} = 2\langle \tilde{\boldsymbol{E}}\ \mathrm{d}\boldsymbol{X}, \mathrm{d}\boldsymbol{X} \rangle. \quad (5.96)$$

Noting that

$$\begin{aligned} \det \tilde{\boldsymbol{C}} &= \det(\tilde{C}^{A}_{.B}) \\ &= \det(\tilde{g}_{\alpha\beta} G^{AC} \tilde{F}^{\alpha}_{.C} \tilde{F}^{\beta}_{.B}) \\ &= (\tilde{g}/G)(\det \tilde{\boldsymbol{F}})^2 \\ &= \tilde{J}^2, \end{aligned} \quad (5.97)$$

and using (2.142), the following identity is derived, holding G_{AB} fixed:

$$\begin{aligned}\frac{\partial \tilde{J}}{\partial \tilde{E}_{AB}} &= 2\frac{\partial \tilde{J}}{\partial \tilde{C}_{AB}}\\ &= \frac{1}{\tilde{J}}\frac{\partial(\tilde{J}^2)}{\partial \tilde{C}_{AB}}\\ &= \frac{1}{\tilde{J}}G^{AC}\frac{\partial(\tilde{J}^2)}{\partial \tilde{C}^{C}_{.B}}\\ &= \frac{1}{\tilde{J}}G^{AC}\frac{\partial \det \tilde{\boldsymbol{C}}}{\partial \tilde{C}^{C}_{.B}}\\ &= \frac{1}{\tilde{J}}G^{AC}(\det \tilde{\boldsymbol{C}})\tilde{C}^{-1B}_{\;\;\;.C}\\ &= \tilde{J}\tilde{C}^{-1BA}\\ &= \tilde{J}\tilde{C}^{-1AB}\\ &= \tilde{J}\tilde{F}^{-1A}_{\;\;\;.\alpha}\tilde{g}^{\alpha\beta}\tilde{F}^{-1B}_{\;\;\;.\beta}.\end{aligned} \tag{5.98}$$

Using (5.71) and (5.85) in (4.1), the total deformation gradient $\boldsymbol{F}$ can be decomposed in a number of ways:

$$\begin{aligned}\boldsymbol{F} &= \bar{\boldsymbol{F}}\tilde{\boldsymbol{F}}\\ &= \bar{\boldsymbol{R}}\bar{\boldsymbol{U}}\tilde{\boldsymbol{R}}\tilde{\boldsymbol{U}}\\ &= \bar{\boldsymbol{R}}\bar{\boldsymbol{U}}\tilde{\boldsymbol{V}}\tilde{\boldsymbol{R}}\\ &= \bar{\boldsymbol{V}}\bar{\boldsymbol{R}}\tilde{\boldsymbol{V}}\tilde{\boldsymbol{R}}\\ &= \bar{\boldsymbol{V}}\bar{\boldsymbol{R}}\tilde{\boldsymbol{R}}\tilde{\boldsymbol{U}}\\ &= \bar{\boldsymbol{V}}\hat{\boldsymbol{R}}\tilde{\boldsymbol{U}},\end{aligned} \tag{5.99}$$

or in indicial notation,

$$\begin{aligned}F^{a}_{.A} &= \bar{F}^{a}_{.\alpha}\tilde{F}^{\alpha}_{.A}\\ &= \bar{R}^{a}_{.\beta}\bar{U}^{\beta}_{.\alpha}\tilde{R}^{\alpha}_{.B}\tilde{U}^{B}_{.A}\\ &= \bar{R}^{a}_{.\beta}\bar{U}^{\beta}_{.\alpha}\tilde{V}^{\alpha}_{.\chi}\tilde{R}^{\chi}_{.A}\\ &= \bar{V}^{a}_{.b}\bar{R}^{b}_{.\alpha}\tilde{V}^{\alpha}_{.\beta}\tilde{R}^{\beta}_{.A}\\ &= \bar{V}^{a}_{.b}\bar{R}^{b}_{.\alpha}\tilde{R}^{\alpha}_{.B}\tilde{U}^{B}_{.A}\\ &= \bar{V}^{a}_{.b}\hat{R}^{b}_{.B}\tilde{U}^{B}_{.A}.\end{aligned} \tag{5.100}$$

In the final expressions of (5.99) and (5.100), $\hat{\boldsymbol{R}} = \bar{\boldsymbol{R}}\tilde{\boldsymbol{R}}$ is a proper orthogonal two-point tensor. The total Jacobian determinant obeys, from (5.63),

(5.75), and (5.89),

$$\begin{aligned} J &= \bar{J}\tilde{J} \\ &= \det \bar{\boldsymbol{U}} \det \tilde{\boldsymbol{U}} \\ &= \det \bar{\boldsymbol{V}} \det \tilde{\boldsymbol{V}} \\ &= \det \bar{\boldsymbol{U}} \det \tilde{\boldsymbol{V}} \\ &= \det \bar{\boldsymbol{V}} \det \tilde{\boldsymbol{U}}. \end{aligned} \tag{5.101}$$

In coordinates, transposes of $\bar{\boldsymbol{F}}$ and $\tilde{\boldsymbol{F}}$ are, respectively,

$$\bar{\boldsymbol{F}}^{\mathrm{T}} = \bar{F}^{\mathrm{T}\,\alpha}_{\ \ \bullet a}\tilde{\boldsymbol{g}}_\alpha \otimes \boldsymbol{g}^a = \bar{F}^b_{\bullet\beta} g_{ab}\tilde{g}^{\alpha\beta}\tilde{\boldsymbol{g}}_\alpha \otimes \boldsymbol{g}^a, \tag{5.102}$$

$$\tilde{\boldsymbol{F}}^{\mathrm{T}} = \tilde{F}^{\mathrm{T}\,A}_{\ \ \bullet\alpha}\boldsymbol{G}_A \otimes \tilde{\boldsymbol{g}}^\alpha = \tilde{F}^\beta_{\bullet B}\tilde{g}_{\alpha\beta}G^{AB}\boldsymbol{G}_A \otimes \tilde{\boldsymbol{g}}^\alpha. \tag{5.103}$$

It follows that determinants of these transpose maps are

$$\det \bar{\boldsymbol{F}}^{\mathrm{T}} = (g/\tilde{g}) \det \bar{\boldsymbol{F}}, \qquad \det \tilde{\boldsymbol{F}}^{\mathrm{T}} = (\tilde{g}/G) \det \tilde{\boldsymbol{F}}. \tag{5.104}$$

Therefore, Jacobian determinants of these mappings obey

$$\bar{J} = \sqrt{g/\tilde{g}} \det \bar{\boldsymbol{F}} = \sqrt{\tilde{g}/g}[(g/\tilde{g}) \det \bar{\boldsymbol{F}}] = \sqrt{\tilde{g}/g} \det \bar{\boldsymbol{F}}^{\mathrm{T}}, \tag{5.105}$$

$$\tilde{J} = \sqrt{\tilde{g}/G} \det \tilde{\boldsymbol{F}} = \sqrt{G/\tilde{g}}[(\tilde{g}/G) \det \tilde{\boldsymbol{F}}] = \sqrt{G/\tilde{g}} \det \tilde{\boldsymbol{F}}^{\mathrm{T}}. \tag{5.106}$$

5.1.4 *Rate equations*

Material time derivatives of terms entering (4.1) are defined as follows [7]:

$$\frac{\mathrm{d}\bar{\boldsymbol{F}}}{\mathrm{d}t} = \frac{\mathrm{d}\bar{F}^a_{\bullet\alpha}}{\mathrm{d}t}\boldsymbol{g}_a \otimes \tilde{\boldsymbol{g}}^\alpha = \left.\frac{\partial \bar{\boldsymbol{F}}}{\partial t}\right|_{X,\,\tilde{\boldsymbol{g}}_\alpha} = \left.\frac{\partial}{\partial t}\right|_X (\bar{F}^a_{\bullet\alpha}\boldsymbol{g}_a) \otimes \tilde{\boldsymbol{g}}^\alpha, \tag{5.107}$$

$$\frac{\mathrm{d}\tilde{\boldsymbol{F}}}{\mathrm{d}t} = \frac{\mathrm{d}\tilde{F}^\alpha_{\bullet A}}{\mathrm{d}t}\tilde{\boldsymbol{g}}_\alpha \otimes \boldsymbol{G}^A = \left.\frac{\partial \tilde{\boldsymbol{F}}}{\partial t}\right|_{X,\,\tilde{\boldsymbol{g}}_\alpha} = \left.\frac{\partial}{\partial t}\right|_X (\tilde{F}^\alpha_{\bullet A})\tilde{\boldsymbol{g}}_\alpha \otimes \boldsymbol{G}^A. \tag{5.108}$$

The notation $\mathrm{d}(\cdot)/\mathrm{d}t$ rather than $\dot{(\cdot)}$ is used preferably here for the material time derivative of quantities with a superposed tilde or bar in order to avoid tower-like constructions with multiple superposed adornments. Notice that intermediate basis vectors $\tilde{\boldsymbol{g}}_\alpha$, and hence their reciprocals $\tilde{\boldsymbol{g}}^\alpha$, are held fixed during these particular material time derivative operations, in addition to coordinates of material particle X which are also held fixed as in the usual material time derivative taken with respect to holonomic coordinates described in Section 3.3.2. Because the total deformation gradient $\boldsymbol{F} = \bar{\boldsymbol{F}}\tilde{\boldsymbol{F}}$ does not depend on the particular choice of intermediate coordinate system, this interpretation of the material time derivative is not inconsistent with (3.161):

$$\frac{\mathrm{d}\boldsymbol{F}}{\mathrm{d}t}=\frac{\partial\boldsymbol{F}}{\partial t}\bigg|_X=\frac{\partial\boldsymbol{F}}{\partial t}\bigg|_{X,\,\tilde{g}_\alpha}=\frac{\partial(\bar{\boldsymbol{F}}\tilde{\boldsymbol{F}})}{\partial t}\bigg|_{X,\,\tilde{g}_\alpha}=\frac{\mathrm{d}(\bar{\boldsymbol{F}}\tilde{\boldsymbol{F}})}{\mathrm{d}t}. \tag{5.109}$$

For the particular case discussed in Section 4.1.3 when $\tilde{\boldsymbol{g}}_\alpha=\tilde{\boldsymbol{g}}_\alpha(X)$, material time derivatives of two-point tensors in (5.107) and (5.108) are identical to the usual material time derivatives in holonomic coordinates:

$$\frac{\partial\tilde{\boldsymbol{g}}_\alpha(X)}{\partial t}\bigg|_X=0\Leftrightarrow\frac{\partial(\cdot)}{\partial t}\bigg|_{X,\,\tilde{g}_\alpha}=\frac{\partial(\cdot)}{\partial t}\bigg|_X. \tag{5.110}$$

For arbitrary intermediate basis vectors of the class $\tilde{\boldsymbol{g}}_\alpha[x(X,t)]$, however,

$$\frac{\partial\tilde{\boldsymbol{g}}_\alpha(x)}{\partial t}\bigg|_X=\frac{\partial\tilde{\boldsymbol{g}}_\alpha}{\partial t}\bigg|_x+\frac{\partial\tilde{\boldsymbol{g}}_\alpha}{\partial x^a}\bigg|_t\frac{\partial x^a}{\partial t}\bigg|_X=(\partial_a\tilde{\boldsymbol{g}}_\alpha)v^a. \tag{5.111}$$

When intermediate coordinates $\tilde{x}^\alpha$ are holonomic, associated basis vectors may be of the general form $\tilde{\boldsymbol{g}}_\alpha[\tilde{x}(X,t)]$, in which case

$$\frac{\partial\tilde{\boldsymbol{g}}_\alpha(\tilde{x})}{\partial t}\bigg|_X=\frac{\partial\tilde{\boldsymbol{g}}_\alpha}{\partial t}\bigg|_{\tilde{x}}+\frac{\partial\tilde{\boldsymbol{g}}_\alpha}{\partial\tilde{x}^\beta}\bigg|_t\frac{\partial\tilde{x}^\beta}{\partial t}\bigg|_X=\partial_\beta\tilde{\boldsymbol{g}}_\alpha\frac{\mathrm{d}\tilde{x}^\beta}{\mathrm{d}t}. \tag{5.112}$$

Definitions (5.107) and (5.108) enable indices corresponding to the intermediate configuration to be treated like those corresponding to the reference configuration when invoking the material time derivative, regardless of choice of intermediate basis vectors.

With such conventions established for material time derivatives, it then follows from application of (3.153) to $\bar{\boldsymbol{F}}(X,t)$ or $\bar{\boldsymbol{F}}(x,t)$ that

$$\begin{aligned}\frac{\mathrm{d}\bar{F}^a_{\cdot\alpha}}{\mathrm{d}t}&=\frac{\mathrm{D}\bar{F}^a_{\cdot\alpha}}{\mathrm{D}t}+\overset{g}{\Gamma}{}^{\cdot\cdot a}_{bc}\bar{F}^c_{\cdot\alpha}v^b\\&=\frac{\partial\bar{F}^a_{\cdot\alpha}}{\partial t}\bigg|_x+\bar{F}^a_{\cdot\alpha;b}v^b\\&=\frac{\partial\bar{F}^a_{\cdot\alpha}}{\partial t}\bigg|_x+\left(\partial_b\bar{F}^a_{\cdot\alpha}+\overset{g}{\Gamma}{}^{\cdot\cdot a}_{bc}\bar{F}^c_{\cdot\alpha}\right)v^b,\end{aligned} \tag{5.113}$$

and from (3.152) for $\tilde{\boldsymbol{F}}(X,t)$,

$$\frac{\mathrm{d}\tilde{F}^\alpha_{\cdot A}}{\mathrm{d}t}=\frac{\partial\tilde{F}^\alpha_{\cdot A}}{\partial t}\bigg|_X. \tag{5.114}$$

Taking the material time derivative of (4.1) and applying the product rule results in

$$\dot{\boldsymbol{F}}=\frac{\mathrm{d}\boldsymbol{F}}{\mathrm{d}t}=\frac{\mathrm{d}\bar{\boldsymbol{F}}}{\mathrm{d}t}\tilde{\boldsymbol{F}}+\bar{\boldsymbol{F}}\frac{\mathrm{d}\tilde{\boldsymbol{F}}}{\mathrm{d}t}. \tag{5.115}$$

Spatial velocity gradient $\boldsymbol{L}$ of (3.162) becomes

$$\begin{aligned}\boldsymbol{L} &= \dot{\boldsymbol{F}}\boldsymbol{F}^{-1} \\ &= \frac{\mathrm{d}\boldsymbol{F}}{\mathrm{d}t}\boldsymbol{F}^{-1} \\ &= \left(\frac{\mathrm{d}\bar{\boldsymbol{F}}}{\mathrm{d}t}\tilde{\boldsymbol{F}} + \bar{\boldsymbol{F}}\frac{\mathrm{d}\tilde{\boldsymbol{F}}}{\mathrm{d}t}\right)\tilde{\boldsymbol{F}}^{-1}\bar{\boldsymbol{F}}^{-1} \\ &= \frac{\mathrm{d}\bar{\boldsymbol{F}}}{\mathrm{d}t}\bar{\boldsymbol{F}}^{-1} + \bar{\boldsymbol{F}}\frac{\mathrm{d}\tilde{\boldsymbol{F}}}{\mathrm{d}t}\tilde{\boldsymbol{F}}^{-1}\bar{\boldsymbol{F}}^{-1} \\ &= \bar{\boldsymbol{L}} + \bar{\boldsymbol{F}}\tilde{\boldsymbol{L}}\bar{\boldsymbol{F}}^{-1},\end{aligned} \tag{5.116}$$

where

$$\bar{\boldsymbol{L}} = \frac{\mathrm{d}\bar{\boldsymbol{F}}}{\mathrm{d}t}\bar{\boldsymbol{F}}^{-1}, \qquad \tilde{\boldsymbol{L}} = \frac{\mathrm{d}\tilde{\boldsymbol{F}}}{\mathrm{d}t}\tilde{\boldsymbol{F}}^{-1}. \tag{5.117}$$

In coordinates, intermediate velocity gradient components of (5.117) are

$$\bar{L}^{a}_{.b} = \frac{\mathrm{d}\bar{F}^{a}_{.\alpha}}{\mathrm{d}t}\bar{F}^{-1\alpha}_{.b}, \qquad \tilde{L}^{\alpha}_{.\beta} = \frac{\mathrm{d}\tilde{F}^{\alpha}_{.A}}{\mathrm{d}t}\tilde{F}^{-1A}_{.\beta}. \tag{5.118}$$

From (4.219) and the second of (5.118), the material time derivative of an intermediate differential line element is

$$\frac{\mathrm{d}}{\mathrm{d}t}(\mathrm{d}\tilde{x}^{\alpha}) = \frac{\mathrm{d}\tilde{F}^{\alpha}_{.A}}{\mathrm{d}t}\mathrm{d}X^{A} = \frac{\mathrm{d}\tilde{F}^{\alpha}_{.A}}{\mathrm{d}t}\tilde{F}^{-1A}_{.\beta}\mathrm{d}\tilde{x}^{\beta} = \tilde{L}^{\alpha}_{.\beta}\mathrm{d}\tilde{x}^{\beta}. \tag{5.119}$$

Material time derivatives of inverses $\bar{\boldsymbol{F}}^{-1}$ and $\tilde{\boldsymbol{F}}^{-1}$ follow from derivations analogous to (3.163), giving

$$\frac{\mathrm{d}}{\mathrm{d}t}(\bar{F}^{-1\alpha}_{.a}) = -\bar{F}^{-1\alpha}_{.b}\bar{L}^{b}_{.a}, \tag{5.120}$$

$$\frac{\mathrm{d}}{\mathrm{d}t}(\tilde{F}^{-1A}_{.\alpha}) = -\tilde{F}^{-1A}_{.\beta}\tilde{L}^{\beta}_{.\alpha}. \tag{5.121}$$

Each of $\bar{\boldsymbol{L}}$ and $\tilde{\boldsymbol{L}}$ can be split into a sum of symmetric part associated with the rate of stretching and a skew part associated with the rate of rotation (*i.e.*, spin):

$$\begin{gathered}\bar{L}_{ab} = \bar{D}_{ab} + \bar{W}_{ab}, \\ \bar{D}_{ab} = \frac{\mathrm{d}\bar{F}_{(a|\alpha|}}{\mathrm{d}t}\bar{F}^{-1\alpha}_{.b)}, \qquad \bar{W}_{ab} = \frac{\mathrm{d}\bar{F}_{[a|\alpha|}}{\mathrm{d}t}\bar{F}^{-1\alpha}_{.b]};\end{gathered} \tag{5.122}$$

$$\begin{gathered}\tilde{L}_{\alpha\beta} = \tilde{D}_{\alpha\beta} + \tilde{W}_{\alpha\beta}, \\ \tilde{D}_{\alpha\beta} = \frac{\mathrm{d}\tilde{F}_{(\alpha|A|}}{\mathrm{d}t}\tilde{F}^{-1A}_{.\beta)}, \quad \tilde{W}_{\alpha\beta} = \frac{\mathrm{d}\tilde{F}_{[\alpha|A|}}{\mathrm{d}t}\tilde{F}^{-1A}_{.\beta]}.\end{gathered} \tag{5.123}$$

Analogously to (3.166), assuming for now that

$$\frac{\mathrm{d}}{\mathrm{d}t}\tilde{g}_{\alpha\beta} = \frac{\mathrm{d}}{\mathrm{d}t}(\tilde{\boldsymbol{g}}_{\alpha} \cdot \tilde{\boldsymbol{g}}_{\beta}) = 0, \tag{5.124}$$

consistent with (5.110), strain rates obey

$$\frac{\mathrm{d}\bar{C}_{\alpha\beta}}{\mathrm{d}t} = 2\frac{\mathrm{d}\bar{E}_{\alpha\beta}}{\mathrm{d}t} = 2\bar{F}^{a}_{.\alpha}\bar{L}_{(ab)}\bar{F}^{b}_{.\beta} = 2\bar{F}^{a}_{.\alpha}\bar{D}_{ab}\bar{F}^{b}_{.\beta}, \tag{5.125}$$

$$\frac{\mathrm{d}\tilde{C}_{AB}}{\mathrm{d}t} = 2\frac{\mathrm{d}\tilde{E}_{AB}}{\mathrm{d}t} = 2\tilde{F}^{\alpha}_{.A}\tilde{L}_{(\alpha\beta)}\tilde{F}^{\beta}_{.B} = 2\tilde{F}^{\alpha}_{.A}\tilde{D}_{\alpha\beta}\tilde{F}^{\beta}_{.B}. \tag{5.126}$$

Notice that material time derivatives of metric tensors introduced in Section 4.1.3 [*i.e.*, (4.50), (4.57), (4.62), and (4.48)] all vanish identically:

$$\tilde{g}_{\alpha\beta} = \delta^{A}_{\alpha}\delta^{B}_{\beta}G_{AB} \Rightarrow \frac{\mathrm{d}\tilde{g}_{\alpha\beta}}{\mathrm{d}t} = \delta^{A}_{\alpha}\delta^{B}_{\beta}\dot{G}_{AB} = 0, \tag{5.127}$$

$$\tilde{g}_{\alpha\beta} = \delta^{a}_{\alpha}\delta^{b}_{\beta}g_{ab} \Rightarrow \frac{\mathrm{d}\tilde{g}_{\alpha\beta}}{\mathrm{d}t} = \delta^{a}_{\alpha}\delta^{b}_{\beta}\dot{g}_{ab} = 0, \tag{5.128}$$

$$\tilde{g}_{\alpha\beta} = \delta^{a}_{\alpha}\delta^{b}_{\beta}\delta_{ab} \Rightarrow \frac{\mathrm{d}\tilde{g}_{\alpha\beta}}{\mathrm{d}t} = \dot{\delta}_{\alpha\beta} = 0, \tag{5.129}$$

$$\tilde{g}_{\alpha\beta} = \partial_{\alpha}\tilde{\boldsymbol{x}} \cdot \partial_{\beta}\tilde{\boldsymbol{x}} \Rightarrow \frac{\mathrm{d}\tilde{g}_{\alpha\beta}}{\mathrm{d}t} = \frac{\partial \tilde{g}_{\alpha\beta}}{\partial t}\bigg|_{\tilde{x}} + \nabla_{\chi}\tilde{g}_{\alpha\beta}\frac{\mathrm{d}\tilde{x}^{\chi}}{\mathrm{d}t} = 0. \tag{5.130}$$

Relations (5.127) and (5.128) follow from (3.169), relation (5.129) is trivial, and vanishing of the covariant derivative in (5.130) follows from definition (4.165). For more general choices of intermediate coordinates, interpretation of the material time derivative as $\mathrm{d}(\cdot)/\mathrm{d}t = \partial(\cdot)/\partial t|_{X,\, \tilde{\boldsymbol{g}}_{\alpha}}$ as introduced in (5.107) and (5.108) will always result in $\mathrm{d}\tilde{g}_{\alpha\beta}/\mathrm{d}t = 0$, as in (5.124).

Material time derivatives of Jacobian determinants are obtained using (5.10), (5.39), and the chain rule as follows, where

$$\frac{\mathrm{d}}{\mathrm{d}t}\tilde{g} = \frac{\mathrm{d}}{\mathrm{d}t}\det\tilde{\boldsymbol{g}} = \tilde{g}\tilde{g}^{\alpha\beta}\frac{\mathrm{d}}{\mathrm{d}t}\tilde{g}_{\alpha\beta} = 0 \tag{5.131}$$

is assumed consistently with (5.127)–(5.130):

$$\begin{aligned}\frac{\mathrm{d}\bar{J}}{\mathrm{d}t} &= \frac{\mathrm{d}}{\mathrm{d}t}\{\bar{J}[\bar{F}^{a}_{.\alpha}(X,t), g(x), \tilde{g}]\}\\ &= \frac{\partial \bar{J}}{\partial \bar{F}^{a}_{.\alpha}}\frac{\mathrm{d}\bar{F}^{a}_{.\alpha}}{\mathrm{d}t}\\ &= \bar{J}\bar{F}^{-1\alpha}_{.a}\frac{\mathrm{d}\bar{F}^{a}_{.\alpha}}{\mathrm{d}t}\\ &= \bar{J}\,\bar{L}^{a}_{.a}\\ &= \bar{J}\bar{D}^{a}_{.a},\end{aligned} \tag{5.132}$$

$$\begin{aligned}\frac{\mathrm{d}\tilde{J}}{\mathrm{d}t} &= \frac{\mathrm{d}}{\mathrm{d}t}\{\tilde{J}[\tilde{F}^{\alpha}_{\boldsymbol{\cdot}A}(X,t),G(X),\tilde{g}]\}\\ &= \frac{\partial \tilde{J}}{\partial \tilde{F}^{\alpha}_{\boldsymbol{\cdot}A}}\frac{\mathrm{d}\tilde{F}^{\alpha}_{\boldsymbol{\cdot}A}}{\mathrm{d}t}\\ &= \tilde{J}\tilde{F}^{-1A}_{\boldsymbol{\cdot}\alpha}\frac{\mathrm{d}\tilde{F}^{\alpha}_{\boldsymbol{\cdot}A}}{\mathrm{d}t}\\ &= \tilde{J}\tilde{L}^{\alpha}_{\boldsymbol{\cdot}\alpha}\\ &= \tilde{J}\tilde{D}^{\alpha}_{\boldsymbol{\cdot}\alpha}.\end{aligned} \tag{5.133}$$

Taking the material time derivative of (5.63) then gives

$$\begin{aligned}\dot{J} &= \frac{\mathrm{d}J}{\mathrm{d}t}\\ &= \frac{\mathrm{d}\bar{J}}{\mathrm{d}t}\tilde{J} + \bar{J}\frac{\mathrm{d}\tilde{J}}{\mathrm{d}t}\\ &= (\bar{J}\bar{L}^{a}_{\boldsymbol{\cdot}a})\tilde{J} + \bar{J}(\tilde{J}\tilde{L}^{\alpha}_{\boldsymbol{\cdot}\alpha})\\ &= J(\bar{L}^{a}_{\boldsymbol{\cdot}a} + \tilde{L}^{\alpha}_{\boldsymbol{\cdot}\alpha})\\ &= JL^{a}_{\boldsymbol{\cdot}a}\\ &= JD^{a}_{\boldsymbol{\cdot}a},\end{aligned} \tag{5.134}$$

which is consistent with (3.172), since, from the trace of (5.116),

$$\begin{aligned}L^{a}_{\boldsymbol{\cdot}a} &= \bar{L}^{a}_{\boldsymbol{\cdot}a} + \bar{F}^{a}_{\boldsymbol{\cdot}\alpha}\frac{\mathrm{d}\tilde{F}^{\alpha}_{\boldsymbol{\cdot}A}}{\mathrm{d}t}\tilde{F}^{-1A}_{\boldsymbol{\cdot}\beta}\bar{F}^{-1\beta}_{\boldsymbol{\cdot}a}\\ &= \bar{L}^{a}_{\boldsymbol{\cdot}a} + \frac{\mathrm{d}\tilde{F}^{\alpha}_{\boldsymbol{\cdot}A}}{\mathrm{d}t}\tilde{F}^{-1A}_{\boldsymbol{\cdot}\alpha}\\ &= \bar{L}^{a}_{\boldsymbol{\cdot}a} + \tilde{L}^{\alpha}_{\boldsymbol{\cdot}\alpha}\\ &= \bar{D}^{a}_{\boldsymbol{\cdot}a} + \tilde{D}^{\alpha}_{\boldsymbol{\cdot}\alpha}\\ &= D^{a}_{\boldsymbol{\cdot}a}.\end{aligned} \tag{5.135}$$

Finally, the material time derivative of an oriented intermediate area element is, from (5.66), (5.121) and (5.133),

$$\begin{aligned}\frac{\mathrm{d}}{\mathrm{d}t}(\tilde{n}_{\alpha}\mathrm{d}\tilde{s}) &= \frac{\mathrm{d}}{\mathrm{d}t}(\tilde{J}\tilde{F}^{-1A}_{\boldsymbol{\cdot}\alpha}N_A\mathrm{d}S)\\ &= \frac{\mathrm{d}}{\mathrm{d}t}(\tilde{J}\tilde{F}^{-1A}_{\boldsymbol{\cdot}\alpha})N_A\mathrm{d}S\\ &= \left(\frac{\mathrm{d}\tilde{J}}{\mathrm{d}t}\tilde{F}^{-1A}_{\boldsymbol{\cdot}\alpha} + \tilde{J}\frac{\mathrm{d}\tilde{F}^{-1A}_{\boldsymbol{\cdot}\alpha}}{\mathrm{d}t}\right)N_A\mathrm{d}S\\ &= \tilde{J}(\tilde{L}^{\beta}_{\boldsymbol{\cdot}\beta}\tilde{F}^{-1A}_{\boldsymbol{\cdot}\alpha} - \tilde{L}^{\beta}_{\boldsymbol{\cdot}\alpha}\tilde{F}^{-1A}_{\boldsymbol{\cdot}\beta})N_A\mathrm{d}S\\ &= (\tilde{L}^{\beta}_{\boldsymbol{\cdot}\beta}\tilde{n}_{\alpha} - \tilde{L}^{\beta}_{\boldsymbol{\cdot}\alpha}n_{\beta})\mathrm{d}\tilde{s}.\end{aligned} \tag{5.136}$$

The material time derivative of intermediate area element $\tilde{n}_\alpha \mathrm{d}\tilde{s}$ can also be written as follows in terms of spatial area element $n_a \mathrm{d}s$ using (3.174) and (5.69):

$$\begin{aligned}
\frac{\mathrm{d}}{\mathrm{d}t}(\tilde{n}_\alpha \mathrm{d}\tilde{s}) &= \frac{\mathrm{d}}{\mathrm{d}t}(\bar{J}^{-1}\bar{F}^a_{.\alpha} n_a \mathrm{d}s) \\
&= \frac{\mathrm{d}}{\mathrm{d}t}(\bar{J}^{-1}\bar{F}^a_{.\alpha}) n_a \mathrm{d}s + \bar{J}^{-1}\bar{F}^a_{.\alpha}\frac{\mathrm{d}}{\mathrm{d}t}(n_a \mathrm{d}s) \\
&= \left(\frac{\mathrm{d}\bar{J}^{-1}}{\mathrm{d}t}\bar{F}^a_{.\alpha} + \bar{J}^{-1}\frac{\mathrm{d}\bar{F}^a_{.\alpha}}{\mathrm{d}t}\right) n_a \mathrm{d}s + \bar{J}^{-1}\bar{F}^a_{.\alpha}\frac{\mathrm{d}}{\mathrm{d}t}(n_a \mathrm{d}s) \\
&= \left(\bar{J}^{-1}\frac{\mathrm{d}\bar{F}^a_{.\alpha}}{\mathrm{d}t} - \bar{J}^{-2}\frac{\mathrm{d}\bar{J}}{\mathrm{d}t}\bar{F}^a_{.\alpha}\right) n_a \mathrm{d}s + \bar{J}^{-1}\bar{F}^a_{.\alpha}\frac{\mathrm{d}}{\mathrm{d}t}(n_a \mathrm{d}s) \\
&= \left(\bar{J}^{-1}\frac{\mathrm{d}\bar{F}^a_{.\alpha}}{\mathrm{d}t} - \bar{J}^{-1}\bar{L}^b_{.b}\bar{F}^a_{.\alpha}\right) n_a \mathrm{d}s \\
&\qquad + \bar{J}^{-1}\bar{F}^a_{.\alpha}(L^b_{.b} n_a - L^b_{.a} n_b)\mathrm{d}s \\
&= \bar{J}^{-1}\bar{F}^a_{.\alpha}(L^b_{.b} - \bar{L}^b_{.b}) n_a \mathrm{d}s + \bar{J}^{-1}\bar{F}^a_{.\alpha}(\bar{L}^b_{.a} - L^b_{.a}) n_b \mathrm{d}s \\
&= \bar{J}^{-1}[\bar{F}^a_{.\alpha}(L^b_{.b} - \bar{L}^b_{.b}) - \bar{F}^b_{.\alpha}(L^a_{.b} - \bar{L}^a_{.b})] n_a \mathrm{d}s.
\end{aligned} \tag{5.137}$$

Material time differentiation and partial (as well as covariant) differentiation with respect to anholonomic coordinates do not generally commute. For example, let $f[x(X,t),t]$ denote a differentiable scalar field. The analog of (3.177) in anholonomic coordinates is

$$\begin{aligned}
\frac{\mathrm{d}}{\mathrm{d}t}(\partial_\alpha f) &= \frac{\mathrm{d}}{\mathrm{d}t}(\tilde{F}^{-1A}_{.\alpha}\partial_A f) \\
&= \tilde{F}^{-1A}_{.\alpha}\partial_A \dot{f} + \frac{\mathrm{d}\tilde{F}^{-1A}_{.\alpha}}{\mathrm{d}t}\partial_A f \\
&= \partial_\alpha \dot{f} - \tilde{F}^{-1A}_{.\beta}\tilde{L}^\beta_{.\alpha}\partial_A f \\
&= \partial_\alpha \dot{f} - \tilde{L}^\beta_{.\alpha}\partial_\beta f.
\end{aligned} \tag{5.138}$$

Similarly, the analog of (3.178) in anholonomic coordinates is

$$\begin{aligned}
\frac{\partial}{\partial t}\bigg|_x(\partial_\alpha f) &= \frac{\partial}{\partial t}\bigg|_x(\bar{F}^a_{.\alpha}\partial_a f) \\
&= \bar{F}^a_{.\alpha}\partial_a\left(\frac{\partial f}{\partial t}\bigg|_x\right) + \frac{\partial \bar{F}^a_{.\alpha}}{\partial t}\bigg|_x \partial_a f \\
&= \partial_\alpha\left(\frac{\partial f}{\partial t}\bigg|_x\right) + \bar{F}^{-1\beta}_{.a}\frac{\partial \bar{F}^a_{.\alpha}}{\partial t}\bigg|_x \partial_\beta f.
\end{aligned} \tag{5.139}$$

Since anholonomic connection coefficients may be time dependent, identities analogous to (3.181) do not generally apply in the intermediate configuration.

5.2 Integral Theorems

Forms of three integral theorems used frequently in continuum mechanics are now described in the context of the intermediate configuration. Recall that Gauss's theorem relates volume and surface integrals, and that Stokes's theorem relates surface and line integrals. In the present context, domains of integration may reside in the possibly anholonomic intermediate configuration. Finally, Reynolds transport theorem is revisited from the perspective of anholonomic deformation.

5.2.1 *Domain transformations*

Because intermediate configuration $\tilde{B}$ is generally an anholonomic space of dimension n, simply connected volumes, areas, and curves generally do not exist in such a configuration. Therefore, in order to apply integral theorems of Gauss and Stokes, domains of integration generally must first be mapped to Euclidean reference (*i.e.*, material) or current (*i.e.*, spatial) configurations. Using (5.4) and (5.34), volume integrals are transformed as

$$\int_{\tilde{V}} (\cdot)\mathrm{d}\tilde{V} = \int_V (\cdot)\tilde{J}\,\mathrm{d}V = \int_v (\cdot)\bar{J}^{-1}\mathrm{d}v. \tag{5.140}$$

Using (5.66) and (5.69), oriented surface integrals are transformed as

$$\int_{\tilde{s}} (\cdot)\tilde{n}_\alpha \mathrm{d}\tilde{s} = \int_S (\cdot)\tilde{J}\tilde{F}^{-1A}_{.\alpha} N_A \mathrm{d}S = \int_s (\cdot)\bar{J}^{-1}\bar{F}^{a}_{.\alpha} n_a \mathrm{d}s. \tag{5.141}$$

Finally, from (4.219), line integrals along intermediate curve $\tilde{c}$ are transformed as

$$\int_{\tilde{c}} (\cdot)\mathrm{d}\tilde{x}^\alpha = \int_C (\cdot)\tilde{F}^{\alpha}_{.A}\mathrm{d}X^A = \int_c (\cdot)\bar{F}^{-1\alpha}_{.a}\mathrm{d}x^a. \tag{5.142}$$

5.2.2 *Gauss's theorem*

A general form of Gauss's theorem on $\tilde{B}$ is obtained by transforming the left and right of (3.133) using (5.140) and (5.141) respectively, here for example mapping from the reference configuration:

$$\int_V \nabla \star \boldsymbol{A}\,\mathrm{d}V = \int_{\tilde{V}} \tilde{J}^{-1}\nabla \star \boldsymbol{A}\,\mathrm{d}\tilde{V}, \tag{5.143}$$

$$\oint_S \boldsymbol{N} \star \boldsymbol{A}\,\mathrm{d}S = \oint_{\tilde{s}} (\tilde{J}^{-1}\tilde{\boldsymbol{F}}\tilde{\boldsymbol{n}}) \star \boldsymbol{A}\,\mathrm{d}\tilde{s}. \tag{5.144}$$

The following anholonomic version of Gauss's theorem is thus obtained:

$$\int_{\tilde{V}} \tilde{J}^{-1}\nabla \star \boldsymbol{A}\, \mathrm{d}\tilde{V} = \oint_{\tilde{s}} (\tilde{J}^{-1}\tilde{\boldsymbol{F}}\tilde{\boldsymbol{n}}) \star \boldsymbol{A}\, \mathrm{d}\tilde{s}. \tag{5.145}$$

As an example, let $\boldsymbol{A} \to \tilde{\boldsymbol{V}}(X) = \tilde{V}^{\alpha}\tilde{\boldsymbol{g}}_{\alpha}$ be a differentiable vector field in coordinates referred to the intermediate configuration, and let $\star \to \langle\ ,\ \rangle$, *i.e.*, the scalar product operator. The divergence theorem for such a vector field is then obtained in intermediate configuration space as

$$\int_{\tilde{V}} \langle \tilde{\nabla}, \tilde{\boldsymbol{V}} \rangle\, \mathrm{d}\tilde{V} = \oint_{\tilde{s}} \langle \tilde{\boldsymbol{n}}, \tilde{\boldsymbol{V}} \rangle\, \mathrm{d}\tilde{s} - \int_{\tilde{V}} \langle \tilde{J}^{-1}\tilde{\boldsymbol{V}}, \langle \overset{G}{\nabla}, \tilde{J}\tilde{\boldsymbol{F}}^{-1} \rangle\rangle\, \mathrm{d}\tilde{V}, \tag{5.146}$$

or derived fully in coordinates using integration by parts with (5.27),

$$\begin{aligned}
\int_{\tilde{V}} \tilde{V}^{\alpha}_{;\alpha}\, \mathrm{d}\tilde{V} &= \int_{\tilde{V}} \tilde{V}^{\alpha}_{:\alpha}\, \mathrm{d}\tilde{V} \\
&= \int_{\tilde{V}} \tilde{V}^{\alpha}_{:A}\tilde{F}^{-1A}_{\ \ \cdot\alpha}\, \mathrm{d}\tilde{V} \\
&= \int_{V} \tilde{V}^{\alpha}_{:A}\tilde{J}\tilde{F}^{-1A}_{\ \ \cdot\alpha}\, \mathrm{d}V \\
&= \int_{V} [(\tilde{V}^{\alpha}\tilde{J}\tilde{F}^{-1A}_{\ \ \cdot\alpha})_{:A} - \tilde{V}^{\alpha}(\tilde{J}\tilde{F}^{-1A}_{\ \ \cdot\alpha})_{:A}]\, \mathrm{d}V \\
&= \oint_{S} \tilde{V}^{\alpha}\tilde{J}\tilde{F}^{-1A}_{\ \ \cdot\alpha}N_A\, \mathrm{d}S - \int_{V} \tilde{V}^{\alpha}(\tilde{J}\tilde{F}^{-1A}_{\ \ \cdot\alpha})_{:A}\, \mathrm{d}V \\
&= \oint_{\tilde{s}} \tilde{V}^{\alpha}\tilde{n}_{\alpha}\, \mathrm{d}\tilde{s} - \int_{V} \tilde{V}^{\alpha}\epsilon_{\alpha\beta\chi}\epsilon^{ABC}\tilde{F}^{\beta}_{\cdot A}\tilde{F}^{\chi}_{\cdot B:C}\, \mathrm{d}V \\
&= \oint_{\tilde{s}} \tilde{V}^{\alpha}\tilde{n}_{\alpha}\, \mathrm{d}\tilde{s} - \int_{\tilde{V}} \tilde{J}^{-1}\tilde{V}^{\alpha}\epsilon_{\alpha\beta\chi}\epsilon^{ABC}\tilde{F}^{\beta}_{\cdot A}\tilde{F}^{\chi}_{\cdot B:C}\, \mathrm{d}\tilde{V}. \\
&= \oint_{\tilde{s}} \tilde{V}^{\alpha}\tilde{n}_{\alpha}\, \mathrm{d}\tilde{s} - \int_{\tilde{V}} \tilde{J}^{-1}\tilde{V}^{\alpha}\epsilon_{\alpha\beta\chi}\epsilon^{ABC}\tilde{F}^{\beta}_{\cdot A}\tilde{F}^{\chi}_{\cdot[B:C]}\, \mathrm{d}\tilde{V}.
\end{aligned} \tag{5.147}$$

The second term (*i.e.*, the volume integral) on the right of the final four equalities in (5.147) vanishes identically when configuration $\tilde{B}$ is holonomic [so that $\tilde{\boldsymbol{F}}(X,t)$ is integrable with respect to X^A over domain V] and when $\Gamma^{\cdot\cdot\alpha}_{[\beta\chi]} = 0$, in which case (4.208) vanishes identically.

5.2.3 *Stokes's theorem*

A general form of Stokes's theorem on configuration $\tilde{B}$ can be obtained by transforming the left and right of (3.143) using (5.141) and (5.142) respectively, here for example mapping from the reference configuration:

$$\int_{S} (\boldsymbol{N} \times \nabla) \star \boldsymbol{A}\, \mathrm{d}S = \int_{\tilde{s}} [(\tilde{J}^{-1}\tilde{\boldsymbol{F}}\tilde{\boldsymbol{n}}) \times \nabla] \star \boldsymbol{A}\, \mathrm{d}\tilde{s}, \tag{5.148}$$

$$\oint_C \mathrm{d}\boldsymbol{X} \star \boldsymbol{A} = \oint_{\tilde{c}} \tilde{\boldsymbol{F}}^{-1} \mathrm{d}\tilde{\boldsymbol{x}} \star \boldsymbol{A}. \tag{5.149}$$

The following anholonomic version of Stokes's theorem is thus obtained:

$$\int_{\tilde{s}} [(\tilde{J}^{-1} \tilde{\boldsymbol{F}} \tilde{\boldsymbol{n}}) \times \nabla] \star \boldsymbol{A} \, \mathrm{d}\tilde{s} = \oint_{\tilde{c}} \tilde{\boldsymbol{F}}^{-1} \mathrm{d}\tilde{\boldsymbol{x}} \star \boldsymbol{A}. \tag{5.150}$$

As an illustrative example, let $\boldsymbol{A} \to \tilde{\boldsymbol{F}}(X,t) = \tilde{F}^{\alpha}_{\cdot A} \tilde{\boldsymbol{g}}_{\alpha} \otimes \boldsymbol{G}^A$, and assign constant intermediate basis vectors $\tilde{\boldsymbol{g}}_{\alpha}$ on $\tilde{B}$ with corresponding vanishing Christoffel symbols. It follows that

$$\begin{aligned}
\oint_{\tilde{c}} \mathrm{d}\tilde{x}^{\alpha} &= \oint_C \tilde{F}^{\alpha}_{\cdot A} \mathrm{d}X^A \\
&= \int_S \epsilon^{ABC} \tilde{F}^{\alpha}_{\cdot A;C} \, N_B \, \mathrm{d}S \\
&= \int_S \epsilon^{ABC} \partial_A \tilde{F}^{\alpha}_{\cdot B} \, N_C \, \mathrm{d}S \\
&= \int_S \epsilon^{ABC} \partial_{[A} \tilde{F}^{\alpha}_{\cdot B]} \, N_C \, \mathrm{d}S.
\end{aligned} \tag{5.151}$$

When the intermediate configuration is holonomic such that $\tilde{F}^{\alpha}_{\cdot A} = \partial_A \tilde{x}^{\alpha}$ is integrable with respect to reference coordinates X^A over domain S, then both the left and right of (5.151) vanish identically.

Analogously, let $\boldsymbol{A} \to \bar{\boldsymbol{F}}^{-1}(x,t) = \bar{F}^{-1\alpha}_{\cdot a} \tilde{\boldsymbol{g}}_{\alpha} \otimes \boldsymbol{g}^a$, and again assign constant intermediate basis vectors $\tilde{\boldsymbol{g}}_{\alpha}$ on $\tilde{B}$ with associated vanishing Christoffel symbols. Mapping the domain of integration to a closed curve c in the current configuration,

$$\begin{aligned}
\oint_{\tilde{c}} \mathrm{d}\tilde{x}^{\alpha} &= \oint_c \bar{F}^{-1\alpha}_{\cdot a} \mathrm{d}x^a \\
&= \int_s \epsilon^{abc} \bar{F}^{-1\alpha}_{\cdot a;c} \, n_b \, \mathrm{d}s \\
&= \int_s \epsilon^{abc} \partial_a \bar{F}^{-1\alpha}_{\cdot b} \, n_c \, \mathrm{d}s \\
&= \int_s \epsilon^{abc} \partial_{[a} \bar{F}^{-1\alpha}_{\cdot b]} \, n_c \, \mathrm{d}s.
\end{aligned} \tag{5.152}$$

When the intermediate configuration is holonomic such that $\bar{F}^{-1\alpha}_{\cdot a} = \partial_a \tilde{x}^{\alpha}$ is integrable with respect to spatial coordinates x^a over domain s, then both the left and right of (5.152) vanish identically.

Integrals in (5.151) and (5.152) can be identified as (in)compatibility conditions for respective mappings to the intermediate configuration $\tilde{\boldsymbol{F}}$ and $\bar{\boldsymbol{F}}^{-1}$, over respective areas S and s in configurations B_0 and B. When these quantities vanish for every possible closed loop C or c on a given simply connected body, then the intermediate configuration $\tilde{B}$ is globally holonomic over that body.

Stokes's theorem can thus be used to demonstrate that the left equality in (4.13) is sufficient for the existence of $\tilde{x}^\alpha(X,t)$. For example, assume the following conditions hold at time t for fields $\tilde{\boldsymbol{F}}(X,t)$ and $\tilde{\boldsymbol{g}}_\alpha(X)$ at all material points X within a simply connected domain of interest S:

$$\partial_{[A}\tilde{F}^{\alpha}_{.B]} = 0, \qquad \partial_A \tilde{\boldsymbol{g}}_\alpha = \Gamma^{..\chi}_{\beta\alpha}\tilde{F}^{\beta}_{.A}\tilde{\boldsymbol{g}}_\chi = 0. \tag{5.153}$$

Substituting into (5.151) yields

$$\Delta\tilde{x}^\alpha(S) = \oint_{\tilde{c}} \mathrm{d}\tilde{x}^\alpha = \int_S \epsilon^{ABC}\partial_{[A}\tilde{F}^{\alpha}_{.B]}\, N_C \,\mathrm{d}S = 0. \tag{5.154}$$

Now consider a simply connected body in the reference configuration covered by a global single valued coordinate system X^A. By definition, any surface S is simply connected in such a body. It follows from (5.154) that $\Delta\tilde{x}^\alpha = 0\ \forall S$ in this body, meaning that $\tilde{x}^\alpha(X,t)$ form a single valued, global coordinate system for the body at time t. Parallel arguments can be used to demonstrate sufficiency of (4.17) for the existence of continuously twice differentiable coordinate functions $\tilde{x}^\alpha(x,t)$ when the body is simply connected in the current configuration.

If a body is not globally simply connected, then Stokes's theorem can only be applied over those local regions that are simply connected, and coordinates $\tilde{x}^\alpha(X,t)$ are single valued functions of X^A necessarily only over those regions [*i.e.*, $\tilde{x}^\alpha(X,t)$ may not form a global covering of the body if it is not simply connected]. Of course, coordinates $X^A(X)$ themselves may not be single valued for each material particle X if the ensemble of such particles comprising the body is not simply connected in the reference configuration.

The following two-point dislocation density tensors can be defined using (4.111) and (4.120) [8]:

$$\tilde{\alpha}^{\alpha A} = \epsilon^{ABC}\partial_B\tilde{F}^{\alpha}_{.C} = \epsilon^{ABC}\tilde{F}^{\alpha}_{.D}\tilde{T}^{..D}_{BC} = \epsilon^{ABC}\tilde{F}^{\beta}_{.B}\tilde{F}^{\chi}_{.C}\tilde{\kappa}^{..\alpha}_{\beta\chi}, \tag{5.155}$$

$$\bar{\alpha}^{\alpha a} = \epsilon^{abc}\partial_b\bar{F}^{-1\alpha}_{\ \ .c} = \epsilon^{abc}\bar{F}^{-1\alpha}_{\ \ .d}\bar{T}^{..d}_{bc} = \epsilon^{abc}\bar{F}^{-1\beta}_{\ \ .b}\bar{F}^{-1\chi}_{\ \ .c}\bar{\kappa}^{..\alpha}_{\beta\chi}. \tag{5.156}$$

Furthermore, a dislocation density tensor referred completely to the intermediate configuration is the Piola transform

$$\alpha^{\alpha\beta} = \tilde{J}^{-1}\tilde{F}^{\beta}_{.A}\tilde{\alpha}^{\alpha A} = \bar{J}\bar{F}^{-1\beta}_{\ \ .a}\bar{\alpha}^{\alpha a}. \tag{5.157}$$

Therefore, according to Nanson's formulae (5.66) and (5.69), line integrals (5.151) and (5.152) become

$$\oint_{\tilde{c}} \mathrm{d}\tilde{x}^{\alpha} = \int_{S} \tilde{\alpha}^{\alpha A}\, N_A \,\mathrm{d}S = \int_{s} \bar{\alpha}^{\alpha a}\, n_a \,\mathrm{d}s = \int_{\tilde{s}} \alpha^{\alpha\beta}\, \tilde{n}_{\beta}\,\mathrm{d}\tilde{s}. \qquad (5.158)$$

Verification of the second equality in (5.157) is straightforward using (3.25), (3.58), and (4.2):

$$\begin{aligned}
\alpha^{\alpha\beta} &= \tilde{J}^{-1}\tilde{F}^{\beta}_{.A}\tilde{\alpha}^{\alpha A} \\
&= \tilde{J}^{-1}\tilde{F}^{\beta}_{.A}\epsilon^{ABC}\partial_B\tilde{F}^{\alpha}_{.C} \\
&= \tilde{J}^{-1}\tilde{F}^{\beta}_{.A}\epsilon^{ABC}(F^{f}_{.C}\partial_B\bar{F}^{-1\alpha}_{.f} + \bar{F}^{-1\alpha}_{.f}\partial_B F^{f}_{.C}) \\
&= \tilde{J}^{-1}\tilde{F}^{\beta}_{.A}\epsilon^{ABC}F^{f}_{.C}\partial_B\bar{F}^{-1\alpha}_{.f} \\
&= (\bar{J}J^{-1})(\bar{F}^{-1\beta}_{.d}F^{d}_{.A})(JF^{-1A}_{.a}F^{-1B}_{.b}F^{-1C}_{.c}\epsilon^{abc}) \\
&\quad \times (F^{f}_{.C}F^{e}_{.B}\partial_e\bar{F}^{-1\alpha}_{.f}) \\
&= \bar{J}\bar{F}^{-1\beta}_{.d}\delta^{d}_{a}\delta^{e}_{b}\delta^{f}_{c}\epsilon^{abc}\partial_e\bar{F}^{-1\alpha}_{.f} \\
&= \bar{J}\bar{F}^{-1\beta}_{.a}\epsilon^{abc}\partial_b\bar{F}^{-1\alpha}_{.c} \\
&= \bar{J}\bar{F}^{-1\beta}_{.a}\bar{\alpha}^{\alpha a}.
\end{aligned} \qquad (5.159)$$

5.2.4 *Reynolds transport theorem*

Consider a differentiable function $f(x,t)$, which may be a scalar, vector, or tensor field of higher order. Recall from (3.173) and (5.133) that material time derivatives of spatial and intermediate volume elements are

$$\begin{aligned}
\frac{\mathrm{d}}{\mathrm{d}t}[\mathrm{d}v(x)] &= \frac{\mathrm{d}}{\mathrm{d}t}[J(X,t)\mathrm{d}V(X)] \\
&= \dot{J}J^{-1}\mathrm{d}v \\
&= v^{a}_{;a}\mathrm{d}v \\
&= L^{a}_{.a}\mathrm{d}v,
\end{aligned} \qquad (5.160)$$

$$\begin{aligned}
\frac{\mathrm{d}}{\mathrm{d}t}[\mathrm{d}\tilde{V}(X,t)] &= \frac{\mathrm{d}}{\mathrm{d}t}[\tilde{J}(X,t)\mathrm{d}V(X)] \\
&= \frac{\mathrm{d}}{\mathrm{d}t}(\tilde{J})\tilde{J}^{-1}\mathrm{d}\tilde{V} \\
&= \tilde{L}^{\alpha}_{.\alpha}\mathrm{d}\tilde{V}.
\end{aligned} \qquad (5.161)$$

Using definition (3.151) of the material time derivative along with (5.135), a version of Reynolds transport theorem can be derived with respect to possibly anholonomic domains as

$$\begin{aligned}\frac{\mathrm{d}}{\mathrm{d}t}\int_{\tilde{V}} f\mathrm{d}\tilde{V} &= \int_{\tilde{V}} \dot{f}\mathrm{d}\tilde{V} + \int_{\tilde{V}} f\frac{\mathrm{d}}{\mathrm{d}t}(\mathrm{d}\tilde{V}) \\ &= \int_{\tilde{V}} \frac{\partial f}{\partial t}\Big|_x \mathrm{d}\tilde{V} + \int_{\tilde{V}} f_{;a}v^a\mathrm{d}\tilde{V} + \int_{\tilde{V}} f\tilde{L}^{\alpha}_{.\alpha}\mathrm{d}\tilde{V} \\ &= \int_{\tilde{V}} \frac{\partial f}{\partial t}\Big|_x \mathrm{d}\tilde{V} + \int_{\tilde{V}} f_{;a}v^a\mathrm{d}\tilde{V} + \int_{\tilde{V}} fv^a_{;a}\mathrm{d}\tilde{V} \\ &\quad + \int_{\tilde{V}} f(\tilde{L}^{\alpha}_{.\alpha} - L^a_{.a})\mathrm{d}\tilde{V} \\ &= \int_{\tilde{V}} \frac{\partial f}{\partial t}\Big|_x \mathrm{d}\tilde{V} - \int_{\tilde{V}} f\bar{L}^a_{.a}\mathrm{d}\tilde{V} + \int_{\tilde{V}} (fv^a)_{;a}\mathrm{d}\tilde{V}. \end{aligned} \tag{5.162}$$

The rightmost term in the final equality can be expressed as follows upon using (5.54), (5.55), (5.69), and the divergence theorem of Gauss:

$$\begin{aligned}\int_{\tilde{V}} (fv^a)_{;a}\mathrm{d}\tilde{V} &= \int_v (fv^a)_{;a}\bar{J}^{-1}\mathrm{d}v \\ &= \int_v (\bar{J}^{-1}fv^a)_{;a}\mathrm{d}v - \int_v fv^a\partial_a(\bar{J}^{-1})\mathrm{d}v \\ &= \oint_s (\bar{J}^{-1}fv^a)n_a\mathrm{d}s - \int_{\tilde{V}} fv^a\partial_a(\bar{J}^{-1})\bar{J}\mathrm{d}\tilde{V} \\ &= \oint_{\tilde{s}} fv^a\bar{F}^{-1\alpha}_{.a}\tilde{n}_\alpha\mathrm{d}\tilde{s} + \int_{\tilde{V}} fv^a\bar{F}^{-1\alpha}_{.b}\bar{F}^b_{.\alpha:a}\mathrm{d}\tilde{V}, \end{aligned} \tag{5.163}$$

where v is a simply connected spatial volume enclosed by surface s with unit outward normal $n_a(x)$. Substituting this result into (5.162) gives

$$\begin{aligned}\frac{\mathrm{d}}{\mathrm{d}t}\int_{\tilde{V}} f\mathrm{d}\tilde{V} &= \int_{\tilde{V}} \frac{\partial}{\partial t}\Big|_x (f)\,\mathrm{d}\tilde{V} + \int_{\tilde{V}} f\bar{F}^{-1\alpha}_{.b}(\bar{F}^b_{.\alpha:a}v^a - \frac{\mathrm{d}}{\mathrm{d}t}\bar{F}^b_{.\alpha})\mathrm{d}\tilde{V} \\ &\quad + \oint_{\tilde{s}} fv^a\bar{F}^{-1\alpha}_{.a}\tilde{n}_\alpha\mathrm{d}\tilde{s}. \end{aligned} \tag{5.164}$$

The rightmost term in equality of (5.164) represents the time rate of change of quantity f associated with the flux of f carried through the boundary $\tilde{s}$ of volume $\tilde{V}$.

When null anholonomic connection coefficients are prescribed such that $\Gamma^{..\alpha}_{\beta\alpha} = 0$, the total covariant derivative reduces to a partial covariant derivative:

$$\bar{F}^{-1\alpha}_{.b}\bar{F}^b_{.\alpha:a} = \bar{F}^{-1\alpha}_{.b}\bar{F}^b_{.\alpha;a} - \Gamma^{..\alpha}_{\beta\alpha}\bar{F}^{-1\beta}_{.a} = \bar{F}^{-1\alpha}_{.b}\bar{F}^b_{.\alpha;a}, \tag{5.165}$$

and the material time derivative in (5.164) becomes

$$\frac{\mathrm{d}}{\mathrm{d}t}\bar{F}^b_{.\alpha} = \frac{\partial}{\partial t}\Big|_x \bar{F}^b_{.\alpha} + \bar{F}^b_{.\alpha:a}v^a, \tag{5.166}$$

such that in this case (5.164) reduces to

$$\frac{\mathrm{d}}{\mathrm{d}t}\int_{\tilde{V}} f\mathrm{d}\tilde{V} = \int_{\tilde{V}} \frac{\partial}{\partial t}\bigg|_x (f)\,\mathrm{d}\tilde{V} - \int_{\tilde{V}} f\bar{F}^{-1\alpha}_{\ \ \ \,\bullet b}\frac{\partial}{\partial t}\bigg|_x (\bar{F}^{b}_{\bullet\alpha})\mathrm{d}\tilde{V}$$
$$+ \oint_{\tilde{s}} fv^a\bar{F}^{-1\alpha}_{\ \ \ \,\bullet a}\tilde{n}_\alpha \mathrm{d}\tilde{s}. \tag{5.167}$$

For the particular case when $\bar{\boldsymbol{F}} = \mathbf{1}$ is uniform in time and space such that $\mathrm{d}v(x) = \mathrm{d}\tilde{V}(x)$, then the second volume integral on the right vanishes and (5.167) degenerates to the standard version of Reynolds transport theorem for holonomic configurations (3.190).

Appendix A

List of Symbols

Item **Definition** (equation of first appearance)

$\boldsymbol{a}, a^a$ spatial acceleration (3.154)

$\mathrm{d}s$ spatial differential surface element (3.130)

$\mathrm{d}S$ referential differential surface element (3.130)

$\mathrm{d}\tilde{s}$ intermediate differential surface element (5.65)

$\mathrm{d}v$ spatial differential volume element (3.37)

$\mathrm{d}V$ referential differential volume element (3.36)

$\mathrm{d}\tilde{V}$ intermediate differential volume element (5.2)

$\mathrm{d}\boldsymbol{x}, \mathrm{d}x^a$ spatial differential line element (3.14)

$\mathrm{d}\boldsymbol{X}, \mathrm{d}X^A$ referential differential line element (3.14)

$\mathrm{d}\tilde{\boldsymbol{x}}, \mathrm{d}\tilde{x}^\alpha$ intermediate differential line element (4.217)

$\mathrm{d}x_a$ covariant components of spatial differential line element (3.33)

$\mathrm{d}X_A$ covariant components of referential differential line element (3.33)

$\mathrm{d}\tilde{x}_\alpha$ covariant components of intermediate differential line element (5.92)

$\boldsymbol{d}X^A$ referential reciprocal basis vector (2.20)

$\boldsymbol{d}\tilde{x}^\alpha$ intermediate reciprocal basis vector (4.64)

$\boldsymbol{e}_a, \boldsymbol{e}^a$ spatial Cartesian basis vector (3.115)

$\boldsymbol{e}_\alpha, \boldsymbol{e}^\alpha$ intermediate Cartesian basis vector (4.62)

e^{abc} contravariant spatial permutation symbols (2.9)

e_{abc} covariant spatial permutation symbols (2.9)

e^{ABC} contravariant referential permutation symbols (2.8)

e_{ABC} covariant referential permutation symbols (2.8)

$e^{\alpha\beta\chi}$ contravariant intermediate permutation symbols (4.5)

$e_{\alpha\beta\chi}$ covariant intermediate permutation symbols (4.5)

g determinant of spatial metric tensor (2.32)

$\tilde{g}$ determinant of intermediate metric tensor (4.30)

g^a_A, g^A_a shifters between reference and current configurations (2.47)

g^a_α, g^α_a shifters between intermediate and current configurations (4.37)
g^α_A, g^A_α shifters between reference and intermediate configurations (4.37)
$\boldsymbol{g}, g_{ab}$ spatial metric tensor (2.23)
$\tilde{\boldsymbol{g}}, \tilde{g}_{\alpha\beta}$ intermediate metric tensor (4.27)
$\boldsymbol{g}_a$ spatial basis vector (2.4)
$\boldsymbol{g}^a$ spatial reciprocal basis vector (2.5)
$\boldsymbol{g}'_A$ convected spatial basis vector (3.349)
$\boldsymbol{g}'^A$ convected spatial reciprocal basis vector (3.349)
$\tilde{\boldsymbol{g}}_\alpha$ intermediate basis vector (4.3)
$\tilde{\boldsymbol{g}}^\alpha$ intermediate reciprocal basis vector (4.3)
$\tilde{\boldsymbol{g}}'_\alpha$ convected intermediate basis vector (4.84)
$\tilde{\boldsymbol{g}}'^\alpha$ convected intermediate reciprocal basis vector (4.84)
$\bar{\boldsymbol{g}}'_\alpha$ convected intermediate basis vector (4.91)
$\bar{\boldsymbol{g}}'^\alpha$ convected intermediate reciprocal basis vector (4.91)
n dimension of Euclidean space (2.82)
$\boldsymbol{n}, n_a$ unit normal to oriented spatial surface (3.130)
$\tilde{\boldsymbol{n}}, \tilde{n}_\alpha$ unit normal to oriented intermediate surface (5.65)
$\tilde{r}^\alpha$ deformation associated with integrable rotation field (4.130)
r, θ, z cylindrical spatial coordinates (3.235)
r, θ, ϕ spherical spatial coordinates (3.290)
t time (2.1)
$\boldsymbol{u}, u^A, u^a$ displacement (3.99)
$\boldsymbol{v}, v^a$ spatial velocity (3.148)
x spatial point (2.1)
x^a spatial coordinates (2.1)
x_a covariant spatial position (3.29)
x^A, x_A shifted spatial position (3.108)
$\boldsymbol{x}$ spatial position vector in Euclidean space (2.4)
$\tilde{x}^\alpha$ intermediate coordinates (4.12)
$\tilde{\boldsymbol{x}}$ intermediate position vector in Euclidean space (4.47)
x, y, z Cartesian spatial coordinates (3.191)
$\boldsymbol{A}, A^a$ material acceleration (3.155)
B spatial configuration
B_0 reference configuration
$\tilde{B}$ intermediate configuration
$\boldsymbol{B}, B^{ab}$ spatial deformation tensor (3.87)
$\bar{\boldsymbol{B}}, \bar{B}^{ab}$ deformation tensor of possibly non-integrable deformation (5.77)
$\tilde{\boldsymbol{B}}, \tilde{B}^{\alpha\beta}$ deformation tensor of possibly non-integrable deformation (5.91)
$\boldsymbol{C}, C_{AB}$ referential deformation tensor (3.86)

$\bar{\boldsymbol{C}}, \bar{C}_{\alpha\beta}$ deformation tensor of possibly non-integrable deformation (5.76)
$\tilde{\boldsymbol{C}}, \tilde{C}_{AB}$ deformation tensor of possibly non-integrable deformation (5.90)
$\boldsymbol{D}, D_{ab}$ spatial deformation rate tensor (3.165)
$\bar{\boldsymbol{D}}, \bar{D}_{ab}$ deformation rate term from non-integrable deformation (5.122)
$\tilde{\boldsymbol{D}}, \tilde{D}_{\alpha\beta}$ deformation rate term from non-integrable deformation (5.123)
$\boldsymbol{E}_A, \boldsymbol{E}^A$ referential Cartesian basis vector (3.220)
$\boldsymbol{E}, E_{AB}$ referential strain tensor (3.92)
$\bar{\boldsymbol{E}}, \bar{E}_{\alpha\beta}$ strain of possibly non-integrable deformation (5.81)
$\tilde{\boldsymbol{E}}, \tilde{E}_{AB}$ strain of possibly non-integrable deformation (5.95)
$\boldsymbol{F}, F^a_{.A}$ total deformation gradient (3.1)
$\bar{\boldsymbol{F}}, \bar{F}^a_{.\alpha}$ possibly non-integrable deformation mapping (4.1)
$\tilde{\boldsymbol{F}}, \tilde{F}^\alpha_{.A}$ possibly non-integrable deformation mapping (4.1)
G determinant of referential metric tensor (2.31)
$\boldsymbol{G}, G_{AB}$ referential metric tensor (2.22)
$\boldsymbol{G}_A$ referential basis vector (2.4)
$\boldsymbol{G}^A$ referential reciprocal basis vector (2.5)
$\boldsymbol{G}'_A$ convected Cartesian basis vector (3.116)
$\boldsymbol{G}'^A$ convected Cartesian reciprocal basis vector (3.116)
$\boldsymbol{G}'_a$ convected referential basis vector (3.358)
$\boldsymbol{G}'^a$ convected referential reciprocal basis vector (3.358)
J Jacobian determinant of total deformation gradient (3.39)
$\tilde{J}$ Jacobian determinant of possibly non-integrable deformation (5.4)
$\bar{J}$ Jacobian determinant of possibly non-integrable deformation (5.34)
$K^{..A}_{BC}$ contortion in reference configuration (4.122)
$\boldsymbol{L}, L^a_{.b}$ spatial velocity gradient (3.160)
$\bar{\boldsymbol{L}}, \bar{L}^a_{.b}$ velocity gradient contribution of non-integrable deformation (5.116)
$\tilde{\boldsymbol{L}}, \tilde{L}^\alpha_{.\beta}$ velocity gradient contribution of non-integrable deformation (5.116)
$M^{..A}_{BC}$ covariant derivative of (metric) tensor (2.89)
$\hat{M}^{..\alpha}_{\beta\chi}$ anholonomic covariant derivative of (metric) tensor (4.79)
$M_{\alpha\beta\chi}$ covariant derivative of intermediate metric tensor (4.178)
$\boldsymbol{N}, N_A$ unit normal to oriented referential surface (3.130)
$\boldsymbol{Q}, Q^\alpha_{.A}$ orthogonal tensor associated with contortion (4.121)
$\boldsymbol{R}, R^a_{.A}$ rotation tensor in polar decomposition (3.74)
$\bar{\boldsymbol{R}}, \bar{R}^a_{.\alpha}$ rotation tensor of possibly non-integrable deformation (5.71)
$\tilde{\boldsymbol{R}}, \tilde{R}^\alpha_{.A}$ rotation tensor of possibly non-integrable deformation (5.85)
$\hat{\boldsymbol{R}}, \hat{R}^a_{.A}$ rotation tensor (5.99)
R Gaussian curvature (2.82)
R, Θ, Z cylindrical reference coordinates (3.257)

R, Θ, Φ spherical reference coordinates (3.312)
R_{AB} Ricci curvature (2.80)
$R^{\cdots A}_{BCD}$ Riemann-Christoffel curvature in reference configuration (2.75)
$\bar{R}^{\cdots A}_{BCD}$ Riemann-Christoffel curvature from part of total connection (2.105)
$\overset{G}{R}{}^{\cdots A}_{BCD}$ Riemann-Christoffel curvature of reference Euclidean metric (2.123)
$\overset{g}{R}{}^{\cdots a}_{bcd}$ Riemann-Christoffel curvature of spatial Euclidean metric (2.151)
$\overset{C}{R}{}^{\cdots A}_{BCD}$ Riemann-Christoffel curvature of deformation tensor (3.127)
$\tilde{R}^{\cdots A}_{BCD}$ Riemann-Christoffel curvature of integrable connection (4.106)
$\bar{R}^{\cdots a}_{bcd}$ Riemann-Christoffel curvature of integrable connection (4.115)
$\hat{R}^{\cdots \alpha}_{\beta\chi\delta}$ arbitrary curvature mapped to intermediate configuration (4.71)
$R^{\cdots \alpha}_{\beta\chi\delta}$ generic curvature of intermediate configuration (4.160)
$S^{\cdot\cdot A}_{BC}$ torsion tensor components of referential contortion (4.123)
$\boldsymbol{T}, T^{\cdot\cdot A}_{BC}$ generic torsion tensor in reference configuration (2.73)
$\overset{C}{T}{}^{\cdot\cdot A}_{BC}$ torsion tensor from connection coefficients of deformation (3.123)
$\hat{T}^{\cdot\cdot \alpha}_{\beta\chi}$ arbitrary torsion tensor mapped to intermediate configuration (4.68)
$\tilde{T}^{\cdot\cdot \alpha}_{\beta\chi}$ reference torsion tensor mapped to intermediate configuration (4.89)
$\bar{T}^{\cdot\cdot \alpha}_{\beta\chi}$ spatial torsion tensor mapped to intermediate configuration (4.96)
$\tilde{T}^{\cdot\cdot A}_{BC}$ torsion tensor of integrable connection coefficients (4.109)
$\bar{T}^{\cdot\cdot a}_{bc}$ torsion tensor of integrable connection coefficients (4.118)
$T^{\cdot\cdot \alpha}_{\beta\chi}$ generic torsion tensor of intermediate configuration (4.159)
$\boldsymbol{U}, U^{A}_{\cdot B}$ right stretch tensor (3.74)
$\bar{\boldsymbol{U}}, \bar{U}^{\alpha}_{\cdot \beta}$ right stretch tensor of possibly non-integrable deformation (5.71)
$\tilde{\boldsymbol{U}}, \tilde{U}^{A}_{\cdot B}$ right stretch tensor of possibly non-integrable deformation (5.85)
$\boldsymbol{V}, V^{a}_{\cdot b}$ left stretch tensor (3.74)
$\bar{\boldsymbol{V}}, \bar{V}^{a}_{\cdot b}$ left stretch tensor of possibly non-integrable deformation (5.71)
$\tilde{\boldsymbol{V}}, \tilde{V}^{\alpha}_{\cdot \beta}$ left stretch tensor of possibly non-integrable deformation (5.85)
$\boldsymbol{V}, V^{a}$ material velocity (3.149)
$\boldsymbol{W}, W_{ab}$ spatial spin tensor (3.165)
$\bar{\boldsymbol{W}}, \bar{W}_{ab}$ spin contribution of possibly non-integrable deformation (5.122)
$\tilde{\boldsymbol{W}}, \tilde{W}_{\alpha\beta}$ spin contribution of possibly non-integrable deformation (5.123)
X reference point or material particle (2.1)
X^{A} referential coordinates (2.1)
X_{A} covariant referential position (3.29)
$\hat{X}^{\hat{A}}$ transformed referential coordinates (2.60)
$\boldsymbol{X}$ referential position vector in Euclidean space (2.4)
X, Y, Z Cartesian reference coordinates (3.211)
$\tilde{\alpha}^{\alpha A}$ contravariant two-point dislocation density tensor (5.155)
$\bar{\alpha}^{\alpha a}$ contravariant two-point dislocation density tensor (5.156)

$\alpha^{\alpha\beta}$ contravariant fully intermediate dislocation density tensor (5.157)
χ_{ABC} partial derivative of referential (metric) tensor (2.93)
$\hat{\chi}_{\alpha\beta\chi}$ partial derivative of anholonomic (metric) tensor (4.80)
δ^a_b spatial Kronecker delta (2.5)
δ^A_B referential Kronecker delta (2.5)
δ^{ab} contravariant Kronecker delta (3.115)
δ_{ab} covariant Kronecker delta (3.115)
δ^α_β intermediate Kronecker delta (4.10)
$\delta^a_\alpha, \delta^\alpha_a$ mixed Kronecker delta (4.55)
$\delta^\alpha_A, \delta^A_\alpha$ mixed Kronecker delta (4.49)
ϵ^{abc} contravariant spatial permutation tensor (2.41)
ϵ_{abc} covariant spatial permutation tensor (2.41)
ϵ^{ABC} contravariant referential permutation tensor (2.40)
ϵ_{ABC} covariant referential permutation tensor (2.40)
$\epsilon^{\alpha\beta\chi}$ contravariant intermediate permutation tensor (4.36)
$\epsilon_{\alpha\beta\chi}$ covariant intermediate permutation tensor (4.36)
φ, φ^a motion (2.1)
κ scalar curvature (2.82)
$\hat{\kappa}_{\beta\chi}^{\cdot\cdot\alpha}$ arbitrary anholonomic object from coordinate transformation (4.68)
$\tilde{\kappa}_{\beta\chi}^{\cdot\cdot\alpha}$ anholonomic object from intermediate connection coefficients (4.89)
$\bar{\kappa}_{\beta\chi}^{\cdot\cdot\alpha}$ anholonomic object from intermediate connection coefficients (4.96)
$\kappa_{\beta\chi}^{\cdot\cdot\alpha}$ generic anholonomic object (4.158)
θ^{AB} Einstein tensor (2.81)
ϑ displacement potential (3.111)
$\tau_{BC}^{\cdot\cdot A}$ torsion components from part of connection (2.102)
ξ^A convected reference coordinates (3.346)
ζ^a convected spatial coordinates (3.355)
Φ, Φ^A inverse motion (2.1)
$\Gamma_{BC}^{\cdot\cdot A}$ arbitrary referential connection coefficients (2.59)
$\hat{\Gamma}_{\hat{B}\hat{C}}^{\cdot\cdot\hat{A}}$ transformed referential connection coefficients (2.62)
$\overset{G}{\Gamma}{}_{BC}^{\cdot\cdot A}$ Levi-Civita connection for Euclidean reference space (2.122)
$\overset{g}{\Gamma}{}_{bc}^{\cdot\cdot a}$ Levi-Civita connection for Euclidean current space (2.150)
$\overset{C}{\Gamma}{}_{BC}^{\cdot\cdot A}$ Levi-Civita connection from deformation tensor (3.120)
$\hat{\Gamma}_{\beta\chi}^{\cdot\cdot\alpha}$ generic connection mapped to intermediate configuration (4.66)
$\bar{\Gamma}_{\beta\chi}^{\cdot\cdot\alpha}$ generic connection mapped to intermediate configuration (4.95)
$\Gamma_{\beta\chi}^{\cdot\cdot\alpha}$ connection from gradients of intermediate basis vectors (4.138)
$\overset{\tilde{g}}{\Gamma}{}_{\beta\chi}^{\cdot\cdot\alpha}$ Levi-Civita connection for intermediate Euclidean space (4.165)
$\Xi_{BC}^{\cdot\cdot A}$ referential connection coefficients in total connection (2.101)

$\Upsilon^{..A}_{BC}$ arbitrary third-order tensor components (2.101)
$\mathbf{1}$ second-order identity tensor or unit tensor (3.77)
$\boldsymbol{\partial}_A$ referential basis vector (2.20)
$\boldsymbol{\partial}_\alpha$ intermediate basis vector (4.64)
$(\cdot)_{(AB)}$ symmetric components (2.12)
$(\cdot)_{[AB]}$ anti-symmetric (skew) components (2.13)
$(\cdot)_{\{ABC\}}$ permutated components (2.92)
$\partial_a(\cdot)$ partial differentiation with respect to spatial coordinates (2.2)
$\partial_A(\cdot)$ partial differentiation with respect to reference coordinates (2.2)
$\partial_\alpha(\cdot)$ partial differentiation with respect to intermediate coordinates (4.15)
$(\cdot)_{;a}$ covariant derivative for spatial Levi-Civita connection (2.152)
$(\cdot)_{;A}$ covariant derivative for reference Levi-Civita connection (2.124)
$(\cdot)_{:a}$ total covariant derivative in spatial coordinates (2.188)
$(\cdot)_{:A}$ total covariant derivative in reference coordinates (2.181)
$(\cdot)_{:\alpha}$ total covariant derivative in intermediate coordinates (4.206)
$\nabla(\cdot)$ generic covariant derivative or gradient operator (2.58)
$\overset{g}{\nabla}_a(\cdot)$ covariant derivative for spatial Levi-Civita connection (2.152)
$\nabla_A(\cdot)$ generic covariant derivative in reference coordinates (2.64)
$\bar{\nabla}_A(\cdot)$ covariant derivative with respect to part of total connection (2.105)
$\overset{G}{\nabla}_A(\cdot)$ covariant derivative for reference Levi-Civita connection (2.124)
$\nabla_\alpha(\cdot)$ covariant derivative with respect to intermediate coordinates (4.138)
$\hat{\nabla}_\alpha(\cdot)$ anholonomic covariant derivative (5.31)
$\dot{(\cdot)}$ material time derivative (3.150)
$\mathrm{d}(\cdot)/\mathrm{d}t$ material time derivative (3.150)
$\mathrm{D}(\cdot)/\mathrm{D}t$ reference partial time derivative with fixed spatial basis (3.153)
$\mathcal{L}_{\boldsymbol{v}}(\cdot)$ Lie derivative with respect to velocity field (3.186)
$\langle\cdot,\cdot\rangle$ scalar product (2.5)
$\otimes$ tensor product (2.19)
$\cdot$ dot product (2.24)
$\times$ cross product (2.43)
$:$ double dot product (2.199)
$\star$ generic operator obeying distributive property (3.133)
$\{_{BC,A}\}$ Christoffel symbols of the first kind (2.85)
$\{^{..A}_{BC}\}$ Christoffel symbols of the second kind (2.86)
$\det(\cdot)$ determinant of square matrix or second-order tensor (2.10)
$\ln(\cdot)$ natural logarithm (2.144)
$\mathrm{tr}(\cdot)$ trace of second-order tensor (2.171)
$(\cdot)^{-1}$ inverse (2.26)

$(\cdot)^{\mathrm{T}}$ transpose of second-order tensor (2.197)
$\int(\cdot)$ generic volume or surface integral (3.133)
$\oint(\cdot)$ integral over closed surface or closed curve (3.133)

Bibliography

[1] Bilby, B., Bullough, R. and Smith, E. (1955). Continuous Distributions of Dislocations: a New Application of the Methods of Non-Riemannian Geometry, *Proceedings of the Royal Society of London A* **231**, pp. 263–273.

[2] Bilby, B., Gardner, L. and Stroh, A. (1957). Continuous Distributions of Dislocations and the Theory of Plasticity, in *Proceedings of the 9th International Congress of Applied Mechanics*, Vol. 8 (Université de Bruxelles, Brussels), pp. 35–44.

[3] Bloom, F. (1979). *Lecture Notes in Mathematics 733: Modern Differential Techniques in the Theory of Continuous Distributions of Dislocations* (Springer-Verlag, Berlin).

[4] Cartan, É. (1926). *La Théorie des Groupes Finis et Continus et la Géométrie Différentielle, Traitées par la Méthode du Repère Mobile: Leçons Professées à la Sorbonne* (Gauthier-Villars, Paris).

[5] Cartan, É. (1928). Sur la Represéntation Géométrique des Systèmes Matériels Non Holonomes, *Proceedings of the International Congress of Mathematicians (Bologna)* **4**, pp. 253–261.

[6] Ciarlet, P. (1988). *Mathematical Elasticity* (North-Holland, Amsterdam).

[7] Clayton, J. (2011). *Nonlinear Mechanics of Crystals* (Springer, Dordrecht).

[8] Clayton, J. (2012). On Anholonomic Deformation, Geometry, and Differentiation, *Mathematics and Mechanics of Solids* **17**, pp. 702–735.

[9] Clayton, J. (2013). Defects in Nonlinear Elastic Crystals: Differential Geometry, Finite Kinematics, and Second-order Analytical Solutions, *Zeitschrift für Angewandte Mathematik und Mechanik* (in press, DOI: 10.1002/zamm.201300142).

[10] Clayton, J. (2014). An Alternative Three-term Decomposition for Single Crystal Deformation Motivated by Non-linear Elastic Dislocation Solutions, *Quarterly Journal of Mechanics and Applied Mathematics* **67**, pp. 127–158.

[11] Eckart, C. (1948). The Thermodynamics of Irreversible Processes. IV. The Theory of Elasticity and Anelasticity, *Physical Review* **73**, pp. 373–382.

[12] Einstein, A. (1928). Riemann-Geometrie mit Aufrechterhaltung des Begriffes des Fernparallelismus, *Preussische Akademie der Wissenschaften, Physikalisch-Mathematische Klasse*, pp. 217–221, 224–227.

[13] Epstein, M. and Elżanowski, M. (2007). *Material Inhomogeneities and their Evolution: A Geometric Approach* (Springer, Berlin).
[14] Ericksen, J. (1960). Tensor Fields, in S. Flügge (ed.), *Handbuch der Physik*, Vol. III/1 (Springer-Verlag, Berlin), pp. 794–858.
[15] Eringen, A. (1962). *Nonlinear Theory of Continuous Media* (McGraw-Hill, New York).
[16] Eringen, A. (1971). Tensor Analysis, in A. Eringen (ed.), *Continuum Physics*, Vol. I (Academic Press, New York), pp. 1–155.
[17] Kondo, K. (1964). On the Analytical and Physical Foundations of the Theory of Dislocations and Yielding by the Differential Geometry of Continua, *International Journal of Engineering Science* **2**, pp. 219–251.
[18] Kondo, K. (1984). Fundamentals of the Theory of Yielding Elementary and More Intrinsic Expositions: Riemannian and Non-Riemannian Terminology, *Matrix and Tensor Quarterly* **34**, pp. 55–63.
[19] Kröner, E. (1960). Allgemeine Kontinuumstheorie der Versetzungen und Eigenspannungen, *Archive for Rational Mechanics and Analysis* **4**, pp. 273–334.
[20] Kröner, E. and Seeger, A. (1959). Nicht-lineare Elastizitätstheorie der Versetzungen und Eigenspannungen, *Archive for Rational Mechanics and Analysis* **3**, pp. 97–119.
[21] Le, K. and Stumpf, H. (1996). On the Determination of the Crystal Reference in Nonlinear Continuum Theory of Dislocations, *Proceedings of the Royal Society of London A* **452**, pp. 359–371.
[22] Malvern, L. (1969). *Introduction to the Mechanics of a Continuous Medium* (Prentice-Hall, Englewood Cliffs, NJ).
[23] Marcinkowski, M. (1979). *Unified Theory of the Mechanical Behavior of Matter* (John Wiley & Sons, New York).
[24] Marsden, J. and Hughes, T. (1983). *Mathematical Theory of Elasticity* (Prentice-Hall, Englewood Cliffs, NJ).
[25] Noll, W., Toupin, R. and Wang, C.-C. (1968). *Continuum Theory of Inhomogeneities in Simple Bodies* (Springer-Verlag, New York).
[26] Ogden, R. (1984). *Non-Linear Elastic Deformations* (Ellis-Horwood, Chichester).
[27] Rakotomanana, L. (2004). *Progress in Mathematical Physics 31: A Geometric Approach to Thermomechanics of Dissipating Continua* (Birkhäuser, Boston).
[28] Schouten, J. (1954). *Ricci Calculus* (Springer-Verlag, Berlin).
[29] Steinmann, P. (2013). On the Roots of Continuum Mechanics in Differential Geometry, in H. Altenbach and V. Eremeyev (eds.), *Generalized Continua–From the Theory to Engineering Applications* (Springer, Udine), pp. 1–64.
[30] Teodosiu, C. (1982). *Elastic Models of Crystal Defects* (Springer-Verlag, Berlin).
[31] Truesdell, C. and Toupin, R. (1960). The Classical Field Theories, in S. Flügge (ed.), *Handbuch der Physik*, Vol. III/1 (Springer-Verlag, Berlin), pp. 226–793.
[32] Vrânceanu, G. (1926). Sur les Espaces Non Holonomes, *Comptes Rendus de*

l'Académie des Sciences **183**, pp. 852–854.

[33] Yavari, A. and Goriely, A. (2012). Riemann-Cartan Geometry of Nonlinear Dislocation Mechanics, *Archive for Rational Mechanics and Analysis* **205**, pp. 59–118.

[34] Zubov, L. (1997). *Lecture Notes in Physics m 47: Nonlinear Theory of Dislocations and Disclinations in Elastic Bodies* (Springer-Verlag, Berlin).

Index

Printed in the United States
By Bookmasters